Electronic Troubleshooting

Fabian J. Lahue
New Hampshire Technical College at Stratham

GLENCOE
Macmillan/McGraw-Hill

New York, New York   Columbus, Ohio   Mission Hills, California   Peoria, Illinois

This book is dedicated to the loving memory of my son, Timothy James, whose life ended too soon. My wife, Marty, and my son, Dale, have always provided loving support. This book is also dedicated to a special friend, Kathleen Boyd.

Lahue, Fabian J.
　　Electronic troubleshooting/Fabian J. Lahue.
　　　　p.　　cm.
　　Includes index.
　　ISBN 0-02-819904-9
　　1. Electronic apparatus and appliances—Maintenance and repair.
　I. Title.
　TK7870.5.L35　1994
　621.3815'4'0288—dc20　　　　　　　　　　　　　93-24452
　　　　　　　　　　　　　　　　　　　　　　　　　　　CIP

Send all inquires to:
GLENCOE DIVISION
Macmillan/McGraw-Hill
936 Eastwind Drive
Westerville, OH 43081

ISBN 0-02-819904-9

Printed in the United States of America

1 2 3 4 5 6 7 8 9 BAW 00 99 98 97 96 95 94 93

CONTENTS

4.

STATIC (DC) TROUBLESHOOTING AT THE SOLID-STATE
COMPONENT LEVEL   51

5.

TROUBLESHOOTING THE POWER SUPPLY   101

6.

AC TROUBLESHOOTING   125

7.

TROUBLESHOOTING UNTUNED AMPLIFIER CIRCUITS   163

8.

TROUBLESHOOTING TUNED AMPLIFIERS   179

9.

TROUBLESHOOTING RADIO FREQUENCY CIRCUITS   201

10.

INTRODUCTION TO DIGITAL CIRCUIT TROUBLESHOOTING 227

APPENDIX A

COMPONENT IDENTIFICATION AND SYMBOLOGY   259

APPENDIX B

INTEGRATED CIRCUITS PACKAGING   271

APPENDIX C

IC TTL DEVICE FAMILIES AND CHARACTERISTICS   273

APPENDIX D

COMPONENT-SELECTION PROCESS   277

PREFACE

Students are taught the fundamentals of electronics theory in the classroom and through their laboratory experience. They begin their study with fundamental theories related to basic components, and complete their training with analyses of complete systems. This pattern of study allows them to see how the components work alone and together with others to form complete circuits. The course of study is normally completed in about two years. While in the classroom, students live in a theoretical world where every situation has a precise, textbook answer. Upon completing their course of study, they enter the real world where they face a myriad of electronic problems that do not have easy and predictable, theoretical answers.

This book is structured to allow students to bridge the gap between the classroom world of precision answers to real situations where components, and even total systems, do not always work according to established theory. Electronic circuits can and do fail, and the technical person must be able to know what to do when failure occurs.

The book is organized to follow a four-semester electronic technology program. Topics for the first two terms are included in Chapters 1, 2, 3, along with appropriate sections in Chapter 6. The third term includes coverage of solid-state topics in Chapters 4, 5, 6, 7, and 8. The concluding fourth term provides digital coverage as found in Chapters 9 and 10. Thus, this single volume can provide troubleshooting insights throughout a student's course of study.

This text is, therefore, a general guide for electronic troubleshooting that fills a void between theoretical and practical circuit operation. The text provides students with the following material:

- How common electronic components and circuits fail.
- How to make proper measurements using common items of test equipment.
- How to analyze symptoms and diagnose problems based on the measurements taken.
- A logical approach to troubleshooting.
- Potential problems due to the interfacing of blocks within a system.
- How to identify parts and select suitable replacements utilizing manufacturers' data sheets.

Throughout the text, the author has kept mathematics to a minimum and presented only the very basic and necessary equations. Likewise, only concise statements of theory are included, as the book's emphasis is troubleshooting. Therefore, instructors are encouraged to tailor their classroom presentations of mathematical and theoretical assignments to pertinent sections, as needed. For ease in making these assignments, the book contains an extremely detailed table of contents with each section identified by page number.

With each assignment, instructors are encouraged to review appropriate safety concerns. The author has found that an initial review of test and measurement equipment prior to assigning troubleshooting exercises facilitates students' abilities in taking the required measurements.

The author would like to acknowledge the recommendations and support provided by Professor Addison Marvin, Department Chairperson of Electronics Technology, New Hampshire Technical College at Stratham, New Hampshire.

Fabian J. Lahue

1

INTRODUCTION

1.1 WELCOME

Welcome to the FL Company. Your position as the only electronic technician at our company is not an easy one. Every day brings new problems to challenge your knowledge. Of course, you have a solid understanding of basic electronic theory; however, that is not enough. It is imperative for you to understand how to recognize and solve problems quickly and how to locate system faults and isolate them fast. In many situations, a data link is critical and cannot be off line (down). Every minute of "down time" costs the company a substantial amount of money. So, when a system does go down, all eyes are on you, the electronic troubleshooter.

1.2 THE BIG PICTURE

Figure 1.1 shows a block diagram of the FL Company's command, control, and communications systems. Take a few minutes to study the various systems. The manufacturing

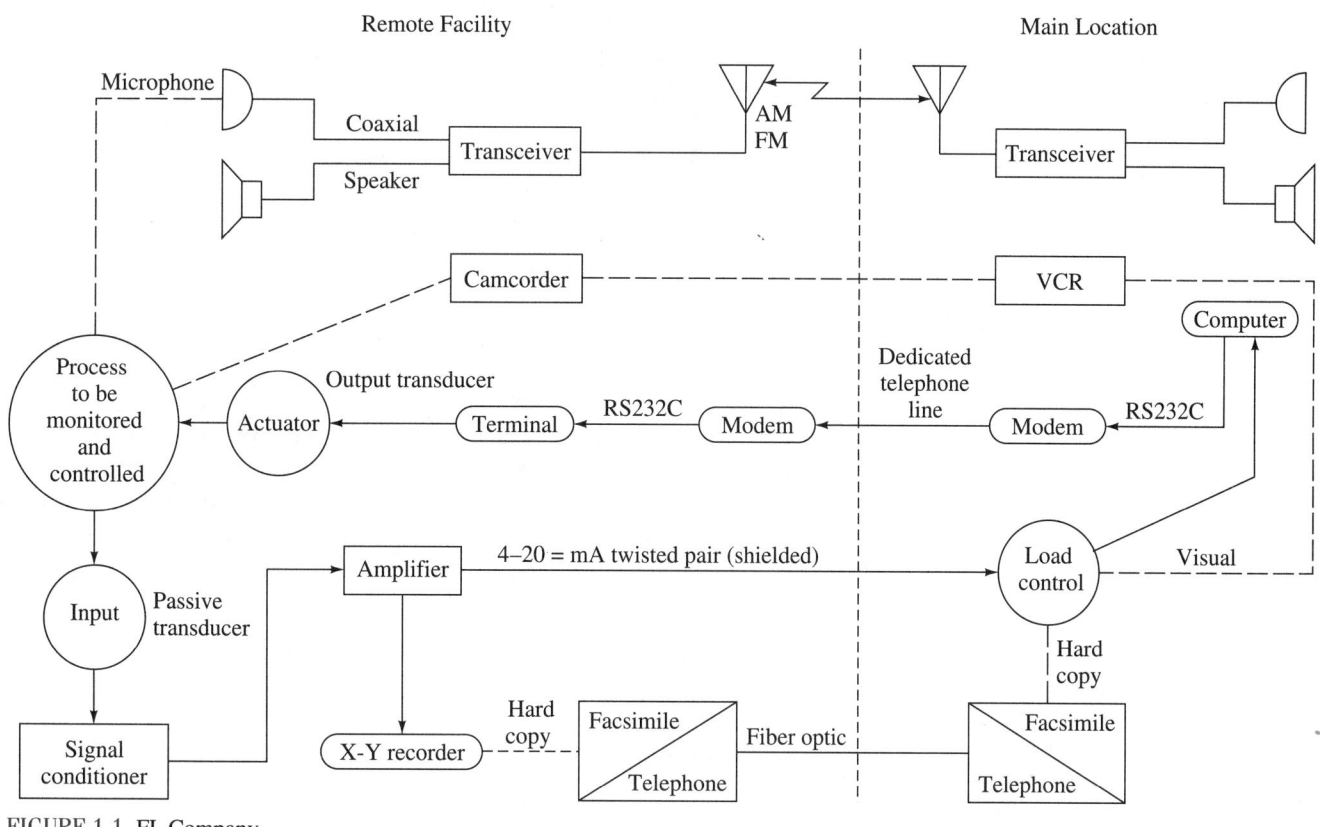

FIGURE 1.1 FL Company

process to be controlled and monitored is located at a remote facility. The process operator provides a (real-time) aural report via the radio link and a (delayed-time) visual report through use of the camcorder. The process is controlled by engineers located in the main facility. Commands entered into the main computer are transferred via a dedicated telephone line to the remote terminal, where they control active transducers (actuators) such as solenoid valves, switches, motors, and heaters to vary such parameters as process temperature, flow rate, pressure, level, and force, to name just a few.

These process parameters are then monitored by various passive transducers, which provide electrical signals that are conditioned and amplified. These amplified signals are sent over a 4-mA to 20-mA twisted-pair loop to provide real time observation and control. They are also sent to the *X-Y* recorder to provide a permanent hard copy of the process parameters. This hard copy can then be sent to the main facility (on demand) via the facsimile and fiber-optic data link.

1.3   THE PROBLEM

Within this big picture, there are five totally different systems:

1. Radio frequency (RF) system
2. Human interface system
3. Digital telemetry system
4. Analog telemetry system
5. Optical data system

These systems mean that you are going to be dealing with both analog and digital signals. It will be necessary to understand test and measurement techniques for both worlds. Here's the problem: You need to understand everything from analog signal conditioning to *X-Y* chart recorders. Do you see the problem? Try counting the number of possible trouble spots — you'll probably be amazed at how many there are.

1.4   TROUBLESHOOTING THE BIG PICTURE

Be familiar with proper troubleshooting techniques.

Actually, the problem is not that devastating, unless lightning destroys several links at once. When a data link does go down, you can isolate the trouble spot quickly with just a few steps.

The goal of the rest of this text is to familiarize you with proper troubleshooting techniques using common items of test equipment.

1.5   COMMON TEST EQUIPMENT

If you are fortunate enough to work in a facility that has the most expensive and sophisticated items of test equipment available, you are indeed extremely lucky. In most cases, however, you will have at your disposal only common items of test equipment. However, this does not mean that you will not be able to do the job properly. Most system problems can be diagnosed and isolated with this equipment. You will be surprised at the amount of information you can obtain simply by making voltage and resistance measurements. Throughout this text, the use of the equipment is discussed where appropriate. Let's break down the equipment into categories of use.

Troubleshooting DC Circuits

Troubleshooting circuits

- Power supplies
- Volt-ohm-milliammeters (VOM)
- Digital multimeters (DMM)
- Oscilloscopes

Troubleshooting Audio Frequency Circuits

- Power supplies
- Volt-ohm-milliameters
- Digital multimeters
- Oscilloscopes
- Audio frequency oscillators
- Signal injectors

Troubleshooting Radio Frequency Circuits

- Power supplies
- Volt-ohm-milliammeters
- Oscillators
- RF probes
- Standing Wave Ratio (SWR) meters
- Dummy loads
- Power meters
- Frequency counters
- Signal injectors
- Audio signal generators
- Spectrum analyzers

Troubleshooting Digital Circuits

- Volt-ohm-milliammeters
- Logic analyzers
- Digital multimeters
- Storage oscilloscopes
- Logic probes
- Logic clips
- Logic pulsers
- Logic comparators
- Current tracers

Troubleshooting Fiber Optic Circuits

- Optical power meter
- Light-Emitting Diode (LED) source and tester
- Oscilloscope
- Microscope

2

SAFETY FIRST

2.1   THE IMPORTANCE OF SAFETY

The importance of *safety* in troubleshooting electronic circuits and systems cannot be overemphasized. You might consider 10 mA to be an insignificant amount of current; however even this small current can cause muscle paralysis, resulting in your inability to let go of the source of current. If your arm, chest, and arm become the circuit, 40 mA can be lethal. A level of 100 mA usually is lethal.

2.2   THE LOW-VOLTAGE TRAP

Don't become careless just because your circuits are low voltage. *Remember:* The habits you learn dealing with low voltages show up again when you are troubleshooting high-voltage circuits. Respect all voltages, especially those greater than 30 V!

 CAUTION
Respect all voltages, especially those greater than 30 V!

2.3   PRACTICING GOOD HOUSEKEEPING

Before you do anything in your work area, you *must* practice good housekeeping and do the following:

1. Remove all jewelry, watches, and metal belt buckles.
2. Wear rubber-soled shoes.
3. Use conductive mats to insulate yourself.
4. Make sure your test bench is grounded.
5. Use wrist straps when handling integrated circuits to protect components from electrostatic charges.
6. Never work on hazardous circuits by yourself.
7. Be familiar with your company's lockout-tagout procedures for remote breakers and switches.
8. Know proper CPR techniques.

 CAUTION
Remember to practice good housekeeping.

2.4   BEFORE TESTING

As you begin to analyze equipment, remember the following:

1. Poor or broken grounds can cause unseen hazards. Consider the situation of a coaxial cable that improperly provides an ac ground for equipment. If you're not careful when disconnecting the cable, you can become part of the circuit. To avoid this, dis-

connect the cable at a connector and hold your hands over insulation before pulling the connector apart.

2. *Turn off all power.*

3. Before removing any covers, check all connections for wires that might be pulled off with the cover.

4. Work with one hand; keep the other behind your back or in your pocket.

5. Always keep metal chassis away from pipes, etc.

6. Don't reach inside energized equipment.

7. Discharge capacitors as follows:

- *Large capacitors:* Drain the charge through resistor of about 100 Ω.
- *Small capacitors:* Short leads together with your screwdriver.

8. Proceed with dead-circuit testing.

2.5 LIVE-CIRCUIT TESTING

When it comes time to perform live-circuit testing, follow these steps.

1. Utilize isolation transformers to avoid the hazard of short-circuiting the ac power line.

2. Never use a solid-state tester in live circuits.

3. Don't destroy your protective grounding. Be sure to keep all ground wires in place. *If the equipment uses a two-wire polarized plug, remember that the wide blade is connected to the neutral of the power line.* Bypassing or trimming the wide blade can make the chassis "hot" and hazardous.

4. Use your meters properly. You can further damage the equipment by using the meters improperly. For example,

- Application of too-high voltage levels, such as pulse voltages
- Use of low-sensitivity dc voltmeters (Such use may upset the dc bias in high-gain dc coupled transistor amplifiers under test.)
- Application of the oscilloscope probe ground lead to a point that is not the circuit common

5. Use circuit-card extenders to make test points more accessible.

6. *Never* allow a laser beam to enter your eyes from any angle.

7. Use the correct probe with your test equipment.

8. Keep high-voltage probes handy, because vacuum tubes are still in use in high voltage and RF applications:

- Magnetrons in radar equipment
- Magnetrons in microwave ovens
- Cathode-ray tube anode potential
- High-power amplifiers in broadcast equipment

If voltage measurements must be made of circuits of 60 V dc or 30 V ac or more, please follow these steps:

1. Turn off all power.
2. Discharge all capacitors.
3. Attach test leads.

 **CAUTION**

Remember that the wide blade is connected to the neutral of the power line if the equipment is using a two-wire polarized plug.

4. Turn on power.
5. Take readings.
6. Turn off power.
7. Discharge capacitors.
8. Remove test leads.
9. Continue with tests.

Use antistatic bags, cases, and sticks for IC circuits. Be sure to keep the leads of all field-effect transistors shorted.

3

DC (STATIC) TROUBLESHOOTING AT THE DISCRETE COMPONENT LEVEL

3.1 BASIC PROCEDURE

Troubleshooting can be performed in many ways. Some technicians use a "shotgun" approach, usually becoming frustrated. A technician using this approach will step up to the plate and swing at anything that moves. In other words, he or she takes measurements randomly with no apparent purpose, substitutes parts randomly for no reason, and amasses a wealth of unrelated data. Sometimes, the technician even stumbles onto the problem. And then there is *your* way — the right way.

Troubleshooting is the sequence of logical and systematic steps taken to diagnose and analyze a problem.

Troubleshooting should be defined as the sequence of logical and systematic steps taken to diagnose a problem, make selective tests and measurements, analyze appropriate data, and, ultimately, isolate the faulty component. These steps can be broken down as follows:

1. *Ask enough questions* to determine all possible symptoms.

   a. What happened?
   b. When did this happen?
   c. What were all the conditions at the time this happened? Was there anything unusual?
   d. Is this a complete failure?
   e. Is this an intermittent failure?
   f. Is this a thermally related failure?
   g. Would you call the output degraded?

2. Define the problem as clearly as possible.

3. Clearly understand how the system is supposed to operate and what voltage levels to look for. This understanding is vital — otherwise any measurements that you make will provide you only with useless data.

4. Study all available literature to determine which functional block might be the problem area. Make use of

   a. Schematic diagrams
   b. Line drawings

c. Blueprints
   d. Block diagrams or layouts

5. Perform selective tests and measurements to eliminate as many functional units as possible (known as the *half-split* method).

6. Use as few tests as possible to identify the functional unit at fault.

7. Use signal-injection and signal-tracing techniques to locate the malfunctioning circuit stage within the functional unit.

8. Isolate the faulty component by

   a. Analyzing output waveforms (using in-circuit testers if available)
   b. Performing voltage measurements with power on but no signal applied
   c. Performing resistance measurements

9. Remove the faulty component, identify it, and locate a replacement part.

10. Repair the equipment.

11. Verify the equipment operation.

3.2   REASONS FOR BREAKDOWN

Why do components fail, causing a system breakdown? Here are a few reasons; you can probably add some of your own.

1. *Operator error.* Improperly applied voltages and currents can contribute to component failures.
2. *Abuse.* Rough handling of equipment can often lead to broken components, lead wires, printed circuit boards, and so on.
3. *Lack of maintenance.* Allowing equipment to become dirty and not replacing frayed wires or broken connectors can cause problems.
4. *Poor installation.* This situation can cause bad solder joints, loose connections, improper grounds, and improper interfacing.
5. *Excessive movement.* Such movement may break connections, cabling, or even printed circuit boards.
6. *Tampering with equipment.* Tampering can cause broken components, boards, cabinets, etc.
7. *Accidental reversal of power supply leads.* This situation can damage equipment that has no polarity protection.
8. *Dropped equipment.* Dropping equipment can cause mechanical damage.
9. *Transients.* Transients are voltage surges that most often exceed the breakdown ratings of components.
10. *Radiation damage.* Radiation is a twentieth-century concern that can damage components or cause degraded performance.
11. *Electromagnetic interference.* This interference in very high levels can cause degraded performance of a system.
12. *Dirty environment.* Excessive levels of dust will cause mechanical problems and even (if extreme) component overheating.
13. *Moisture.* Short circuits and even open circuits can occur due to the formation of oxides.
14. *Extreme temperatures.* Resistances increase, and materials dry out, expand, and eventually crack under extreme temperatures.
15. *Rodent damage.* Rodents often nibble at the wires, causing open circuits.
16. *Spilled coffee or other substances on the keyboard.*

3.3 EFFECTS OF BREAKDOWN

Breakdown effects may be numerous. Let's look at some common ones.

1. *Short circuits* normally result in excessive current flow. The most evident signs of a short-circuit condition are

- Blown fuses
- Hot components
- Low voltage levels
- High amperage levels
- Smoke

2. *Open circuits* are due to broken connectors, wires, or printed circuit boards or burned components. Look for

- Very high resistance values
- Zero current levels
- Higher than normal voltage levels
- Dead circuits

3. *Grounds* where there should not be grounds can cause loss of ac signals and dc voltages. Look for

- Abnormal current levels
- Abnormal voltage levels
- Abnormal resistance levels
- Shocks
- Poor performance
- Hum problems

4. *Mechanical breakage* causes the following:

- Noisy operation
- Poor operation
- Degraded performance
- Complete failures

5. *Dead equipment* is the easiest problem to diagnose.

6. *Degraded performance* is very difficult to diagnose. It may be due to an amplifier stage whose gain has degraded, often due to prolonged high temperatures.

7. *Erratic intermittent problems* are the worst. The symptoms always seem to disappear when the technician arrives.

8. *Hum problems* are most often caused by one of several different situations:

- Bad filter capacitors in the power supply
- Missing ground connections
- Ground loops causing circulating currents
- Bad amplifier circuits

3.4 REQUIRED TEST EQUIPMENT

If you are proficient with the following items of test equipment, you will be able to perform most static tests and measurements necessary. Make certain that you have the following available:

1. *Volt-ohm-milliammeter.* This multimeter is simple, inexpensive, and easy to use. It is an analog device and performs ammeter, ohmmeter, and voltmeter functions. The meter uses a d'Arsonval movement (a moving coil inside a magnetic field), which draws

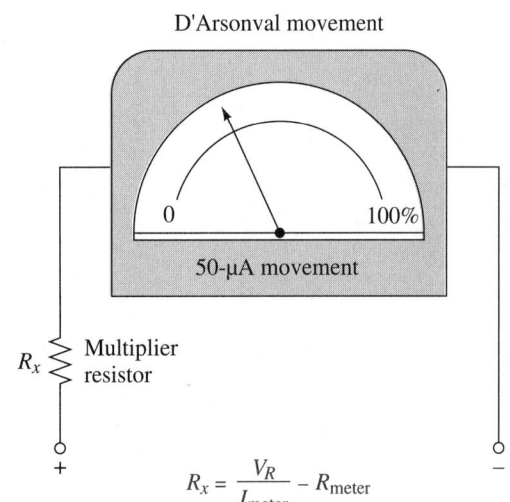

FIGURE 3.1 Basic voltmeter circuit.

$$R_x = \frac{V_R}{I_{meter}} - R_{meter}$$

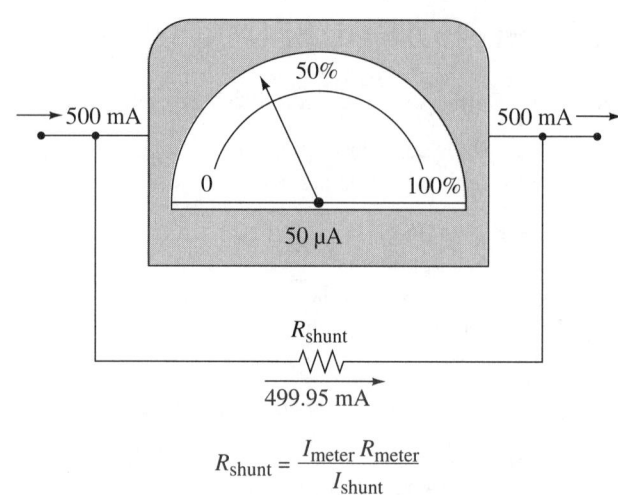

$$R_{shunt} = \frac{I_{meter} R_{meter}}{I_{shunt}}$$

FIGURE 3.2 Basic ammeter circuit.

current from the circuit under test. Because it draws current from the circuit under test, the meter can act as a load and give false indications when making measurements on low-voltage and/or high-impedance circuits. The meter therefore represents a low value of input impedance to the circuit. The meter movement does, however, offer the advantage of being able to view slowly varying voltages.

a. *Voltmeter.* When used to measure voltages, the meter selector places a resistor in series with the meter movement. This multiplier resistor is used to limit the current through the coil; higher range settings select larger values of resistance. The current through the resistor is in direct proportion to the level of voltage being measured. See Figure 3.1. The voltmeter is, therefore, polarized. It must be placed in parallel with the circuit under test and must never be placed in series. The sensitivity of the voltmeter is directly related to the meter-movement sensitivity expressed in ohms/volt. From Ohm's law, $R = V/I$, a 50-μA meter movement would represent 20,000 Ω/V. If you are using the 10-V range, the meter represents a total resistance of 20,000 Ω/V multiplied by 10 V, or 200,000 Ω. Therefore, the accuracy of the voltage measurement depends on the range that the meter is on, the voltmeter sensitivity, and the resistance of the circuit under test. *This is known as the loading effect.*

 CAUTION

Never place the ammeter in parallel with a circuit component. It can be destroyed.

b. *Ammeter.* The ammeter is used to measure circuit current. It must always be placed in series and should have a very low resistance value in order to not disturb the circuit resistance. It is polarized and must be placed in the circuit properly. *Never place the ammeter in parallel with a circuit component. It can be destroyed.* Inside the ammeter, a shunt resistor bypasses most of the circuit current around the meter. Thus, a 50-μA movement receives a maximum of only 50 μA (see Figure 3.2) and the meter value is proportional to the actual circuit current. In order to calculate the value of the shunt resistor, it is necessary to know the sensitivity and internal resistance of the meter coil:

$$R_{shunt} = \frac{I_{meter} \times R_{meter}}{I_{shunt}}$$

c. *Ohmmeter.* The ohmmeter is used to measure resistance, continuity, opens, shorts, etc. It can also be used to indicate the condition of components such as

- Capacitors
- Inductors

- Transformers
- Transistors
- Diodes and various other devices.

Never use the ohmmeter in a live circuit.

The ohmmeter uses its own internal dc voltage source, and you should always isolate the component being tested to avoid any parallel paths that might change the result.

The basic ohmmeter (Figure 3.3) has a sensitive dc meter movement, series current limiter, zero-adjust rheostat, and a dc source. The battery is used to activate the meter movement.

FET multimeters use field-effect-transistor amplifiers to increase the input resistance to a value in excess of 10 MΩ.

d. *Accuracy.* The overall accuracy of any reading is, therefore, dependent on

- The accuracy of the movement
- The care taken in reading the scale (watch out for the parallax error)
- The condition of the battery

As a final note, the analog meter is excellent for trend observations and is less susceptible to interference from electromagnetic fields than its digital counterpart (depending on the proximity and strength of the EMF).

CAUTION

Never use the ohmmeter in a live circuit.

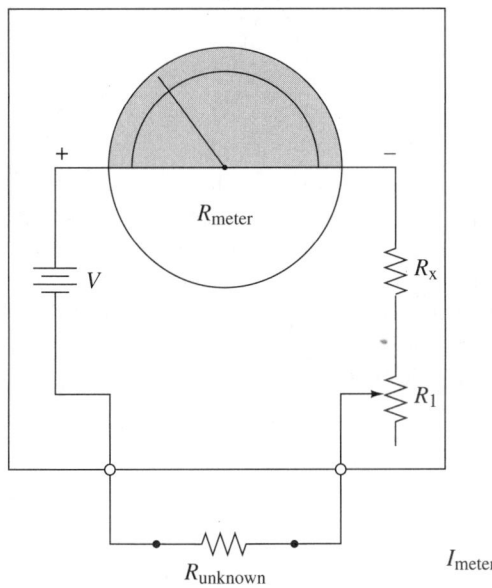

Series circuit of R_{meter}
$+ R_x + R_1 + R_{unknown}$

$$I_{meter} = \frac{V}{R_m + R_x + R_1 + R_{unknown}}$$

FIGURE 3.3 Basic ohmmeter.

2. *Digital multimeter (DMM).* A typical DMM (Figure 3.4) is built around a custom-designed CMOS LSI multiprocessor chip. With most of the electronic circuitry on the chip, there are very few, if any, discrete parts to degrade the lifetime of the meter. Peripheral components are used with the chip to provide a variety of measurements. Usually, the input signal to be measured is conditioned and converted to a dc voltage in the millivolt range. This millivolt signal is sent to a filter, where any interference such as power-line noise or miscellaneous audio frequency signals is removed. A clean, filtered signal is then digitized by the analog to digital converter and used to drive the individual segments of the digital display. Because the input signal is essentially averaged, the DMM cannot follow slowly varying analog signals. Another disadvantage is that a DMM is not useful for tuning receiver circuits because digital meters require a delay time to update the display and cannot easily be used to find null points or peaks. There are, however, many advantages to a DMM. Here are a few:

- Easy to read
- High degree of accuracy
- Extremely high input impedance
- Consistent readings
- Automatic display of polarity
- Autoranging often available
- Display even of microvolt levels
- Battery operated
- Storage capabilities often available

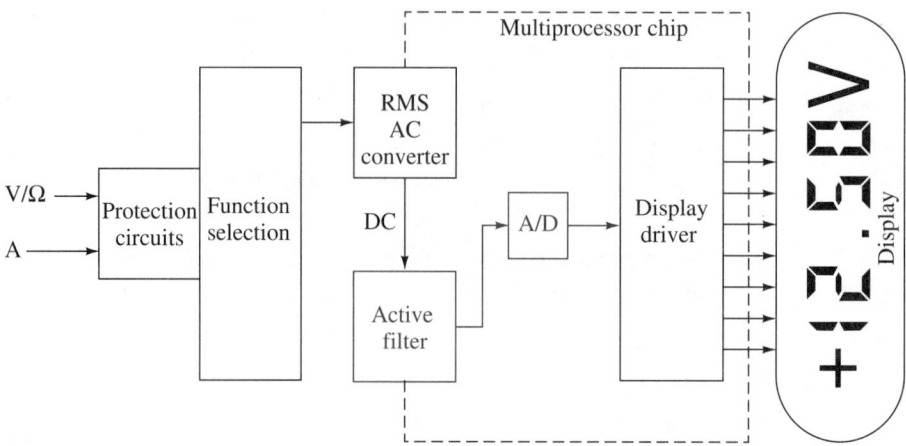

FIGURE 3.4 Typical DMM.

3. *Oscilloscope.* You will soon find the oscilloscope to be invaluable. It is advisable to have at least a 100-MHz bandwidth on the vertical system and dual-channel capability. Be sure that the input can be dc coupled. A separate ground reference button makes reference levels easy to follow.

Keep in mind that the input is usually grounded to the power line; this can present problems by creating improper ground points (known as ground loops) within the circuit under test. This is discussed more fully later in the text.

4. *Power supply.* Your power supply should have at least a 25-V capability and be able to source 5 A.

3.5   STATIC (DC) TROUBLESHOOTING — DISCRETE COMPONENTS

Resistors, capacitors, and inductors are the three basic components used in electronic circuits.

There are three basic components used in electronic circuits: resistors, capacitors, and inductors. All three are very complex, with each one exhibiting characteristics of the other two when the operating frequency is high enough. This concept is discussed again in the ac section of this book. These three basic elements are shown in Figure 3.5 with ideal schematic symbols. When studying the dc characteristics of any electronic circuit, you must remember the following:

1. *Resistors* present their rated ohmic value to the circuit and will establish a level of current established by Ohm's law, $I = V/R$.

2. *Capacitors* are frequency dependent and offer reactance to a circuit determined by the equation $X_C = 1/(6.28FC)$ where F is the frequency in hertz and C is the capacitance in farads. Therefore, for dc (frequency of 0 Hz), X_C approaches infinity and the capacitor looks like an open circuit.

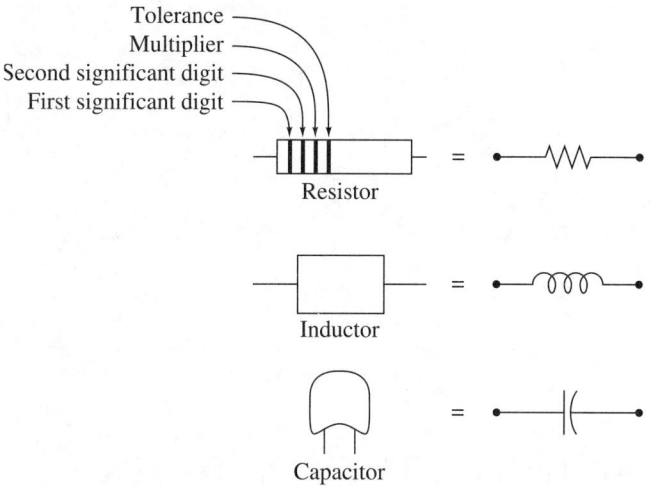

FIGURE 3.5 Ideal component symbols.

3. *Inductors* are also frequency dependent and establish a level of inductive reactance determined by the equation $X_1 = 6.28FL$, with *L* expressed in henrys. This shows that for dc, X_1 is equal to 0 Ω. The inductor looks like a short circuit. However, all practical inductors do have some small amount of inherent resistance determined by their physical composition. Therefore, for dc the inductor looks like this inherent resistance and can be measured directly.

When making dc measurements, visualize the equivalent dc circuit as follows:

1. Leave all resistors as they are.
2. Replace all capacitors with open circuits.
3. Replace all inductors with short circuits.

The circuit in Figure 3.6 is a typical class A audio amplifier in two stages.

In order to perform a static (dc) analysis, the circuit can be redrawn to give the dc equivalent circuit shown in Figure 3.7.

3.6 RESISTANCE MEASUREMENTS: A CAUTION

Never perform resistance measurements in live circuits. This can damage your ohmmeter and even introduce new failures.

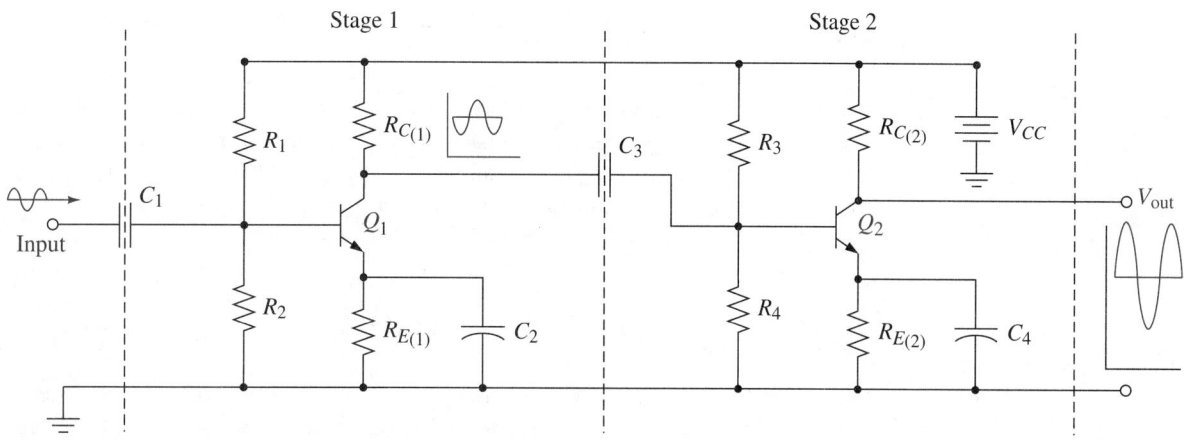

FIGURE 3.6 Typical two-stage class A amplifier.

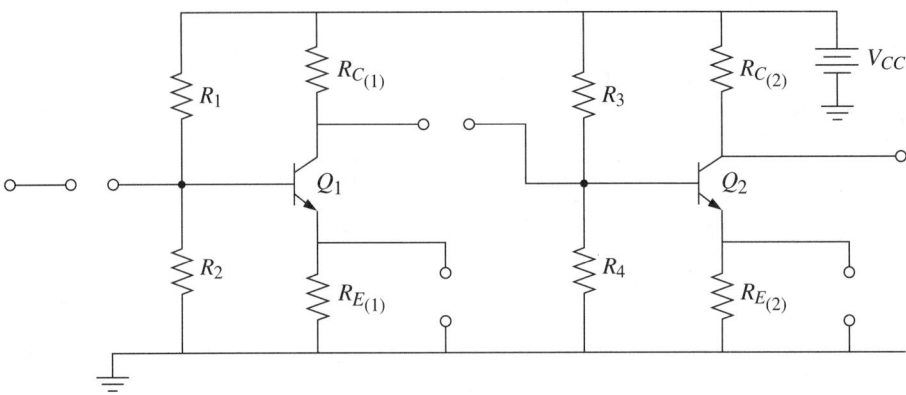

FIGURE 3.7 DC equivalent circuit.

3.7 TROUBLESHOOTING WITH YOUR OHMMETER

Your ohmmeter is a very valuable piece of test equipment and will readily identify the three basic trouble conditions — open, short, and out-of-tolerance circuits.

Open Circuit

Open circuits occur when a break interrupts the normal current flow. This interruption may be due to a break in the printed circuit board, broken wires, poor solder joints, blown fuses, faulty switches, faulty connections, or burned components. Solid-state components will usually short internally and then burn open. Placing the test leads of your ohmmeter across the component will result in an infinite ohms indication.

Short Circuit

Short circuits have components that are bypassed (internally or externally) by a low-resistance path. An internal short may be caused by a voltage transient, which is a sudden surge in voltage or current. Excessive current flowing through a component will overheat it, damaging it internally. The component may be bypassed externally by a broken wire, a solder bridge between two circuit points, bent terminals, or even worn insulation on circuit conductors.

Out-of-Tolerance Circuit

An out-of-tolerance circuit can be detected by measuring the actual resistance value and comparing with the manufacturer's data or the resistor color code.

3.8 VOLTAGE MEASUREMENTS

The concept of voltage measurement is vital to any troubleshooting situation. Most problems are found by voltage measurements. Be sure that your voltmeter test leads are in good condition and the probe tips are clean and sharp. It is usually most convenient to attach the black (negative) voltmeter lead to the circuit ground. This allows you to move easily through the circuit with one hand on the red (positive) lead.

3.9 POLARITY OF VOLTAGE DROPS

It is very important to understand the polarity of voltage drops within a circuit. Refer to Figure 3.8. Electron current leaves the negative battery terminal and proceeds through the circuit back to the positive battery terminal. The end of the resistor at which electron

Ohmmeter tests for three basic trouble conditions

Your voltmeter test leads should be in good condition and the probe tips clean and sharp.

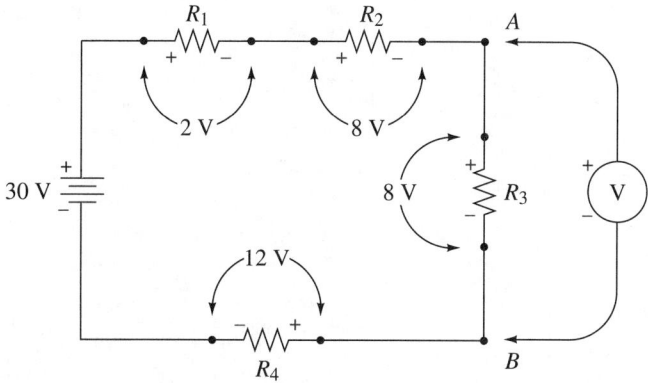

FIGURE 3.8 Polarity of voltage drops.

current enters is the negative end. The other end is positive. Be careful here. The resistor is not actually polarized. When considering the voltage drop across any resistor, the positive end is the end at the higher potential. The (−) and (+) symbols merely represent the relative polarity of voltage drop. See Figure 3.9. The potential at a point (A) is always measured with reference to some other point in the circuit. For example, the voltage dropped across R_1 in Figure 3.9 (labeled V_{R_1}, V_{AB}, etc.) is measured by connecting the voltmeter in parallel with R_1 as shown. *Voltmeters are always connected in parallel to the circuit component being tested.* Point A is the most positive and point C is the most negative. It can be said that A is very positive relative to C or that C is very negative relative to A. The voltage from point A to B (represented by V_{AB}) is +6 V, whereas the voltage from B to A, V_{BA} is – 6 V. Likewise, V_{BC} equals +12 V and V_{CB} equals − 12 V.

As a result, it is possible to obtain both positive and negative voltages from the circuit relative to the circuit ground.

Voltmeters are always connected in parallel to the circuit component being tested.

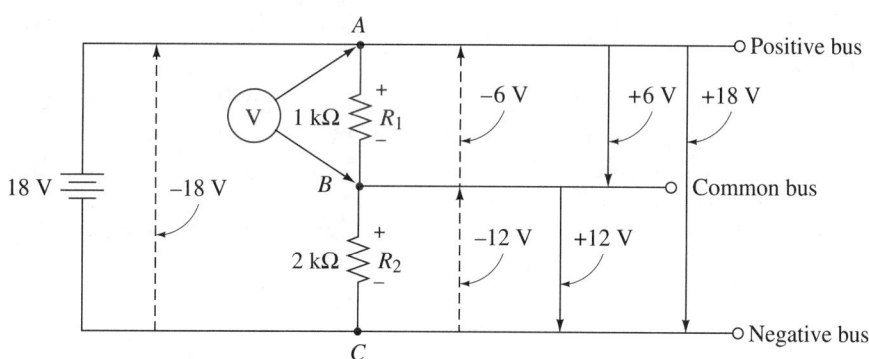

FIGURE 3.9 Relative voltage levels.

3.10   COMMON GROUND

Simple circuits have two primary conductors carrying current to and from the components involved. Most circuits, however, are far more complicated. Including all the wires leading to and from the power source quickly becomes confusing. Therefore, common-ground symbols are used to reduce the confusion. Let's discuss the concept of a ground (Figure 3.10). When you use a voltmeter, you always measure the voltage level at one point in the circuit relative to another. A resistor voltage drop of 10 V means that one end of the resistor is 10 V more positive or negative than the other end. To make this measurement, the technician must hold both probes of the voltmeter across the resistor in the

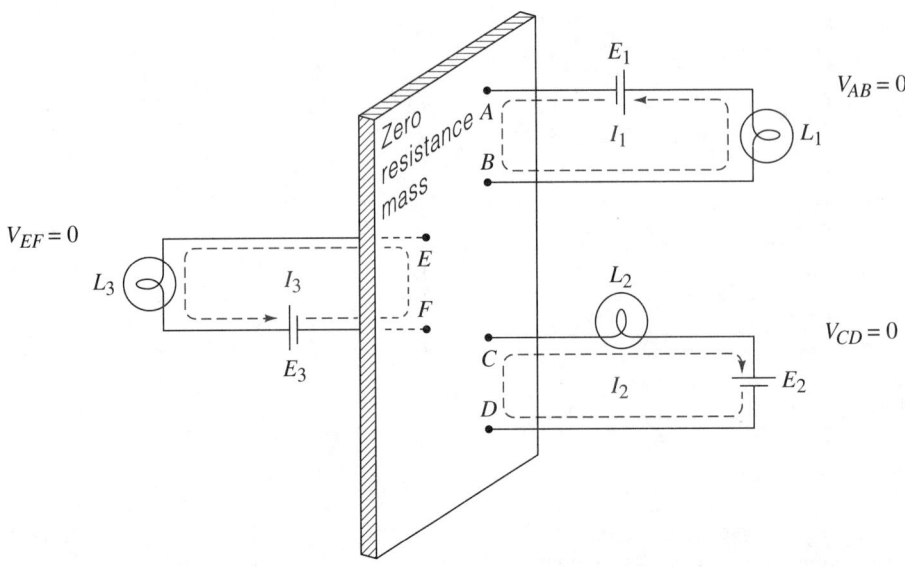

FIGURE 3.10 The "ground" concept.

live circuit. This procedure involves the risk of slipping and causing short circuits or even danger to the technician. For safety reasons, the technician should have only one hand in the circuit under test at any time. Thus, one probe of the voltmeter needs to be located at some common circuit point; the other probe is used to measure the voltage level on either end of the resistor. The difference, of course, will be the voltage drop. Usually, when troubleshooting, the voltage measurements are taken relative to 0 V, or *ground*. There are considered to be two types:

1. *Earth ground.* Your local electric company delivers voltage that is referenced to earth ground. One side of the source at the power plant is directly connected to the earth. The earth is considered to be a large conducting mass with virtually zero resistance between any two points. Figure 3.11(a) represents the commonly used symbol for earth ground.

2. *Circuit or chassis ground.* In your automobile battery, the negative post is connected to the frame or engine block. Therefore, the entire metal body is used as the common ground (reference) system. This is known as a negative ground. In electronic equipment, the chassis is used as the reference or ground. One side of the voltage source is connected to it and all measurements can be taken relative to it. The accepted symbol for chassis ground is shown in Figure 3.11(b). This symbol is also used to represent the

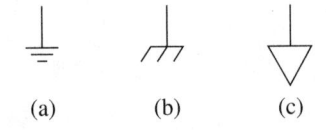

(a)   (b)   (c)

FIGURE 3.11 Commonly used ground symbols.

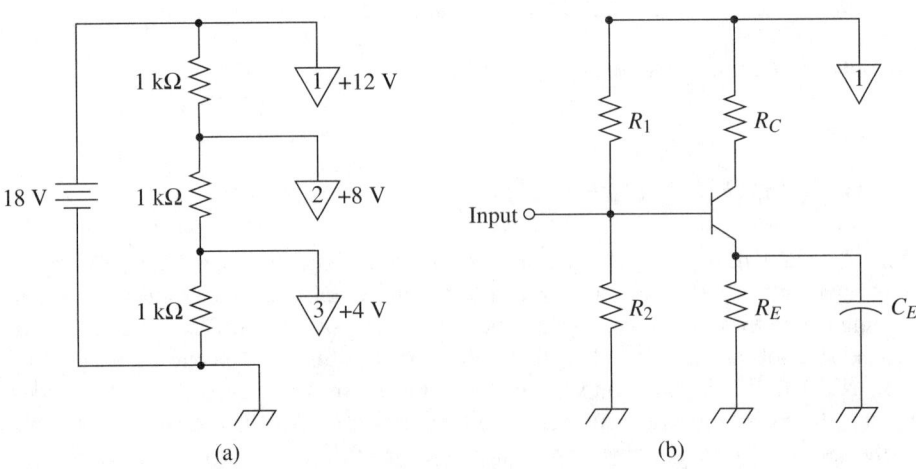

(a)   (b)

FIGURE 3.12 Designating points of common potential.

large-area copper foil on printed circuit boards. The concept of a zero-resistance ground reference path is very important. The circuits in Figure 3.11 are all connected to a zero-resistance mass. This means that no voltage exists between any of the terminal locations *A–F*. Thus all three circuits can operate independently of the other circuits and still have a common current return path.

Although Figure 3.11(c) is used to designate a ground return path, it was primarily intended to indicate a common connection. These may be points of common voltage, as in Figure 3.12.

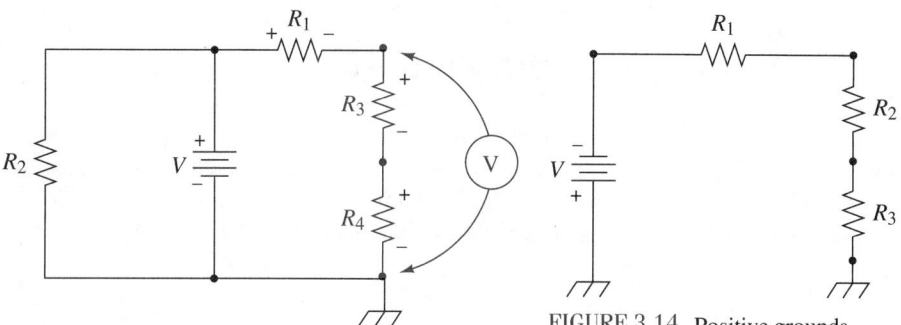

FIGURE 3.13 Negative grounds.

FIGURE 3.14 Positive grounds.

3.11   POSITIVE OR NEGATIVE GROUND

As shown in Figure 3.13, the return paths of all components in a negative ground are connected to the negative side of the supply. All voltages are positive relative to this ground. Some circuits, such as that in Figure 3.14, require use of negative voltages and, therefore, employ positive grounds.

In dual-polarity circuits such as the one in Figure 3.15, both positive and negative voltages are available relative to a circuit ground. Remember this when making voltage measurements.

3.12   DETECTING OPEN CIRCUITS

Whatever the cause, in an open circuit we know that circuit current has been stopped. An open circuit can be treated as an infinite resistance and, therefore, as the largest resistance in the circuit. Ohm's law tells us that the largest resistance in a circuit will have the largest voltage drop. We can expect all the supply voltage to appear across the open circuit and no voltage drops to occur across other components due to the lack of current flow.

In summary, referring to Figure 3.16,

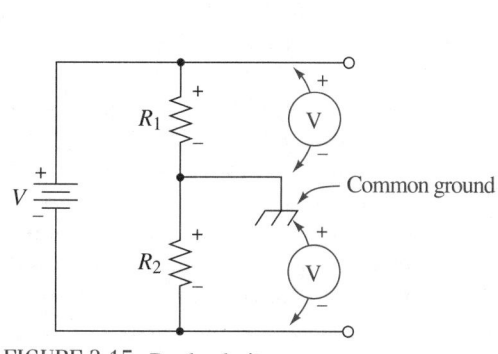

FIGURE 3.15 Dual polarity.

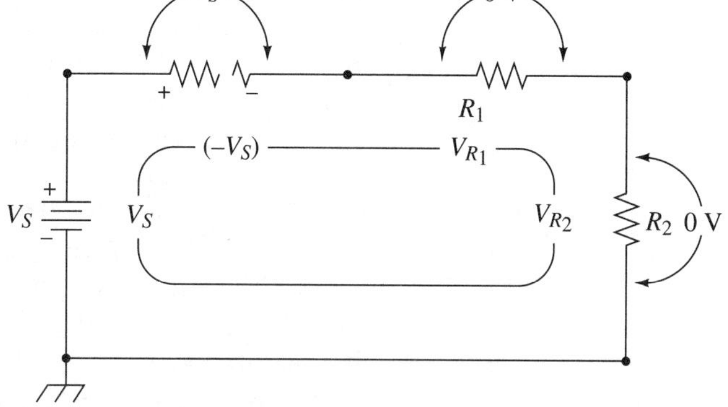

FIGURE 3.16 Open circuit.

1. When an open occurs in a circuit, the current is zero.

2. The entire source voltage shows up across the open.

3. All other circuit components will have no voltage drops. Kirchhoff's voltage law tells us that the sum of all the voltage drops around a closed loop must equal zero. Starting at point X and proceeding clockwise around the loop, we have

$$+V_S - V_S - V_{R_1} - V_{R_2} = 0$$
$$+V_S - V_S - 0 - 0 = 0$$

3.13 DETECTING SHORT CIRCUITS

The short circuit in Figure 3.17 caused an increase in current level due to lower resistance values. Regardless of the cause, a short creates a zero resistance path from one point (A) to another point (B). Therefore, the voltage from point A to point B should be zero. This loss of voltage will cause the voltage distribution throughout the circuit to change (increase). In summary,

1. Total circuit current increases because $R(T)$ decreases.
2. Voltage across the short drops to zero because $R = 0$.
3. Voltage across other components increases due to the increasing current.

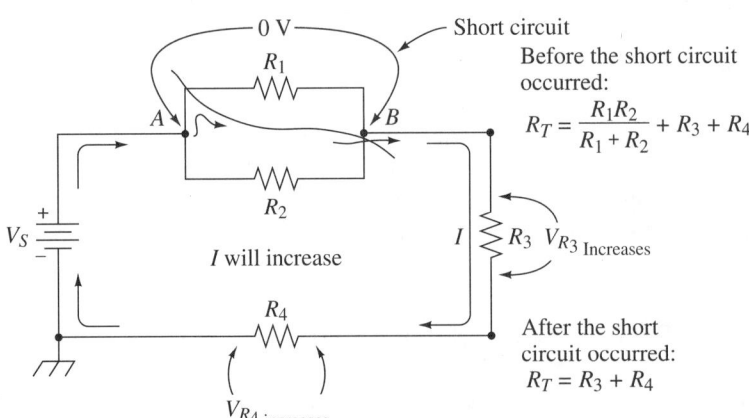

FIGURE 3.17 Short circuits.

3.14 VOLTMETER LOADING EFFECTS

Because the voltmeter draws current from the circuit under test, placing a voltmeter across a component to measure its voltage has the same effect as adding a shunt (parallel) resistor. The voltmeter becomes a load on the circuit and changes the values of I and V. Therefore, as shown in Figure 3.18, the voltmeter will give false readings. The amount of loading is determined by the internal resistance of the meter and expressed as the *sensitivity* of the meter. Lower-priced analog multimeters may have sensitivity ratings of 20,000 Ω/V, 30,000 Ω/V, or 50,000 Ω/V. Thus, if a 20,000-Ω/V meter is set on the 10-V range, the total voltmeter resistance will be 200,000 Ω and would be best used on low-resistance circuits. The true voltage V_{AB} in Figure 3.18 is 5 V.

Let's look at the readings obtained by three different meters. The data are compiled in Table 3.1. As you can see, the higher the voltmeter sensitivity is, the more accurate will be your measurement. Not all analog meters have such a low resistance. Most meters have field-effect transistors (FETs), which give the meter 10 MΩ of resistance or more, resulting in more accuracy.

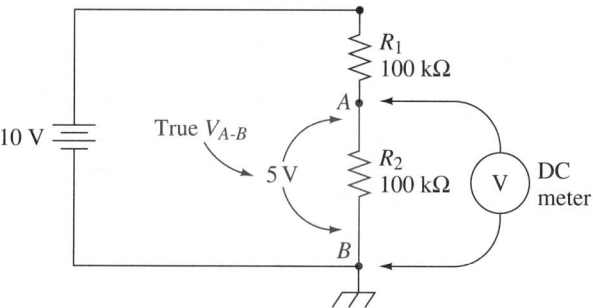

FIGURE 3.18 Voltmeter loading effects.

TABLE 3.1 Effects of Meter Sensitivity on DC
Voltage Readings

Meter	A	B	C
Sensitivity	1 kΩ/V	20 kΩ/V	100 kΩ/V
Scale setting	5 V	5 V	5 V
Meter resistance	5 kΩ	100 kΩ	500 kΩ
Current drain	0.09 mA	0.034 mA	0.0094 mA
Meter reading	0.45 V	3.33 V	4.54 V

3.15 TROUBLESHOOTING WITH A DMM

The DMM has become the most popular meter in use today. Current designs utilize operational amplifiers in their front end and, therefore, have no loading effects on the circuit at all. The DMM presents many advantages compared to a VOM:

Advantages of a DMM to a VOM

1. Very high input impedances
2. Measurement even of microvolts with a high degree of accuracy
3. Better repeatability between successive readings
4. Very accurate
5. Autoranging capabilities available
6. Easy to read

It is necessary, however, to change your DMM's battery periodically.

3.16 DC TROUBLESHOOTING WITH
AN OSCILLOSCOPE

Most oscilloscopes can also be used to measure dc voltages. Make sure the vertical amplifier section is dc coupled. The scope also allows you easily to look at the ground reference. Usually, it's best to adjust your vertical position so that the ground position is at the center of the screen. Thus positive voltages will be reflected by a scope trace above ground; similarly, negative voltages will be represented by a trace below ground. Be careful here and observe the proper ground connection to the circuit under test. The oscilloscope input circuit common is usually connected to the power-line ground. If you connect the probe ground lead to the wrong circuit location, you change the circuit ground reference and might damage the circuit. It may be necessary to connect the oscil-

Observe the proper ground connection when testing circuits.

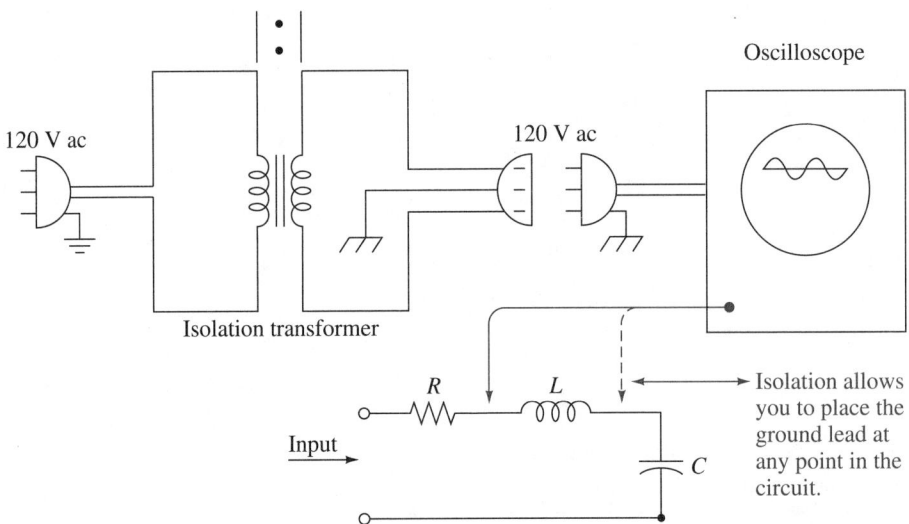

FIGURE 3.19 Using the isolation transformer.

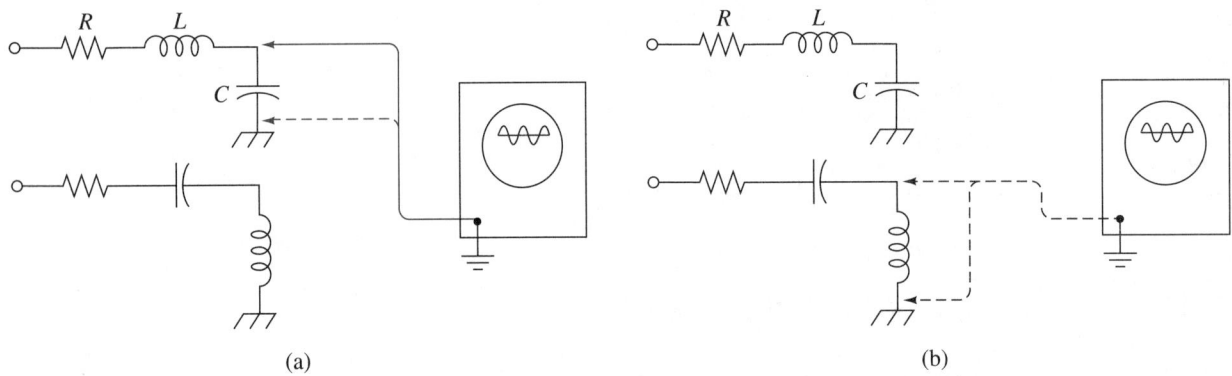

(a)   (b)

To avoid ground loops:

Step 1: Check waveform across *C*.

Step 2: Relocate component position.

Step 3: Check waveform across *L*.

FIGURE 3.20 Procedure for avoiding ground loops.

loscope to an isolation transformer first. (See Figure 3.19.) If an isolation transformer is not available, then you will have to reposition each component physically relative to the circuit ground before making your test with the oscilloscope. This procedure is shown in Figure 3.20.

Always use an appropriate probe for your measurements in order to reduce noise and hum pickup.

One advantage of using the oscilloscope for dc measurements is that any unwanted ac signals, such as 60-Hz hum, will be visible on the trace.

3.17 MAKING DC VOLTAGE MEASUREMENTS

Comparing actual voltage measurements with those found in a normal circuit

In troubleshooting a circuit the technician compares actual voltage measurements with those found in a normal circuit. To do this, the technician must visualize the schematic as it would appear to a dc signal. Study Figure 3.21. Remember, a capacitor looks like an open circuit to the dc signal and an inductor looks like a short circuit (or very small value of resistance due to the winding). After the dc circuit is drawn, voltage levels can be esti-

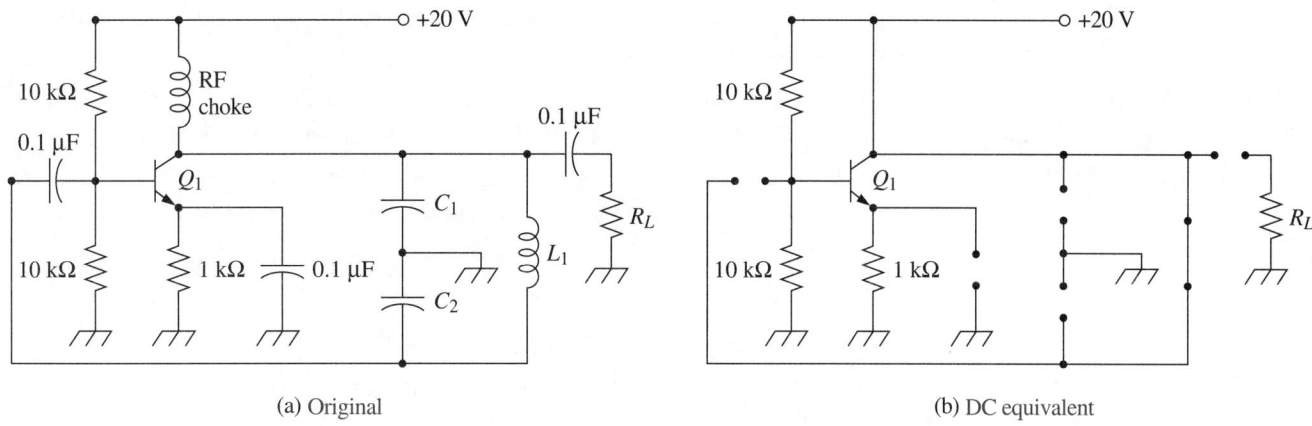

(a) Original   (b) DC equivalent

FIGURE 3.21

mated by applying simple series, parallel, and series-parallel circuit-analysis techniques.

It is very important for the technician to be able to estimate circuit voltages rapidly. This will make troubleshooting the circuit much easier. Refer again to Figure 3.21. The static voltage at point B should be +5 V dc. Remember that all resistors have tolerance factors, so a reading within $\pm X$ volts could be considered correct unless extreme precision is expected due to the use of precision resistors.

A Voltage Measurement That Is Too Low

Let's assume that the voltage level measured is much lower than you expected. Here are some of the possible reasons for the condition:

1. An improper source voltage exists.
2. The value of resistance whose voltage drop is being measured is too low.
3. The circuit is being loaded down by the meter.
4. The meter is faulty.
5. The circuit is faulty.

A Voltage Measurement That Is Too High

If, on the other hand, the voltage level measured is much higher than you expected, suspect the following:

1. The source voltage is too high.
2. The resistance is too high.
3. The source resistance is too low.
4. The meter is faulty.
5. The circuit is faulty.

Remember that voltage measurements are the most common and useful method available to the technician for circuit troubleshooting.

3.18 TROUBLESHOOTING WITH AN AMMETER

Current measurements are not routinely taken during troubleshooting. However, we need to discuss a few basic facts.

1. Refer to Figure 3.22. Remember that the ammeter must be inserted in the conductive path; therefore, it is important to observe proper polarity of circuit connections. The ammeter is a polarized instrument. This is especially important when using analog

Some basic facts about ammeters

meters. When using digital meters, a polarity reversal will be indicated by a negative sign on the indicator.

 **CAUTION**

Never place an ammeter in parallel with the circuit.

2. *Never place an ammeter in parallel with the circuit.* Because of the very low input impedance, severe damage can occur to the meter if it is not protected.

3. It is always best to start your readings on the highest range and switch to lower ranges as needed. Many DMMs have autoranging, which makes this unnecessary.

4. Always turn off the circuit power before connecting and disconnecting the ammeter.

In summary, when troubleshooting with the ammeter,

1. Turn off the power before connecting the ammeter.
2. Observe proper polarity.
3. Place the ammeter in series.
4. Start on the highest range setting first.
5. Turn off the power before disconnecting the ammeter.

Testing Open Circuits With the Ammeter

An open circuit will not allow a current to flow. A circuit branch with an open resistor will indicate a zero current level. If R_1 in Figure 3.22 is open, meter M_2 will indicate 0 A and the total circuit current indicated by meter M_1 will be lower than normal.

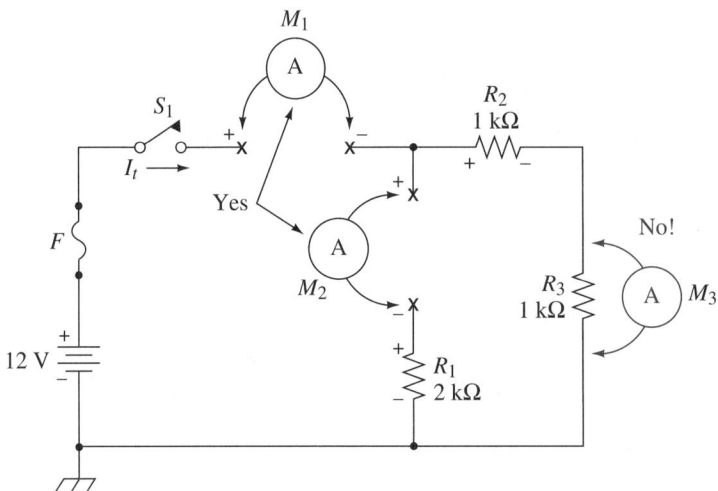

Open S_1 when inserting meters.
Close S_1 when taking readings.

$$R_T = 2\ k\Omega\ ||\ 2\ k\Omega = 1\ k\Omega$$

$$I_T = \frac{12\ V}{1\ k\Omega} = 12\ mA$$

If R_2 is shorted:
$$I_T = 18\ mA$$

FIGURE 3.22 Inserting instruments.

Testing Short Circuits With an Ammeter

A shorted component will cause the total circuit current to increase. In Figure 3.22, if R_2 were shorted, the normal circuit current of 12 mA would increase to 18 mA. At this point, it would be necessary to take voltage readings to isolate the defective component. Remember, the voltage across a short will be zero.

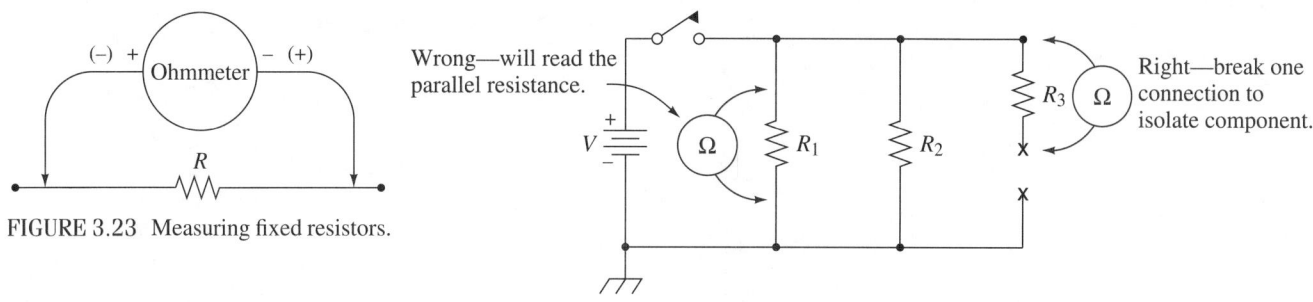

FIGURE 3.23 Measuring fixed resistors.

FIGURE 3.24 Resistance measurements in circuit.

3.19 TESTING A RESISTOR

Caution: Never perform resistance measurements in a live circuit. This procedure can damage your meter.

 CAUTION

Never perform resistance measurements in a live circuit.

Fixed Resistors

With your meter set on the ohms range, you can measure fixed resistors, as shown in Figure 3.23. Since resistors are not polarity sensitive, you do not need to worry about positive or negative terminations. If you are making in-circuit measurements, be aware of false readings due to shunt-resistance paths (which are often hidden). It is always best to take one lead of the resistor out of the circuit, if possible, as shown in Figure 3.24. Figure 3.24 illustrates the importance of isolating a parallel component from the rest of the circuit. As you can see, R_1 has not been isolated, whereas R_3 has. Therefore, meter M_1 is reading the parallel values of both R_1 and R_2, or R_1/R_2. In most cases, it is not evident which parallel paths exist.

If your measured value does not agree with the value you calculated from the color codes on the resistor, remember to check the resistor tolerance factor. The fourth band will give you this as a percentage, and you can determine the range within which the value of the resistor should fall. For example, a 1000-Ω resistor with 20% tolerance is good if its value falls between 1000 Ω ±20%, or between 800 Ω and 1200 Ω.

3.20 VOLT-AMMETER METHOD

The volt-ammeter method is a very simple method of measuring the dc resistance of a component in a dc circuit by measuring the voltage and current. Ohm's law ($V = IR$) allows you to calculate the resistance directly. Slight inaccuracies may result from the voltmeter loading effect on the circuit. See Figure 3.25.

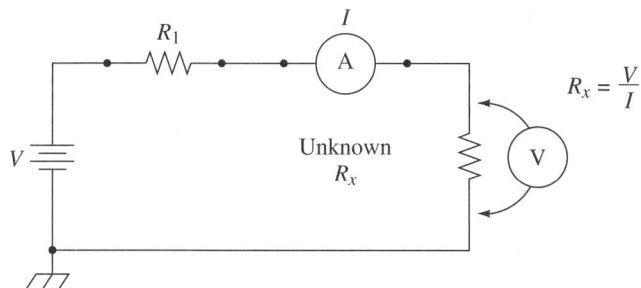

$$R_x = \frac{V}{I}$$

FIGURE 3.25 Volt-ammeter method.

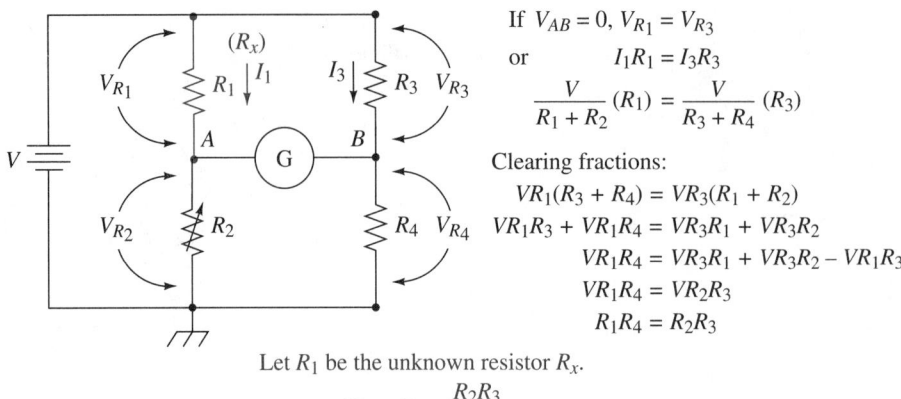

If $V_{AB} = 0$, $V_{R_1} = V_{R_3}$

or $\qquad I_1R_1 = I_3R_3$

$$\frac{V}{R_1 + R_2}(R_1) = \frac{V}{R_3 + R_4}(R_3)$$

Clearing fractions:

$VR_1(R_3 + R_4) = VR_3(R_1 + R_2)$
$VR_1R_3 + VR_1R_4 = VR_3R_1 + VR_3R_2$
$VR_1R_4 = VR_3R_1 + VR_3R_2 - VR_1R_3$
$VR_1R_4 = VR_2R_3$
$R_1R_4 = R_2R_3$

Let R_1 be the unknown resistor R_x.

Thus $R_x = \dfrac{R_2R_3}{R_4}$

FIGURE 3.26 Wheatstone bridge.

3.21 WHEATSTONE BRIDGE METHOD

The Wheatstone bridge is a parallel, precision resistor network, as shown in Figure 3.26, with a galvanometer connected as shown. If the voltage across R_1 is equal to the voltage across R_3 and the voltage across R_2 equals that across R_4, the potential difference, V_{AB}, will be zero; the circuit is then in the null, or balanced, condition. The general bridge equation is

$$R_1R_4 = R_2R_3 \tag{3.1}$$

R_3 and R_4 are known, standard values and R_2 is a known potentiometer. If we replace R_1 by the unknown resistor R_x and balance the bridge by varying R_2 until the meter indicates zero current, the value of R_2 will be equal to the value of R_x. R_x can be calculated from the equation

$$R_x = \frac{R_2R_3}{R_4} \tag{3.2}$$

3.22 TESTING POTENTIOMETERS

Testing the total resistance value of a potentiometer (variable resistor) is shown in Figure 3.27(a) and testing the variable resistance value is shown in Figure 3.27(b).

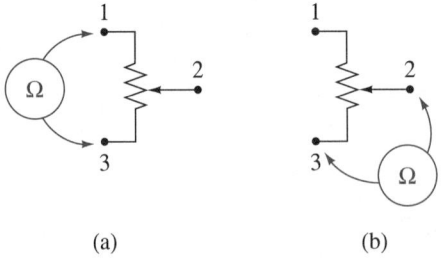

(a)　　　　　　　(b)

FIGURE 3.27 Measuring the potentiometer.

3.23 COMMON RESISTOR FAILURE MODES

Reasons for resistor failure

The most common cause of resistor failure is excessive current through it. As the level of current increases, the power dissipated from the resistor in the form of heat increases according to the following law:

$$P = I^2R \qquad (3.3)$$

where

$$R = p(L/A) + a(dt) \qquad (3.4)$$

First, the resistance value increases beyond its tolerance limits. Eventually, if the high level of current continues, the resistor burns out or becomes an open circuit. If you use your meter to check this component, the meter will indicate a reading close to the infinite ohms position. Very often variable resistors get dirty inside and become noisy when the rotor is turned.

3.24 HOW RESISTORS AFFECT DC MEASUREMENTS

You need to remember how voltage distribution changes across resistors in series and parallel. When measuring voltages referenced to ground, you must take into consideration all voltage drops. See Figure 3.28.

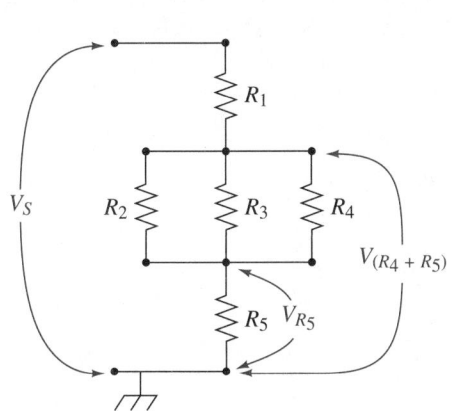

FIGURE 3.28 Checking voltage drops.

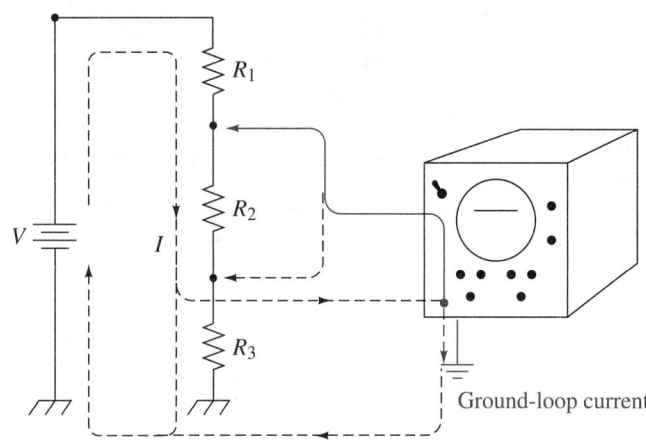

FIGURE 3.29 Oscilloscope ground loop.

Be careful when using the oscilloscope because the probe ground lead is probably connected to the power-line ground, which could damage some circuits or result in erroneous readings. See Figure 3.29.

The voltage drop across a resistor may be misleading. Resistors in parallel with other components can be open, yet show an almost normal voltage. The voltage drop across R_3 in Figure 3.30 may be misleadingly considered correct. Confusing situations often arise when resistors are in parallel with capacitors to force division of voltages, as in Figure 3.31. If C_1 has leakage of 10 MΩ, and C_2 has leakage of 40 MΩ, then C_2 will fail with 80 V across it. By placing two parallel resistors as shown, the voltage is distributed evenly and protects C_2 from failure.

 CAUTION

Be careful when using the oscilloscope.

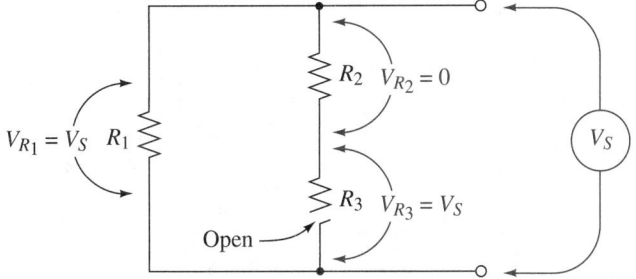

FIGURE 3.30 Misleading voltages.

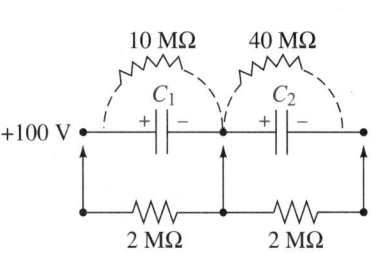

FIGURE 3.31 Confusing situations.

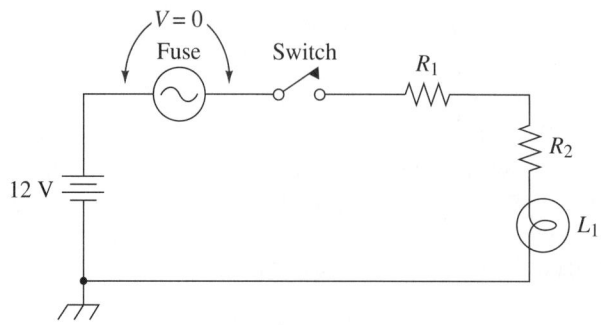

FIGURE 3.32 Series circuit with overload protection.

3.25 TROUBLESHOOTING THE SERIES RESISTANCE CIRCUIT

Troubleshooting short circuits

Figure 3.32 shows a series circuit with overload protection. The fuse has a resistance of just a few ohms and has virtually no voltage dropped across it. It is, therefore, a short circuit.

If the light L_1 does not operate when the switch S_1 is closed, the following basic troubleshooting procedures are in order:

1. Identify the problem: The light does not operate.
2. List all possible troubles:
    a. Bad light bulb
    b. Blown fuse
    c. Broken connection or conductor
    d. Open resistor
    e. Bad switch
    f. Bad power supply
3. Inspect all circuit components for broken terminals or wires.
4. Check the condition of the fuse; an ommeter should indicate 0 Ω.
5. Measure the supply voltage.
6. Check the light bulb; if it is incandescent, the ohmic value should be small.
7. Make voltage measurements.
8. Isolate and identify the bad component.

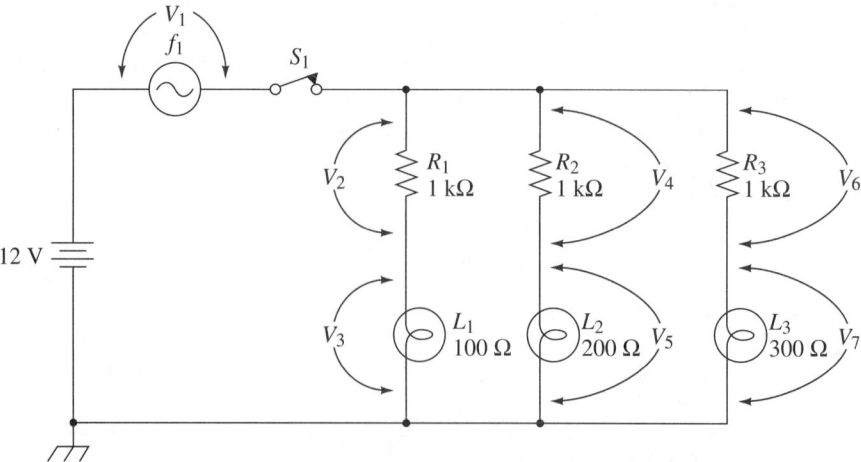

FIGURE 3.33 Troubleshooting the parallel resistive circuit.

3.26 TROUBLESHOOTING THE PARALLEL RESISTANCE CIRCUIT

The situation in Figure 3.33 is this: L_1 is too bright, L_2 is all right, and L_3 is off. Following the troubleshooting steps, we obtain the voltage measurements listed in Table 3.2. Follow these measurements and see if you agree with the conditions stated.

Troubleshooting steps in voltage measurements

TABLE 3.2 Measurements for Figure 3.26

Voltage	Should Be	Measured	Condition
V_1	0	0	O.K.
V_2	10.91	8	Wrong res.
V_3	1.09	4	Light O.K.
V_4	10	10	O.K.
V_5	2	2	O.K.
V_6	9.23	0	O.K.
V_7	2.77	12	Open L_3

3.27 TROUBLESHOOTING THE LOADED VOLTAGE DIVIDER

Short Circuits

Let's start by examining the simplest and most common voltage-divider circuit shown in Figure 3.34. The symptoms of trouble are as follows:

1. R_1 is unusually warm.
2. The light L_1 is off.

Troubleshooting short circuits and open circuits

These symptoms suggest the presence of a short-circuit condition. Excessive current levels increase the temperature of R_1. Our assumption is that either R_2 is shorted or L_1 is shorted. Since these components do not normally short internally, such conditions would have to be caused externally. A quick voltage measurement across R_1 indicates a drop of 18 V. We also measure 0 V across R_2, which reinforces our short-circuit theory. Disconnect L_1 and remeasure V_{R_2}. If V_{R_2} is still 0 V, then R_2 must be shorted externally in some way. Visually inspect the circuit board and components surrounding the terminals of R_2.

On the other hand, if V_{R_2} increased to 7.36 V when L_1 was removed, we have to check the connections to the light for short circuits.

Now, let's change the symptoms as follows:

1. L_1 is brighter than normal.
2. Supply voltage is normal level.

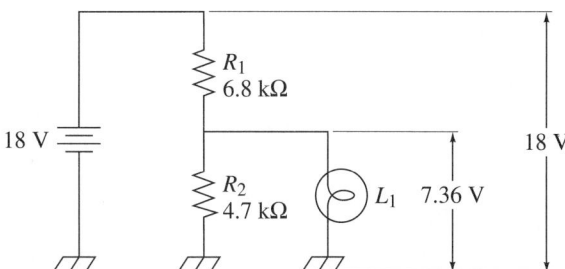

FIGURE 3.34 Troubleshooting the loaded voltage divider.

Our voltage measurements show that the total supply voltage is dropped across R_2, with 0 V across R_1. The indications are that R_1 is shorted; resistance measurements will prove this.

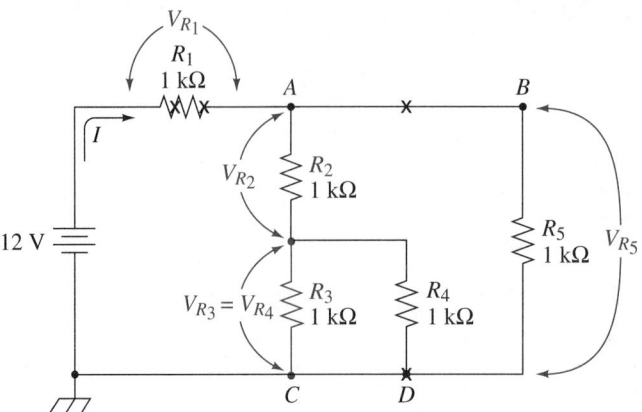

FIGURE 3.35 Open-circuit condition.

Open Circuits

The next example, Figure 3.35, contains an open circuit somewhere. The open circuit might be a broken resistor, broken or loose connection, or even a broken conductor. Refer to Figure 3.36 and Table 3.3. If R_1 were opened, V_{R_1} would equal the source voltage, or 12 V dc. All other voltage drops would be zero, as shown in Figure 3.36(a). If R_2 were

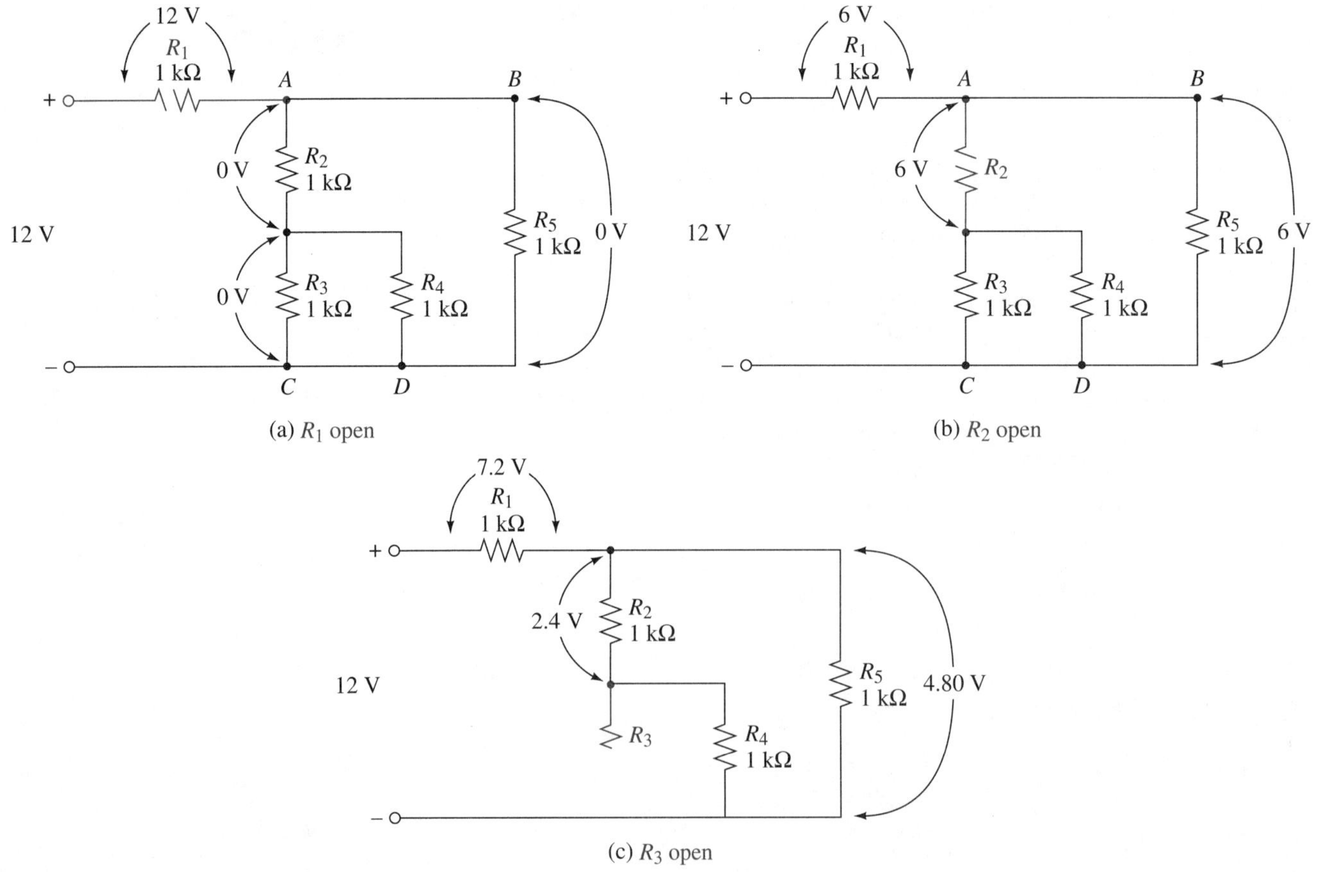

(a) R_1 open

(b) R_2 open

(c) R_3 open

FIGURE 3.36 Open-circuit effects.

opened, V_{R_2}, V_{AC}, and V_{BC} would all be equal. V_{R_3} and V_{R_4} would be zero. See Figure 3.36(b). With R_3 open (Figure 3.36(c)), all circuit voltage drops are abnormal. First, measure all voltage drops and compare them to normal readings. V_{R_1} is lower than normal, suggesting an open circuit, decreasing the circuit current. If R_2 were opened, V_{R_3} and V_{R_4} would be zero. So R_2 is all right. This leaves R_3, R_4, and R_5 as suspects. Quick calculations show that V_{R_1} equals V_{R_2} when R_5 is opened. Thus the problem is in R_3 or R_5. A resistance check is used to isolate R_3.

An open circuit is often the result of a poor solder joint or mechanical connection. Let's assume a loose connection at point D. Voltage measurements show $V_{R_4} = 0$, whereas $V_{R_3} = 2$ V. If R_4 were opened, we would measure $V_{R_4} = 2$ V. This points to a loose connection at either points D or E. An ohmic test will show continuity (or $0\ \Omega$) at these connections if they are all right but a high resistance (possibly infinite) if they are broken. Also, a voltage will be dropped across the junction when it is broken.

Other common causes of open circuits are a broken conductor, broken PC board land (trace), and cracked foils. These conditions are very often difficult to locate. Invariably, a broken wire is hidden inside a bundle or insulation and, therefore, is not visible. The technician can very easily overlook this problem.

Let's assume that conductor A-B is broken at the $\times$ (Figure 3.35). From Table 3.3 the voltmeter shows $V_{R_5} = 0$ and voltage across R_2, R_3, and R_4. If R_5 were opened, V_{R_5} would not equal 0, and if R_5 were shorted, $V_{R_2} = V_{R_3} = V_{R_4} = 0$. Therefore, the problem has to be in the conductor.

TABLE 3.3 Open-Circuit Effects

Voltage	Normal Values (V)	R_1 Open	R_2 Open	R_3 Open	R_4 Open	R_5 Open	Broken Terminal at D	A–B Broken
V_{R_1}	7.5	12	6	7.2	7.2	4.8	8	4.8
V_{R_2}	3.0	0	6	2.4	2.4	4.8	2	4.8
V_{R_3}	1.5	0	0	2.4	2.4	2.4	2	2.4
V_{R_4}	1.5	0	0	2.4	2.4	2.4	0	2.4
V_{R_5}	4.5	0	6	4.8	4.8	7.2	7.2	0

3.28 CRITICAL PARAMETERS

When selecting a resistor, you must determine the following:

1. Type (fixed, variable, temperature-sensitive, etc.)
2. Resistance value in ohms, kilohms, or megohms
3. Wattage ratings ($\frac{1}{8}$, $\frac{1}{2}$, 1, 2 W, etc.)

3.29 FUSIBLE LINKS

The only component that is good if it is a short circuit is the fusible link. A good fuse measures about $0\ \Omega$. A blown fuse measures infinite ohms across its terminals.

3.30 TROUBLESHOOTING THE TRANSFORMER

The primary and secondary windings of transformers have internal resistance due to their physical composition. This dc resistance is what an ohmmeter will indicate. Transformer leads are identified by their color codes. Power transformers usually have black or black

Testing transformers for faults

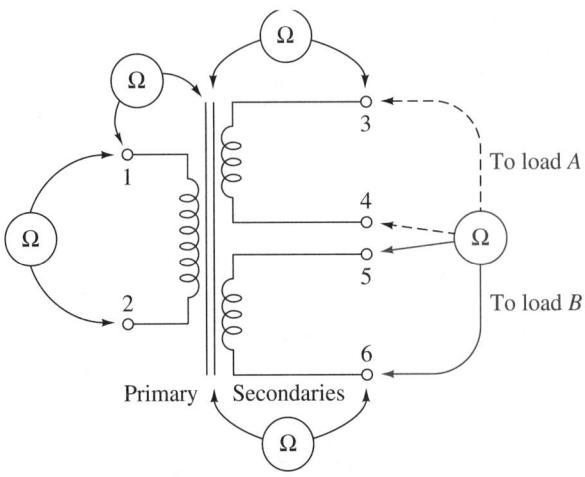

FIGURE 3.37 Transformer resistance measurements.

and white primary leads and red, green, or yellow secondary leads. Remember that the dc resistance of the wires is usually very low, so you need to use the R × 1 range setting on your meter and connect to the circuit as shown in Figure 3.37. Once you have identified the leads,

1. Test each winding for continuity.
2. Windings showing infinite resistance indicate a defective transformer.
3. Check for continuity between individual windings — there should not be any.
4. Check for continuity between the windings and the core — there should not be any.

If the main fuse has blown,

5. Turn off all power and disconnect all secondary leads.
6. Apply power to the primary. If the fuse blows, replace the transformer.
7. If the fuse didn't blow, reconnect the secondary loads one at a time until the fuse blows. This will identify the faulty stage.

3.31   TRANSFORMER FAILURE MODES

If a component in the secondary circuits fails or becomes shorted and demands an excessive level of current from the transformer windings, the heat generated can be sufficient enough to melt wire insulation, fuse the windings together, and cause a short circuit or cause breaks in the coil windings, resulting in open circuits. It is more common for a transformer to burn up than to develop open circuits.

3.32   TROUBLESHOOTING CAPACITORS IN DC CIRCUITS

Charge in a capacitor

Let's assume that a capacitor is placed in a dc circuit as shown in Figure 3.38 and switch S_1 is closed. Electrons will move from the negative terminal of the battery and distribute themselves within the surface of the dielectric at the end of the capacitor, which then becomes the negative end. (The electrons do not accumulate on the plate itself but are actually in the dielectric.) Because of the electrostatic induction, the accumulating negative charge at B induces an equal but opposite (+) charge at surface A, thereby creating the positive end of the capacitor. Initially, a large number of electrons accumulate at the negative end of the capacitor, with a corresponding positive charge at the positive end. The ability to store an electric charge is called the *capacitance* of the capacitor. The factors contributing are the following:

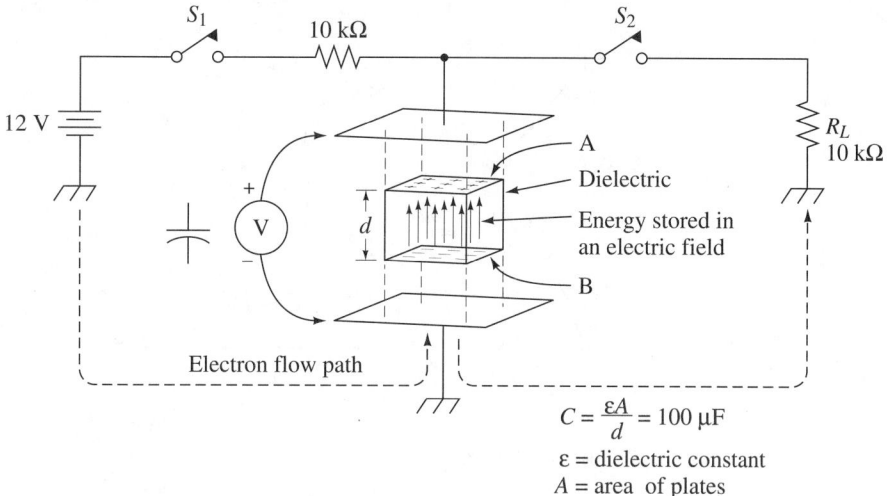

FIGURE 3.38 Capacitor in a dc circuit.

$$C = \frac{\varepsilon A}{d} = 100 \ \mu F$$

ε = dielectric constant
A = area of plates

1. Area of the plates
2. Absolute permittivity of the dielectric
3. Thickness of the dielectric

The capacitance is given by the equation

$$C = \frac{eA}{d} \qquad (3.5)$$

where C is in farads, microfarads, etc., e is the permittivity in farads per meter, A is in meters squared, and d is in meters.

Although no electrons actually move through the dielectric, there is an apparently large current flow through the circuit. A voltage (potential) difference is developed across the plates of the capacitor, producing an electric field. The capacitor is said to be charging. As more electrons move into the surface of the dielectric, the surface fills and the rate of charge decreases to zero, at which time the capacitor is fully charged and acts like a battery. The amount of charge is determined by

$$Q = CV$$

where Q is the charge in couloumbs, C is the capacitance in farads, and V is the voltage across the capacitor.

3.33 CAPACITOR TIME CONSTANTS

How long does this process take? The circuit resistance limits the current flow (rate of charge) and the size of the capacitor determines the amount of charge (Q) needed to become fully charged. Thus the length of time required is determined by both R and C. In fact, the product of R and C determines a time period known as the time constant (symbolized by the Greek letter τ (tau)), during which the capacitor charges to 63% of its final value. Thus

$$1\tau = RC \qquad (3.6)$$

After a time period equal to 5τ, the capacitor is considered to be fully charged. In Figure 3.39, $1\tau = RC = 1$ s. The charge curve is shown in Figure 3.40.

At time t_0, as the capacitor begins to charge, Kirchhoff's voltage law tells us that the entire source voltage is across the resistor. The circuit current is maximum and is equal to

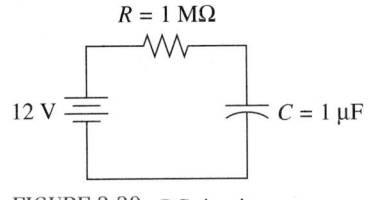

FIGURE 3.39 *RC* circuit.

$1\tau = RC$
$1\tau = (1 \times 10^{+6})(1 \times 10^{-6})$
$1\tau = 1$ s

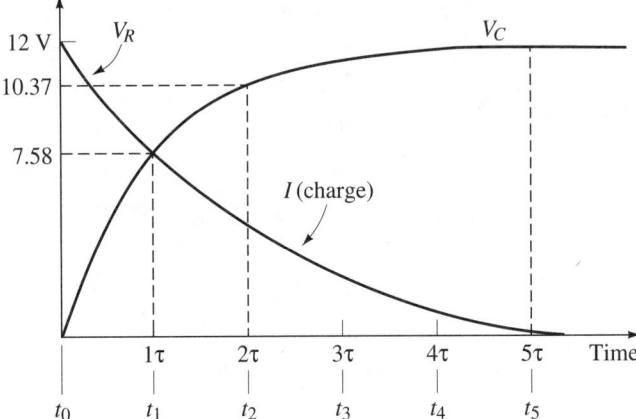

FIGURE 3.40 *RC* charge curves.

1.2 mA. As the capacitor charges, the voltage redistributes across the capacitor until the capacitor is fully charged and the entire source voltage is across its plates, with 0 V across the resistor. The charge current also drops to zero. At the end of the first time constant, V_C increases to 63% of 12 V, or 7.56 V. This leaves 4.44 V across the resistor. The circuit current decreases by 63% to 444 µA.

If we now open switch S_1, the capacitor will retain its charge and act like a 12-V battery. Theoretically, the capacitor will remain charged indefinitely. Practically, the capacitor will discharge in time due to leakage (but this may be a very long time). *Remember this when troubleshooting: A charged capacitor can produce a strong shock.* You can rapidly discharge the capacitor by closing switch S_2. If the capacitor is discharged through the 10-kΩ resistor, the discharge time will equal the charge time. The positive and negative charges will redistribute until the current decreases to zero, as shown in Figure 3.41.

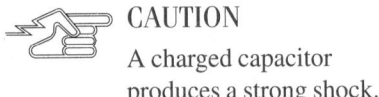

CAUTION

A charged capacitor produces a strong shock.

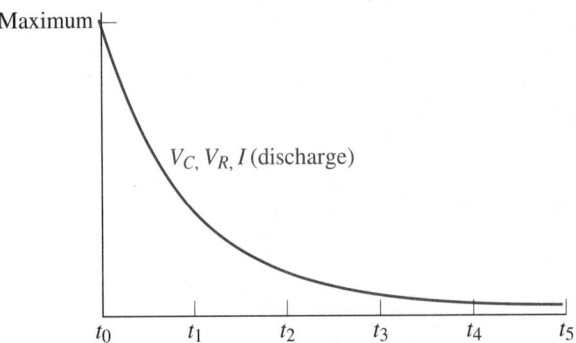

FIGURE 3.41 *RC* discharge curves.

3.34 CAPACITOR LEAKAGE MEASUREMENTS

Theoretically, an ideal capacitor, fully charged to its rated operating dc voltage, will have zero current through it. Practically, there is always a small value of leakage current. See Figure 3.42. The capacitor is said to have leakage resistance and the value of leakage resistance can vary significantly depending on the capacitance type. For example,

1. Mica, ceramic, paper, and glass capacitors have a typical leakage resistance in excess of thousands of megohms.
2. Electrolytic capacitors may have leakage resistance values less than 1 MΩ.

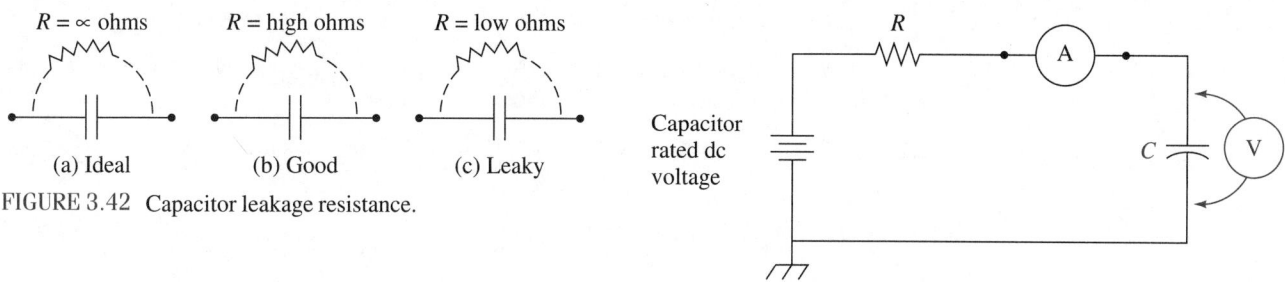

FIGURE 3.42 Capacitor leakage resistance.

FIGURE 3.43 Leakage resistance test circuit.

Excessive leakage will cause a capacitance error. A simple circuit for measuring the leakage resistance of a capacitor is shown in Figure 3.43.

3.35 TESTING FIXED CAPACITORS

Ohmmeter Tests

The following test provides a quick method of checking the health of electrolytic capacitors with ratings of 1 μF and above.

1. Turn off the power.

2. Discharge the capacitor by shorting the leads through a resistor, such as one with 100 Ω. Some capacitors can be damaged by dead-shorting techniques.

3. Remove one lead of the capacitor from the circuit.

4. Connect the ohmmeter leads across the capacitor observing correct polarity — positive lead to the positive terminal and negative lead to the negative terminal, as shown in Figure 3.44.

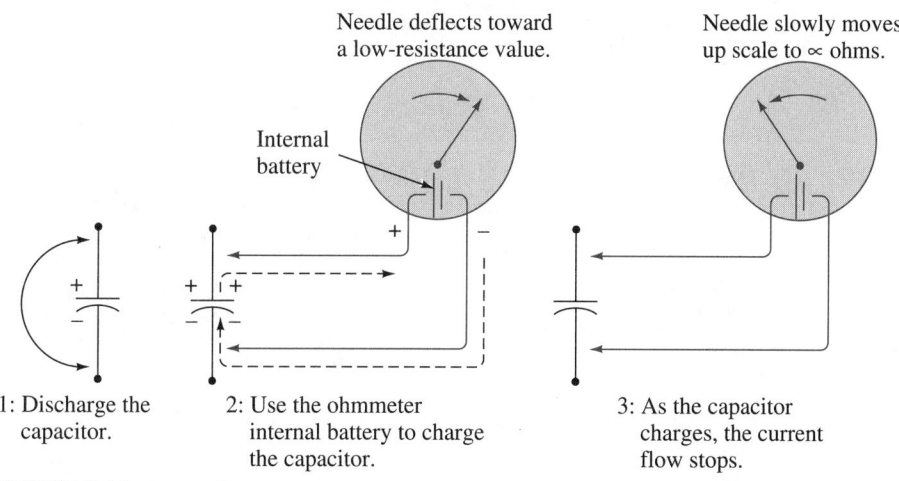

FIGURE 3.44 Testing fixed capacitors.

5. Set the ohmmeter range to R × 1K.

6. The meter should indicate a deflection from infinity toward 0 Ω as the initial surge of current starts to charge the capacitor. As the capacitor charges, the needle should move back up scale toward the infinite ohms position.

7. Nonpolarized capacitors with ratings less than 1 μF charge so fast that the preceding steps should show only a high resistance value for a good capacitor. Of course, you do not have to observe polarities when testing these small capacitors.

8. If there is no deflection of the meter, the capacitor is probably open.

9. If there is a deflection toward a low value of resistance and the meter does not return to a high-resistance reading, the capacitor is shorted.

10. If the meter deflects and returns partially, the capacitor is leaky.

3.36   LIVE VOLTAGE TESTS

A capacitor can be checked while in a live circuit by using a voltmeter as follows:

1. Break the ground (common) lead and insert the voltmeter as shown in Figure 3.45. If the capacitor is good, the meter will move to 12 V momentarily while the capacitor is charging and then back toward 0 V as the capacitor becomes charged.

2. If the capacitor is open, the voltmeter will show no indication.

3. A shorted capacitor will give a full 12-V reading on the meter.

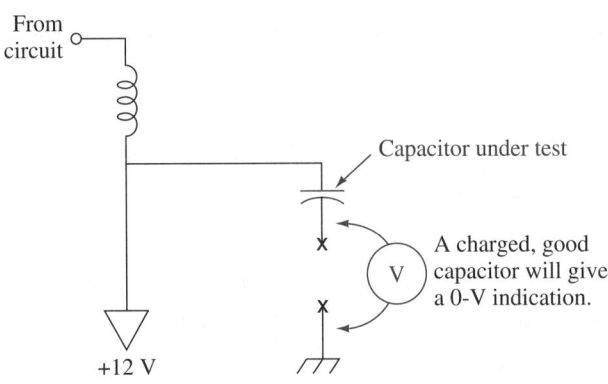

FIGURE 3.45 Capacitor live circuit test.

3.37   CAPACITANCE MEASUREMENTS

A capacitance checker can be used to verify the capacitance value. Some checkers also indicate when a capacitance is shorted, open, or leaky.

The Shering bridge circuit in Figure 3.46 can also be used to calculate the value of an unknown capacitor.

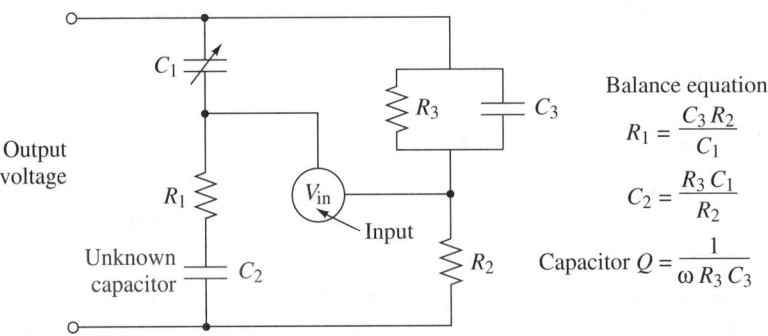

FIGURE 3.46 The Shering bridge.

3.38 THE SPARK TEST

Exercise caution with the spark test, and do it only with low-voltage applications. If the capacitor is good and able to hold a charge, a shorting stick will produce a spark. Absence of a spark indicates a poor capacitor.

CAUTION

Only perform a spark test with low-voltage applications.

3.39 BRIDGING METHOD

When you suspect a capacitor in a circuit of being faulty, a quick test is to take another of equal capacitance and shunt the suspected one. Any change in circuit performance indicates correct thinking.

Make sure that you use a capacitor with a breakdown voltage rating equal to or greater than the one in the circuit.

Only use a capacitor with a breakdown voltage rating equal to or greater than the one in the circuit.

3.40 SUBSTITUTION METHOD

The substitution method involves simply replacing the suspected capacitor with a good one while observing circuit performance.

3.41 CAPACITOR FAILURE MODES

1. *Shorted.* Leakage resistance is close to 0 Ω and the capacitor will not charge.

2. *Open.* Resistance is close to infinity, but the capacitor will not charge.

3. *Leaky.* The capacitor cannot hold a charge due to high leakage resistance. Polarized capacitors tend to be more leaky than nonpolarized types.

Reasons why capacitors fail

In capacitors other than electrolytics and tantalum, the failure rate is very low. Paper, mica, ceramic, and plastic can be assumed to be good unless the ohmmeter test shows zero resistance.

3.42 CRITICAL PARAMETERS

When replacing a capacitor you must consider the following:

1. Type: polarized, nonpolarized, mica, polyester, etc.
2. Capacitance value in farads, microfarads, or picofarads
3. Working voltage rating in volts

3.43 INDUCTORS IN DC CIRCUITS

In an inductor, when the operating frequency is dc (or zero), the only opposition to current flow is the wire resistance of the inductor's coil. The coil's wire resistance (known as the dc resistance of the coil) is usually very low and is insignificant in most cases. Coils are therefore considered to be short circuits to dc current. If we divide the inductance value (in henries) by the dc resistance (in ohms), the result is a characteristic known as the Q of the coil.

When inductors are used in circuits with dc voltages, a stable magnetic field is established surrounding the coil turns. A significant amount of potential energy becomes stored within this field. This can create unexpected problems. Consider the circuit of Figure 3.47. When switch S is closed, the current in the circuit (I) reaches its maximum

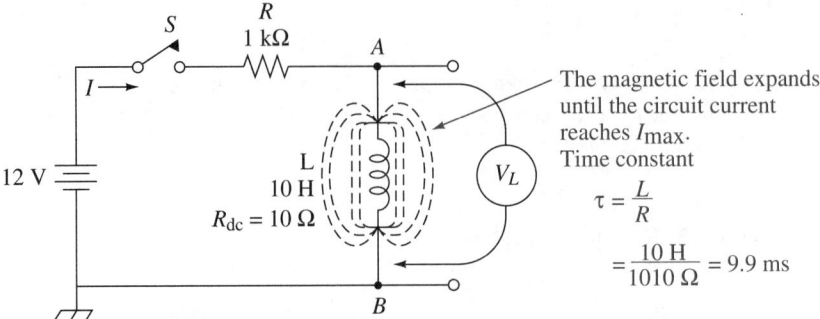

FIGURE 3.47 Inductors in dc circuits.

value after 5 time constants, during which time a counterelectromotive force is developed whose polarity opposes the 12-V supply.

A coil does not like change and initially blocks the flow of current. At this time the voltage across the coil is maximum and is equal to the supply voltage, or 12 V. After 5 time constants, maximum current flows and the coil voltage approaches zero, with most of the supply dropped across the resistor, R. See Figure 3.48.

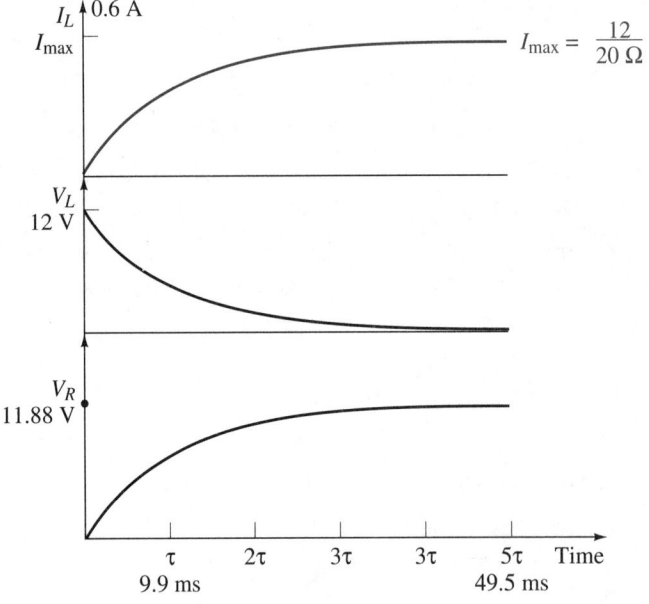

FIGURE 3.48 *RL* circuit waveforms.

If the switch is opened, the circuit current (I) is interrupted. However, remember the inductor does not like change, and the collapsing magnetic field (whose lines of force are now cutting the coil turns and inducing additional current flow) attempts to maintain circuit current across the gap. This results in an extremely large voltage spike across the gap, witnessed by a spark across the switch terminals. This spike, commonly referred to as a voltage transient, is known as the CEMF (counterelectromotive force), or inductor kickback voltage. The arcing across the switch terminals can eventually cause switch failure or pitted terminals, whereas the voltage spike can cause false triggering of digital circuits or even destruction of solid-state components.

A well-designed inductive circuit will have protection like that provided by the diode in Figure 3.49.

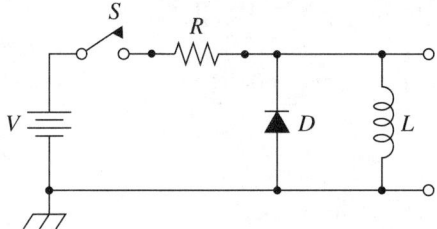

FIGURE 3.49 Inductive circuit with protection.

When the switch is closed, the diode looks like an open circuit. However, when the switch is opened, the CEMF forward-biases the diode and the voltage spike is eliminated. Inductors are extremely useful in electromechanical dc relays.

3.44 FAILURE MODES OF COILS

Basically only two conditions cause failure in a coil:

1. The coil wire can break, causing an open circuit.
2. The windings can short together.

3.45 TROUBLESHOOTING THE COIL

To troubleshoot an opened coil, make a continuity measurement with your ohmmeter. The dc resistance is not really important, but continuity is. It is more difficult to troubleshoot a shorted coil. Proceed as follows:

Troubleshooting a shorted coil

1. Measure the dc resistance with your ohmmeter and compare with any available data.
2. Conduct a visual inspection, looking for discoloration or any evidence of overheating.
3. Check for a burned odor.
4. Listen for any crackling sounds when power is applied.
5. Check by substitution of another coil.

3.46 MEASURING INDUCTANCE

A typical method of measuring inductance is by using the Maxwell bridge circuit shown in Figure 3.50.

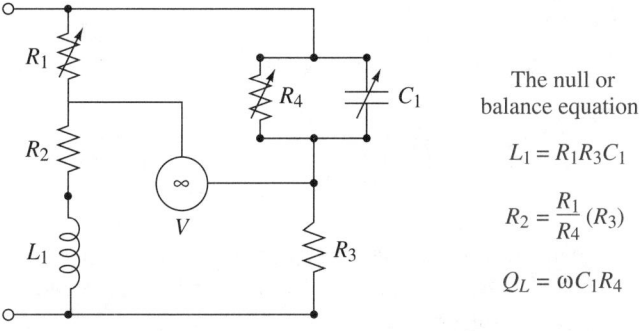

The null or balance equation:

$$L_1 = R_1 R_3 C_1$$

$$R_2 = \frac{R_1}{R_4}(R_3)$$

$$Q_L = \omega C_1 R_4$$

FIGURE 3.50 The Maxwell bridge for measuring induction.

3.47 TROUBLESHOOTING THE DC-*RC* CIRCUIT

The behavior of capacitors is different in series than in parallel circuits.

It is important to remember how capacitors act in series and parallel circuits.

Series

The total capacitance of *n* capacitors in series is found by the equation:

$$C_T = \frac{1}{\sum \frac{1}{C_n}}$$

or

$$\frac{1}{C_T} = \frac{1}{C_1} + \frac{1}{C_2} + \cdots + \frac{1}{C_n} \tag{3.7}$$

Therefore, capacitors in series behave like resistors in parallel, and the total capacitance is less than that of the smallest value of capacitance.

Example 1

The circuit in Figure 3.51 has four capacitors in series. Find the total capacitance.

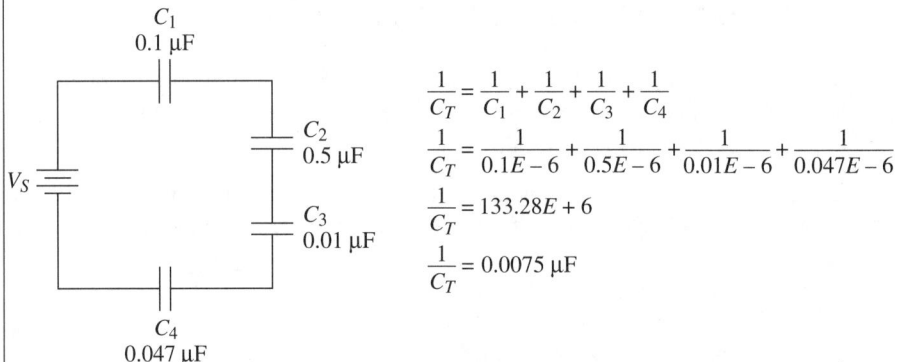

$$\frac{1}{C_T} = \frac{1}{C_1} + \frac{1}{C_2} + \frac{1}{C_3} + \frac{1}{C_4}$$

$$\frac{1}{C_T} = \frac{1}{0.1E-6} + \frac{1}{0.5E-6} + \frac{1}{0.01E-6} + \frac{1}{0.047E-6}$$

$$\frac{1}{C_T} = 133.28E+6$$

$$\frac{1}{C_T} = 0.0075 \ \mu F$$

FIGURE 3.51 Series capacitors (Example 1).

In the series circuit the source voltage, V_S, is distributed throughout the circuit so that the lowest value of capacitance receives the greatest proportion of the total voltage. The current flows in the circuit during the charging process.

Solution: We know that the dc current is equal to the rate of change of electrons (*Q* or charge), or

$$I = \frac{dQ}{dt} \tag{3.8}$$

Solving for the total charge gives

$$dQ = I dt$$

Using integration, we find that

$$Q = \int_o^t \frac{I}{dt} \qquad \text{← wrong} \qquad Q \int_o^T I \cdot d\tau \tag{3.9}$$

which means that the charge available is equal to the total current accumulated over a period of time (*t*). Since this charge is fixed and since $Q = CV$, or $V = Q/C$, the smaller value of *C* receives a larger value of voltage.

Example 2

Three capacitors are connected in series across a 12-V source, as shown in Figure 3.52. Calculate the total capacitance, total circuit charge, and all voltage drops.

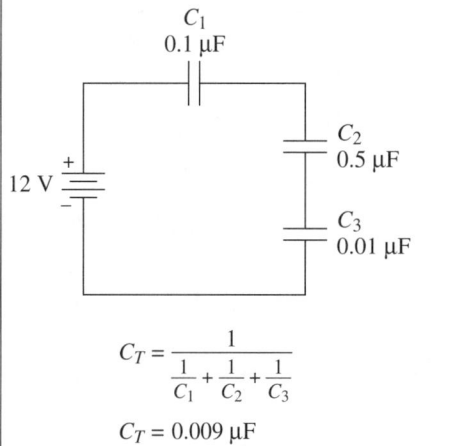

Total circuit charge

$$Q = C_T V$$
$$Q = (0.009^{-6})(12) = 0.108^{-6} \text{ C}$$

Since $V = \dfrac{Q}{C}$,

$$V_{C_1} = \frac{0.108^{-6}}{0.100^{-6}} = 1.08 \text{ V}$$

$$V_{C_2} = \frac{0.108^{-6}}{0.500^{-6}} = 0.216 \text{ V}$$

$$V_{C_3} = \frac{0.108^{-6}}{0.010^{-6}} = 10.8 \text{ V}$$

$$V_T = V_{C_1} + V_{C_2} + V_{C_3} = 12.096 \text{ V}$$

$$C_T = \cfrac{1}{\dfrac{1}{C_1} + \dfrac{1}{C_2} + \dfrac{1}{C_3}}$$

$$C_T = 0.009 \text{ } \mu F$$

FIGURE 3.52   Examples 2 and 3.

Solution: The voltage-divider rule applies here also. In the cirucit of Figure 3.52, $Q_T = Q_1 = Q_2 = Q_3$.

$$Q_T = V_T C_T = V_1 C_1 = V_2 C_2 = V_3 C_3.$$

So,

$$V_T C_T = I = V_2 C_2$$

and

$$V_2 = \frac{V_T C_T}{C_2}$$

In general,

$$V_x = \frac{V_T C_T}{C_x}$$

Equation (3.10) can be very helpful when estimating voltage drops in a capacitive circuit.

Example 3

Refer to Figure 3.52 and use the voltage-divider rule to calculate the voltage drop across capacitor C_3.

If the switch in Figure 3.53 is opened, the light will remain on as capacitors C_2 and C_3 discharge.

Problem 1: The light does not function but tests all right; S_1 is closed and V_S is all right.
Solution: From Table 3.4, it is clear that there are numerous possible causes. In fact the only ones that are not probable causes are C_1 shorted and R_1 shorted. This means that you must make all voltage measurements in order to identify the cause.

Problem 2: The light glowed very bright and then went out.
Solution: Here we may have a multiple problem. We test the light and find that it is open. So far, so good, but why did it go out? Did the light bulb filament fail from fatigue or from excessive current? If the latter is true, then we have a second problem. Look at Table 3.4 again. The only condition that causes a higher-than-normal voltage for V_2 is if C_1 is shorted.

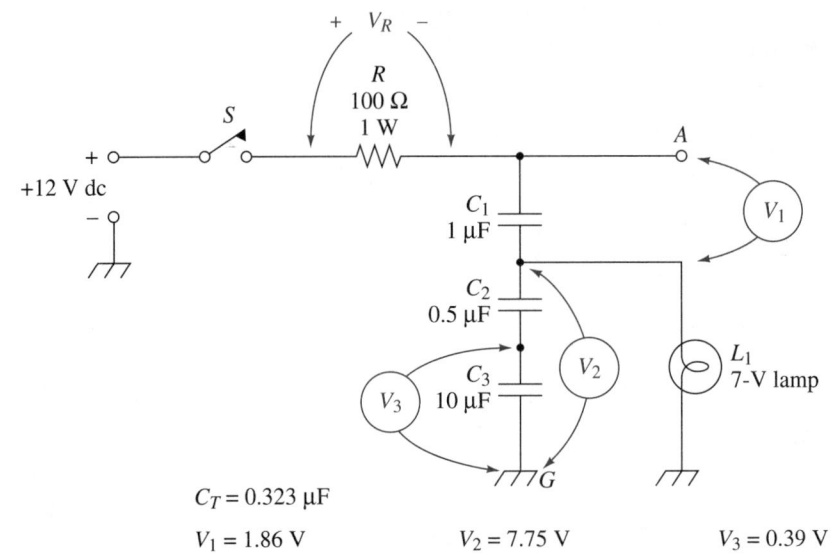

$$C_T = 0.323\ \mu F$$

$$V_1 = 1.86\ V \qquad V_2 = 7.75\ V \qquad V_3 = 0.39\ V$$

FIGURE 3.53 Problem 1.

TABLE 3.4

V_R	V_1	V_2	V_3	V_{AG}	Cause
0	1.86	7.75	0.39	12	Normal
0	12	0	0	12	C_1 open
0	0	11.43	.57	12	C_1 short
0	12	0	0	12	C_2 open
0	10.92	1.08	1.08	12	C_2 short
0	9.96	2.04	0	12	C_3 short
0	12	0	0	12	C_3 open
12	0	0	0	0	A–G short
0	12	0	0	12	B–G short
0	1.86	7.75	0.39	12	L_1 open
0	1.86	7.75	0.39	12	R_1 short
12	0	0	0	0	R_1 open

Problem 3: The resistor burned open.
Solution: The first question you should ask is Why? The answer has to be excessive current and, therefore, excessive voltage drop across R. Check the table. There is only one condition that causes this, a short from A to G. It's possible that all three capacitors shorted, but it is not likely. Look for a broken wire, solder bridge, etc., that caused the condition.

Parallel

Capacitors in parallel tend to act like an equivalent capacitance with a larger plate area and behave like series resistors according to the formula

$$C_T = C_1 + C_2 + \cdots + C_n \tag{3.11}$$

In Figure 3.54, each capacitor has equal voltage across it. Since $Q = CV$, the amount of

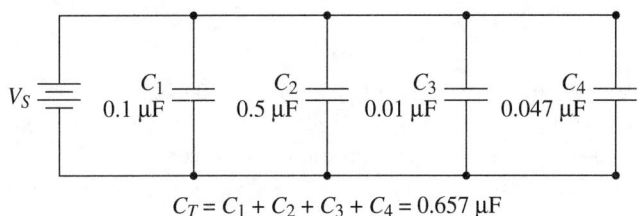

$$C_T = C_1 + C_2 + C_3 + C_4 = 0.657\ \mu F$$

FIGURE 3.54 Parallel capacitors.

charge varies for each path. If any one capacitor becomes shorted, the voltage source will discharge and there will be no voltage dropped across the good capacitors.

3.48 TROUBLESHOOTING AN *RLC* CIRCUIT

Study the circuit in Figure 3.55 and the corresponding voltage curves shown in Figure 3.56. Assume that switch S_1 has been closed for a long time, $T \gg 5\tau_c$. Table 3.5 lists various sets of measurements and the possible faults.

We can see that if the light goes out but the bulb tests good, the problem could be

Finding a solution when the light goes out but the bulb tests good

1. Open resistor
2. Open inductor
3. Shorted capacitor

An open resistor is usually visible, but you should perform an ohmic continuity test even if the resistor looks good. A few quick checks should isolate the faulty component.

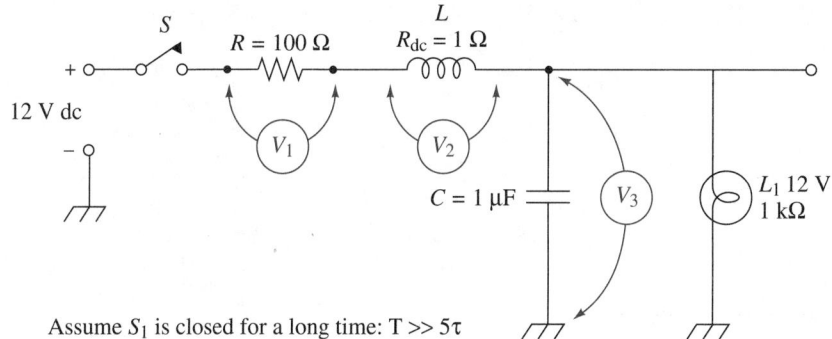

Assume S_1 is closed for a long time: $T \gg 5\tau$

FIGURE 3.55 *RLC* circuit.

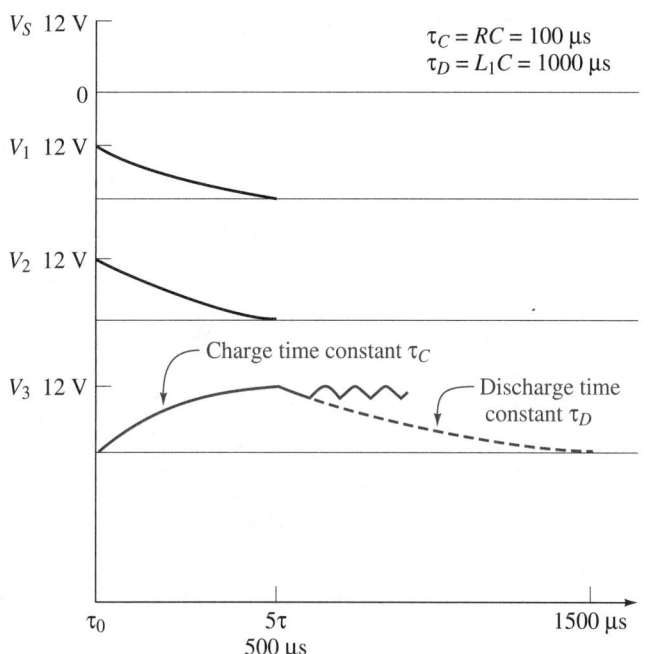

$$\tau_C = RC = 100\ \mu s$$
$$\tau_D = L_1 C = 1000\ \mu s$$

FIGURE 3.56 Voltage curves.

TABLE 3.5

V_1	V_2	V_3	L_1	Condition
0	0	12	On	Normal
0	0	12	On	L shorted
0	12	0	Off	L open
12	0	0	Off	C shorted
1.1	0	10.9	Dim	C open
0	0	12	On	R shorted
12	0	0	Off	R open
12	0	0	Off	L_1 shorted
0	0	12	Off	L_1 open

TABLE 3.6

Capacitance	Leakage Resistance
10 μF	R =
100 μF	R =

3.49   TESTING CAPACITORS

For this procedure, select two electrolytic capacitors: 100 μF and 10 μF. First short the capacitor terminals to discharge the capacitor. Set your VOM on the R × 10K scale. Connect the positive lead to the positive terminal of the capacitor and the negative lead to the negative terminal of the capacitor. You should observe a needle kick upscale toward a low resistance value and then gradually move back toward the infinite ohms end as the capacitor begins to charge to the value of the meter's internal battery. When the needle stops moving, read the ohmic value and record it in Table 3.6 as the value of leakage resistance.

3.50   TESTING THE TRANSFORMER

Select a transformer as follows: 120-V primary and 12.6-V secondary. With no voltage applied to the primary, measure the resistance of both windings and enter the values in Table 3.7. Also measure the resistance from secondary back to the primary and the resistance from *A* to the transformer case (see Figure 3.57) and enter the values in Table 3.7.

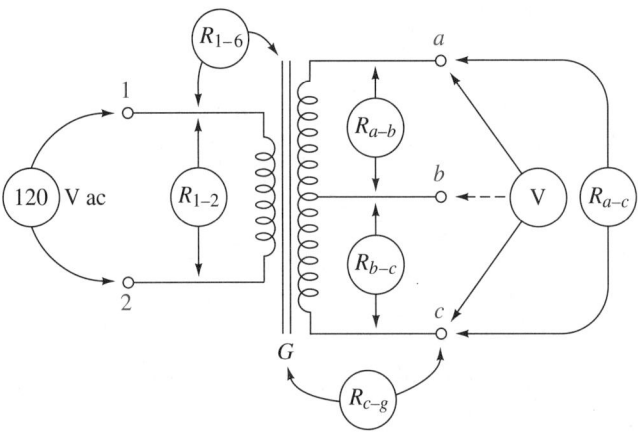

FIGURE 3.57   Transformer testing.

TABLE 3.7

Static Tests	
Resistance of primary winding:	$R_{1-2}=$
Resistance between primary and case:	$R_{1-6}=$
Resistance of secondary:	$R_{ac}=$
	$R_{ab}=$
	$R_{bc}=$
Resistance between secondary and case:	$R_{cg}=$
Dynamic Tests	
Primary voltage:	$V_{1-2}=$
Secondary voltages:	$V_{ac}=$
	$V_{ab}=$
	$V_{bc}=$

Next plug the transformer in and complete the voltage checks requested in Table 3.7. Enter these data in the table.

3.51 TESTING INDUCTORS

Select various inductors as indicated in Table 3.8 and measure their dc resistances. Enter the measurements in the space provided.

TABLE 3.8 Testing Inductor dc Resistance

Inductance	DC Resistance
1 µH coated	
10 µH coated	
100 µH coated	
10 µH coiled	
100 µH coiled	
1.5 mH coiled	
4.7 mH coiled	
10 mH coiled	
22 mH coiled	
47 mH coiled	
100 mH coiled	

3.52 EXERCISES

Exercise 1: Static Troubleshooting the *RC* Filter

See Figure 3.58 and Table 3.9.

Static troubleshooting the *RC* filter

Step 1 Let $R = 2.2$ kΩ
$C = 10.0$ µF

Step 2 Breadboard the filter only.

Step 3 Measure the leakage resistance of the capacitor: R_{BC}.

Step 4 Measure R_{AB}.

Step 5 Disconnect point *B*. Measure R_{AC}.

Step 6 Using a jumper wire, short *B* to *C*.

Step 7 Measure R_{AC}.

Step 8 Measure R_{BC}.

Step 9 Short *A* to *B*. Measure R_{AC}.

TABLE 3.9

Step	Measurement
3	$R_{BC} =$
4	$R_{AB} =$
5	$R_{AC} =$
7	$R_{AC} =$
8	$R_{BC} =$
9	$R_{AC} =$

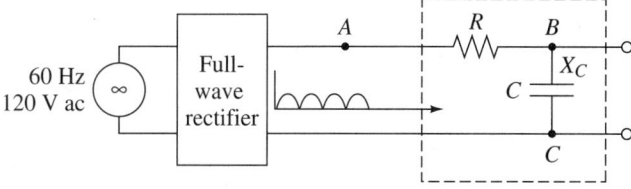

The signal into this *RC* filter has a ripple frequency of 120 Hz. By design, the value of *R* is usually 10 times the value of X_C. Thus the ripple voltage appears across *R*.

FIGURE 3.58 Exercise 1.

Exercise 2: Static Troubleshooting the Wheatstone Bridge

See Figure 3.59.

Step 1 Connect the bridge circuit shown.
Step 2 Zero the voltmeter.
Step 3 Select a 470-Ω resistor for R_x.
Step 4 Set the voltmeter on a high range.
Step 5 Close switch S.
Step 6 Adjust R_3 for a zero indication on V.
Step 7 Change the voltmeter setting. Adjust R_3 for a zero on the lowest possible range.
Step 8 Open switch S.
Step 9 Change to a higher voltmeter setting.
Step 10 Read the resistance of R_3 from its dial or measure it.
Step 11 Calculate the resistance of R_8 from $R_x = (R_3) R_1/R_2$.
Step 12 Measure R_x.
Step 13 Complete Table 3.10.

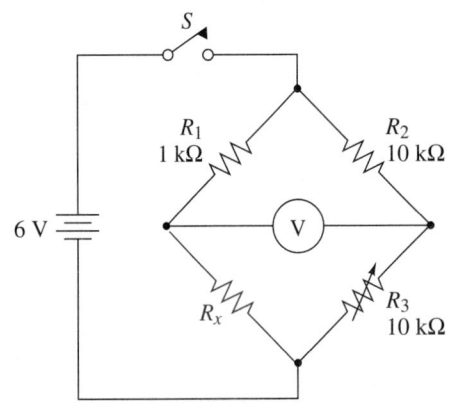

FIGURE 3.59 Exercise 2.

TABLE 3.10

Step	Value
10	
11	
12	

3.53 PROBLEMS

1. Draw the dc equivalent circuit for Figure 3.60.

2. Indicate the polarity of each resistor in Figure 3.61.

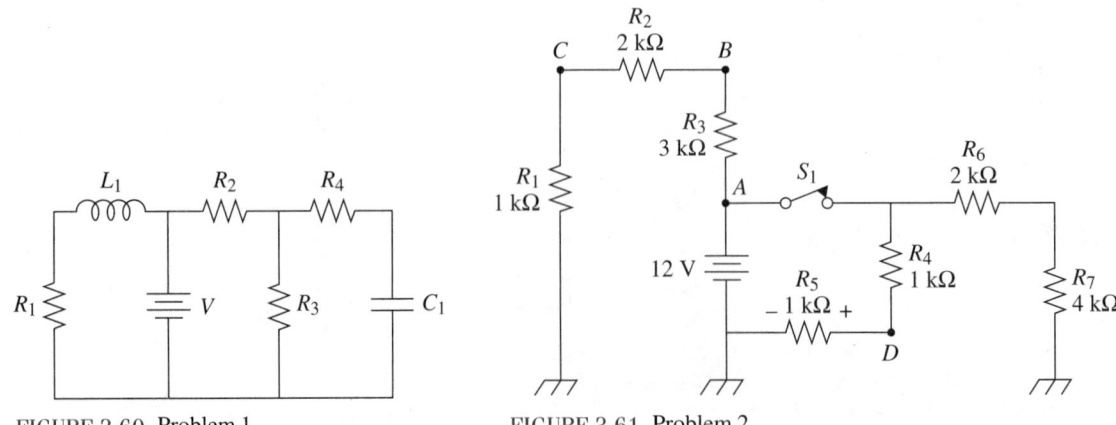

FIGURE 3.60 Problem 1.   FIGURE 3.61 Problem 2.

3. Referring to Figure 3.61, calculate the voltages listed in Table 3.11 .

4. Identify the use of the symbols in Figure 3.62.

5. Given the circuit in Figure 3.61, explain the step-by-step procedure for measuring the resistance of R_4.

6. In the Wheatstone bridge circuit of Figure 3.63, find all voltage drops, label each resistor polarity, and calculate the value of R_x. The bridge is balanced.

TABLE 3.11

$V_{AG} =$	$V_{AD} =$
$V_{AB} =$	$V_{FD} =$
$V_{BA} =$	$V_{GF} =$
$V_B \; =$	$V_{AF} =$
$V_{CA} =$	$V_{FC} =$
$V_{BG} =$	$V_{BD} =$

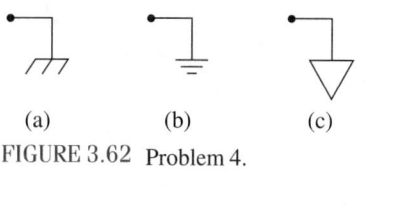

(a)   (b)   (c)

FIGURE 3.62 Problem 4.

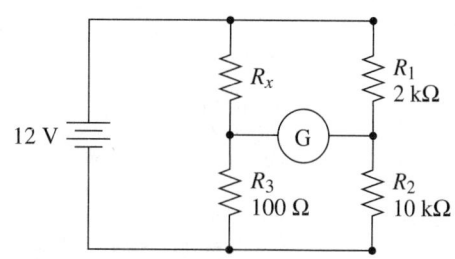

FIGURE 3.63 Problem 6.

7. An iron-core transformer with a 10 : 1 step-down ratio has 120 V applied to the primary winding. The secondary voltage equals (a) 24 V; (b) 10 V; (c) 12 V; (d) 1200 V.

8. Calculate C_T for the series parallel combination of capacitors in Figure 3.64.

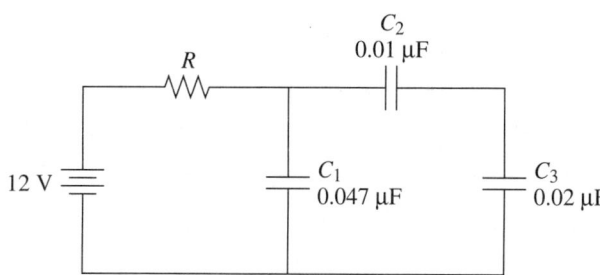

FIGURE 3.64 Problem 8.

9. At time $t = t_0$, switch S_1 in Figure 3.65(a) is closed. Calculate the charge time constant. At time t_1 ($6\tau_c$), switch S_1 is opened and S_2 is closed. Sketch the complete charge-discharge response curves in Figure 3.65(b).

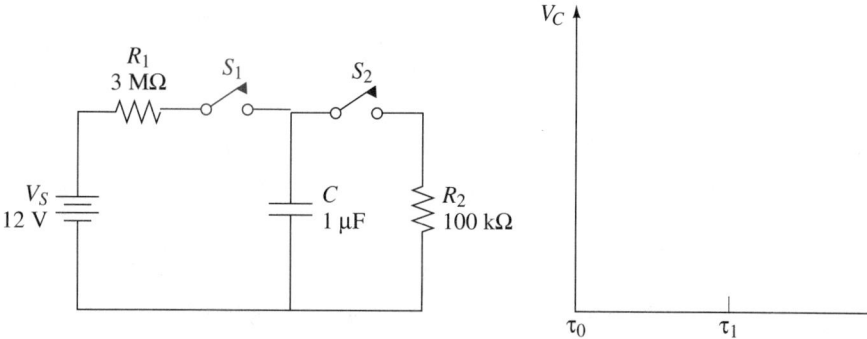

FIGURE 3.65 Problem 9.

10. In Figure 3.66 a 75-V square wave at a frequency of 20 Hz is applied. Sketch the curves for V_C, I, and V_R.

11. In Figure 3.67 what is the value of circuit current 0.1 s after the switch is closed?

12. What is the most likely trouble in the circuit of Figure 3.68?

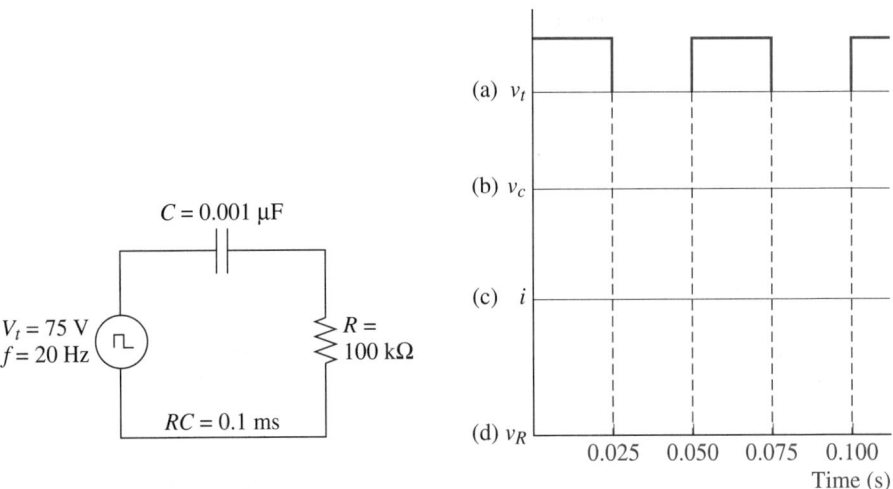

FIGURE 3.66 Problem 10.

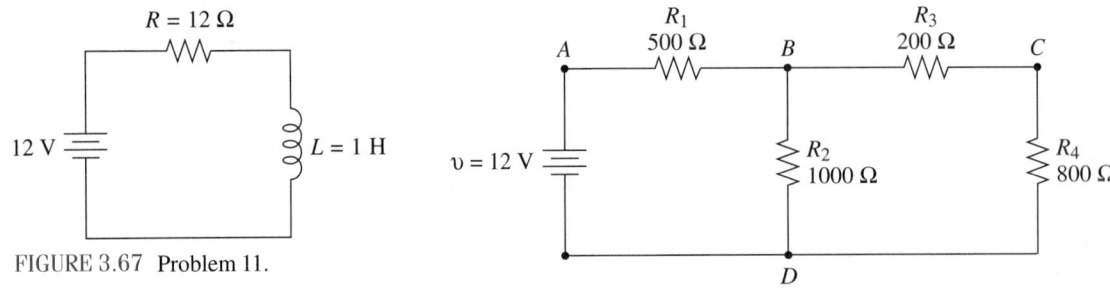

FIGURE 3.67 Problem 11.

FIGURE 3.68 Problem 12.

13. If V_{BC} measured 8 V, what's the most likely problem?

14. A 20-mH inductor and a 40-mH inductor are connected in series. Calculate the total inductance.

15. Refer to Figure 3.69. Calculate the parameters listed and complete Table 3.12.

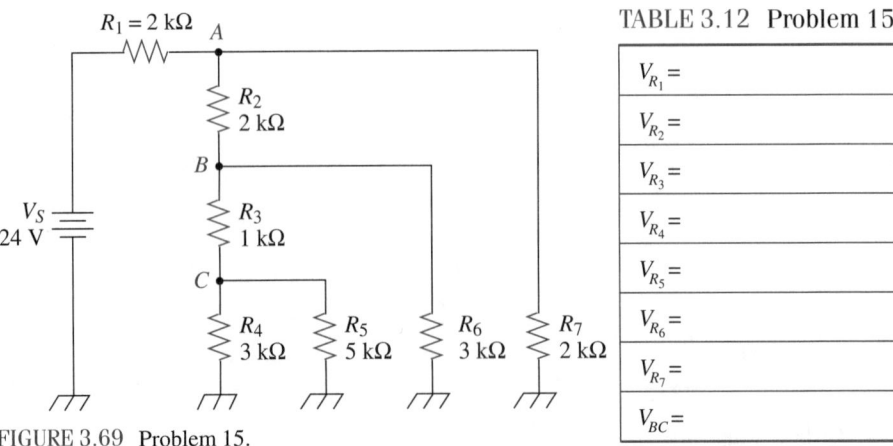

FIGURE 3.69 Problem 15.

TABLE 3.12 Problem 15

$V_{R_1} =$
$V_{R_2} =$
$V_{R_3} =$
$V_{R_4} =$
$V_{R_5} =$
$V_{R_6} =$
$V_{R_7} =$
$V_{BC} =$

16. *Troubleshooting the loaded voltage divider.* Voltage measurements were taken for the circuit in Figure 3.70 for various trouble situations. In Table 3.13, identify the trouble for each set of measurements.

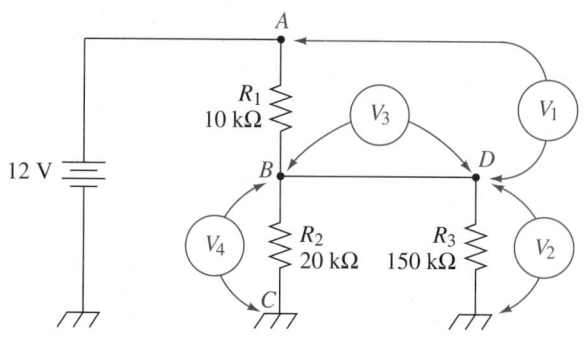

FIGURE 3.70 Problem 16.

TABLE 3.13 Problem 16

V_1	V_2	V_3	V_4	Trouble
0.75	11.25	0	11.25	
4.00	8.00	0	8.00	
4.00	0.00	8	8.00	
0.00	12.00	0	12.00	
12.00	0.00	0	0.00	

17. *Troubleshooting RC circuits.* Complete Table 3.14 for Figure 3.71 by identifying the possible problem areas.

TABLE 3.14 Problem 17

Measured Voltages				Troubles
V_{AB}	V_{BC}	V_{CD}	V_{BD}	
24	0	24	24	
48	0	0	0	
0	0	0	0	

$$V_{CX} = \frac{C_T}{C_X} V_T$$

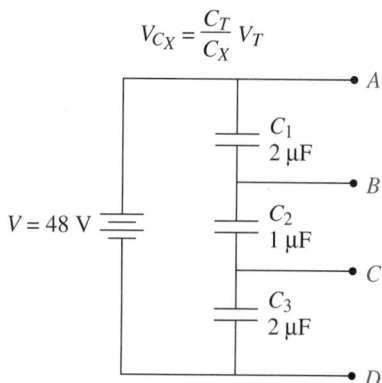

FIGURE 3.71 Problem 17.

18. You have been asked to repair the switch in the circuit in Figure 3.72. You were also asked to explain the following:

a. Why does the switch fail so often?
b. How can you correct the problem?

What would you say?

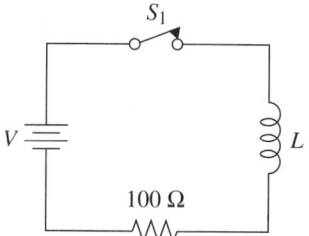

FIGURE 3.72 Problem 18.

19. You have examined the waveform across the resistor of Figure 3.73(a) and it appears as shown in Figure 3.73(b).

  a. What could be the reason?
  b. Explain your troubleshooting procedure, step by step.

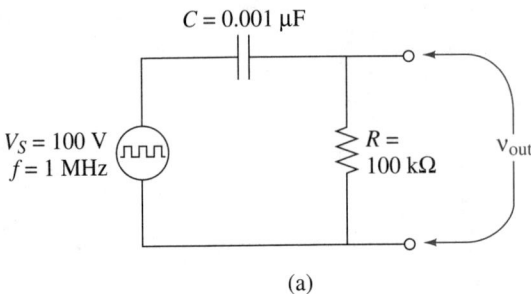

(a)

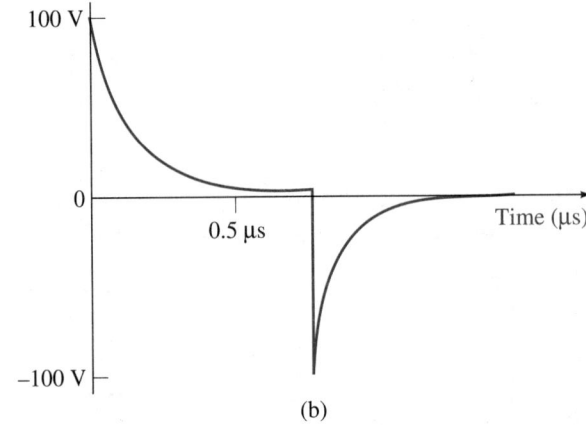

(b)

FIGURE 3.73 Problem 19.

4

STATIC (DC) TROUBLESHOOTING AT THE SOLID-STATE COMPONENT LEVEL

4.1   THE DIODE

We begin with the most fundamental of solid-state components—the *pn* junction diode. In a very simplistic sense, let's start with two blocks of pure silicon atoms (Figure 4.1) and dope them (add impurity atoms) as follows: one with atoms of boron, gallium, or indium, which have an excess of positive charges (these are referred to as *acceptor atoms*) and the other with atoms of antimony, arsenic, or phosphorus, which have an excess of negative charges (known as *donor atoms*).

The silicon block with an excess of positive charges is referred to as *p-type material.* The silicon block with an excess of negative charges is referred to as *n-type material.*

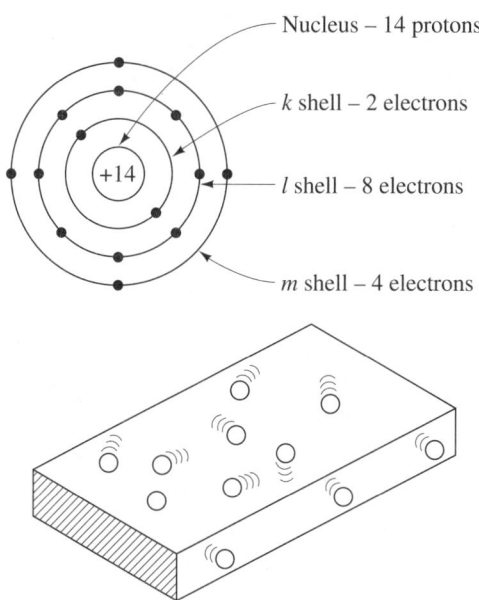

Nucleus – 14 protons

k shell – 2 electrons

+14

l shell – 8 electrons

m shell – 4 electrons

Figure 4.1 Silicon atoms.

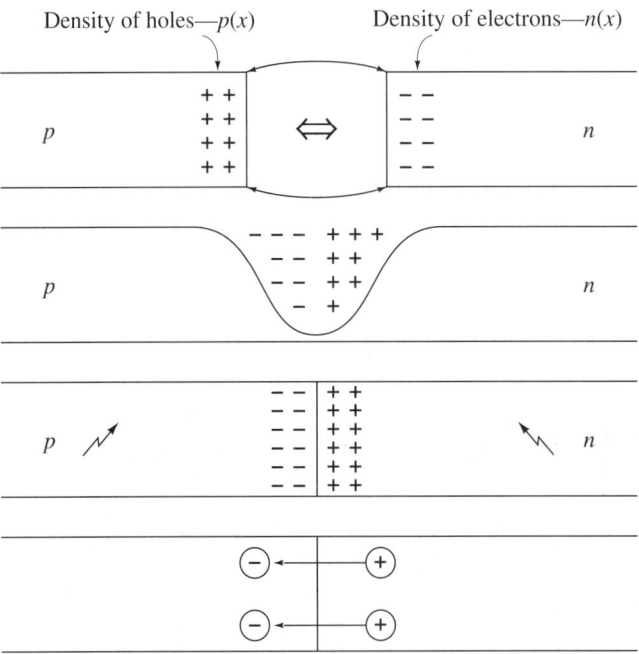

Density of holes—$p(x)$ Density of electrons—$n(x)$

FIGURE 4.2 The pn junction.

Imagine bringing these two blocks together and joining them instantaneously to form one unit, as shown in Figure 4.2.

For a very brief amount of time, the free electrons from the *n*-type material fall into holes in the *p*-type material adjacent to the boundary, leaving a net positive charge behind. At the same time, the *p* side obtains a net negative charge due to the additional electrons. The result is an internal dipole layer with an associated electric field, resulting in a potential difference of about 0.7 V for silicon and 0.3 V for germanium.

The space within the layer (known as the *depletion region*) is now stable and represents a barrier to the movement of additional electrons from the *n*-type material into the *p*-type material. As a result, the diode is able to conduct only when forward-biased (Figure 4.3) because the external positive voltage pulls electrons over the barrier and the negative voltage pushes electrons from the *n*-type material. (The positive voltage must exceed the barrier potential.) The depletion region will, however, change with a change in temperature. Devices based on this characteristic are known as *thermistors*.

When reverse-biased (Figure 4.4) the external voltage creates a wider barrier by attracting negative charges to the positive voltage and positive charges to the negative voltage.

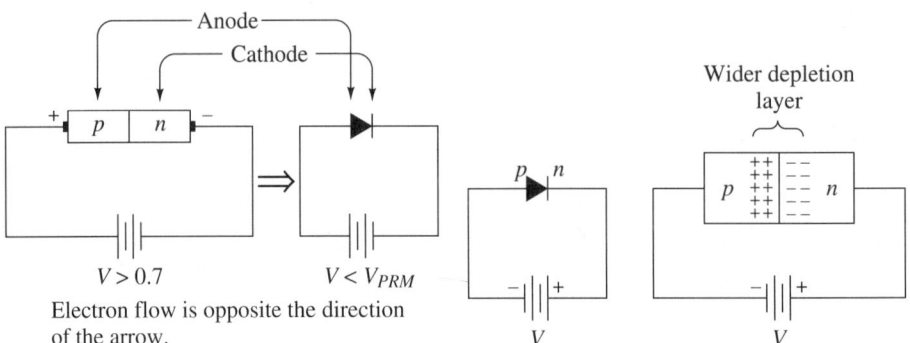

FIGURE 4.3 Forward-biased diode.   FIGURE 4.4 Reverse-biased diode.

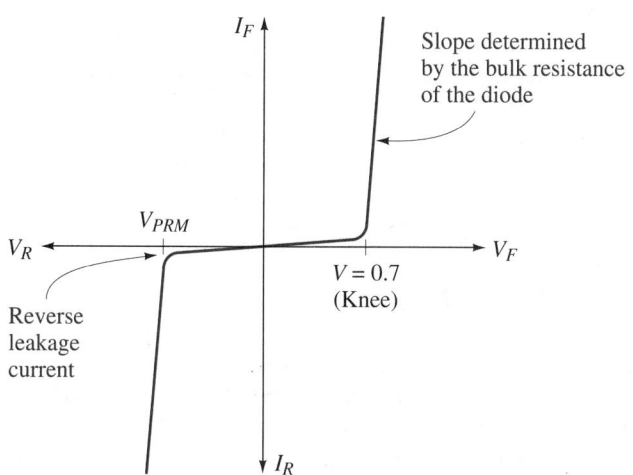

FIGURE 4.5 Diode characteristic curve.

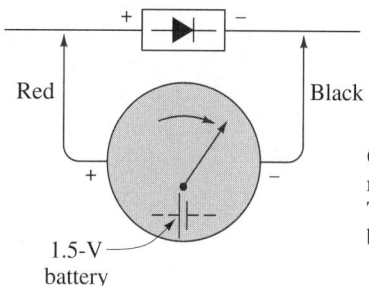

Ohmmeter indicates low resistance.
The internal battery forward-biases the diode.

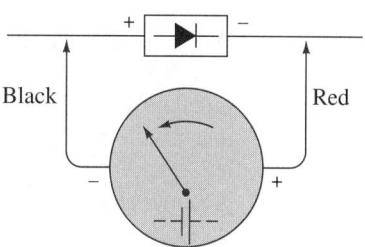

Ohmmeter indicates a high resistance.

FIGURE 4.6 Testing diodes.

The diode characteristic curve of forward voltage (V_F) versus forward current (I_F) is shown in Figure 4.5.

It is important not to exceed the forward maximum power dissipation rating, or $P_{D(max)}$, of the diode. You can calculate the actual power dissipation by using the equation

$$P = V_F I_F \qquad (4.1)$$

4.2 DIODE FAILURE MODES

Diode failures are typically due to excessive current, component age, or application of reverse voltages exceeding the maximum reverse voltage rating, specified as V_{RRM} on a data sheet. You may also find the maximum reverse voltage listed as PIV (peak inverse voltage), PRV (peak reverse voltage), VRWM (voltage reverse working maximum), or VBR (voltage breakdown). After exceeding the power dissipation rating, the diode will short internally and, eventually, if current continues flowing, crack and fall apart.

4.3 TESTING DIODES OUT OF THE CIRCUIT

When performing out-of-circuit checks on diodes, there are four basic tests to perform.

Ohmic

1. **a.** The diode should conduct when forward-biased and should not conduct when reverse-biased. See Figure 4.6.
   b. The ohmmeter should indicate a low resistance, perhaps a few hundred ohms, when the internal battery of the meter forward-biases the diode.
   c. The ohmmeter should indicate a very high resistance when the diode is reverse-biased.

Be careful! Some ohmmeters can supply levels of current sufficient to destroy a low-signal diode. It is best to keep your meter on the higher R × scales (R × 100, R × 1000). Use the lower scales (R × 1, R × 10) for larger rectifier diodes.

 CAUTION

Ohmmeters can supply too much current.

Dynamic

2. For a given value of reverse voltage (V_R), the reverse current (I_R) should not exceed a given value.

3. For a given value of forward current (I_F), the forward voltage (V_F) should not exceed a given value.

4. The switching times should be fast enough for a given digital application.

4.4   USING DIGITAL MULTIMETERS

When using DMMs to test diodes, it is important to remember that due to the inherent design of the typical DMM, the low-voltage ohmmeter section cannot be used to test *pn* junctions. Many DMMs have a special diode setting for this purpose. On this setting the meter indicates the voltage drop across the diode's *pn* junction. If the voltage is much less than 0.7 V (silicon diodes), the diode is probably shorted. If the voltage is much greater than the 0.7 V, the diode is probably open.

4.5   USING CURVE TRACERS

Curve tracers are very useful for dynamic testing. The curve tracer uses an oscilloscope to display the characteristic V_F–I_F curve of the diode. Typical displays are shown in Figure 4.7.

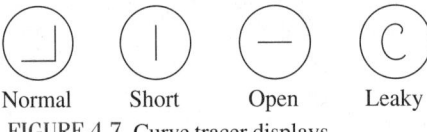

Normal　　　Short　　　Open　　　Leaky

FIGURE 4.7 Curve tracer displays.

4.6   TESTING DIODES IN LIVE CIRCUITS

When checking a diode's performance in a circuit, look at the voltage drop across its terminals. The diode should have 0.3 V (typical for germanium) or 0.7 V (typical for silicon) across its terminals when forward-biased (turned on).

　　If the diode drop is much less than 0.3 V, consider the diode to be shorted. If it is much greater, consider the diode to be open. In Figure 4.8 diodes D_1 and D_2 are normal, but D_3 and D_4 are faulty. What about D_5? Do you see that D_4 is shorted, D_3 is open, and D_5 is reverse-biased and normal?

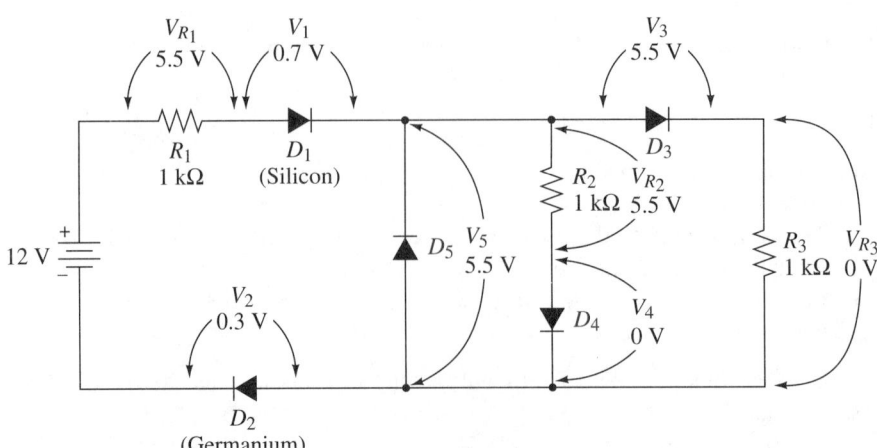

FIGURE 4.8 Complex diode circuit.

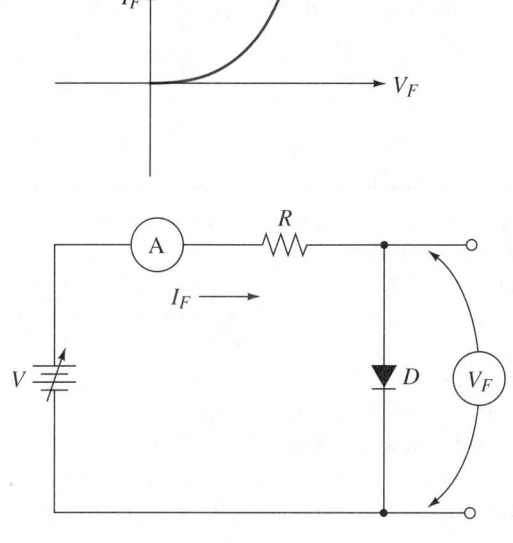

Vary V until I_F equals the specified value. Read the value of V_F.

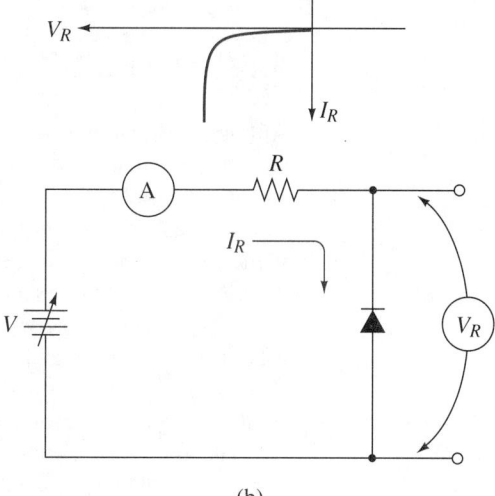

Vary V_R while observing I_R.

(a)

(b)

FIGURE 4.9 Diode test circuits.

4.7 DIODE STATIC (DC) CHARACTERISTICS

In many situations the forward voltage drop across the diode and the reverse leakage current must be known. Simple test circuits to determine these values are shown in Figure 4.9.

4.8 CRITICAL DIODE PARAMETERS

When it becomes necessary to replace a diode in a circuit,

1. You must specify whether the diode is a small-signal diode or a rectifier diode. Rectifier diodes must handle larger currents and will have higher forward-voltage drops. If the application is an extremely small-signal one, the germanium diode may be the better choice due to its lower barrier potential. The AM receiver circuit in Figure 4.10 is a good example.

Replacing a diode in a circuit

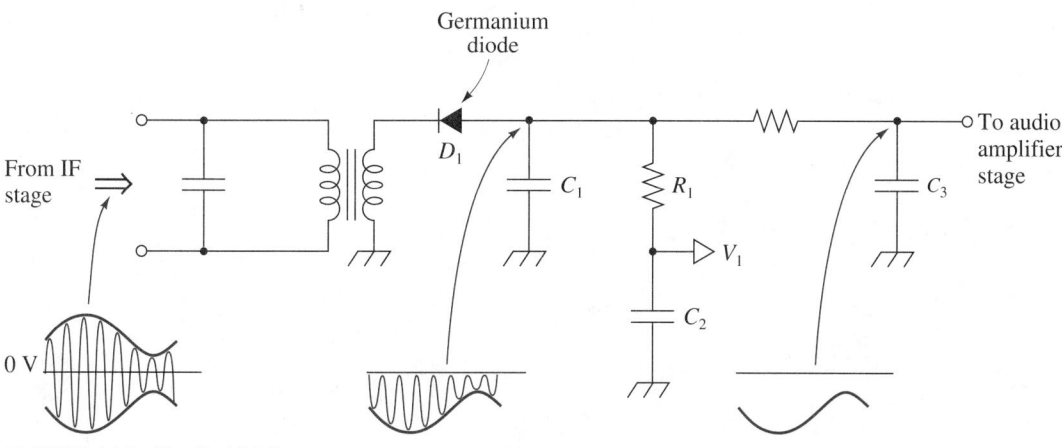

FIGURE 4.10 Simple AM detector stage.

2. You need to ensure that the forward current rating and maximum power dissipation rating are not exceeded.

3. You must check to be sure that the reverse breakdown voltage will not be exceeded.

4.9 ZENER DIODES

Although normal diodes operate in the forward conduction region, zener diodes are designed to operate in the breakdown region and are normally reverse-biased. The zener can also be used in the forward-biased mode and will operate like a normal diode.

The value of reverse voltage that causes breakdown is called the *zener voltage*. Once in the breakdown region, the zener will maintain a nearly constant voltage (referred to as the zener voltage) across its terminals.

A typical complete volt-ampere characteristic curve of a zener is shown in Figure 4.11. The circuit used to determine the curve is shown in Figure 4.12.

V_Z ranges in ratings from 1.8 V to hundreds of volts. This voltage will be nearly constant as long as the reverse current is between the knee value and the maximum value.

The manufacturer will list in the data sheet a given value of V_Z for a particular value of current, called the zener test current (I_{ZT}).

It is possible to calculate the zener impedance from the equation

$$Z_Z = \frac{\Delta V_Z}{\Delta I_Z} \qquad (4.2)$$

I_R is the reverse current through the zener when $V < V_Z$. Because the zener conducts in both directions, you cannot perform ohmmeter tests properly and have to measure the voltage drop across it.

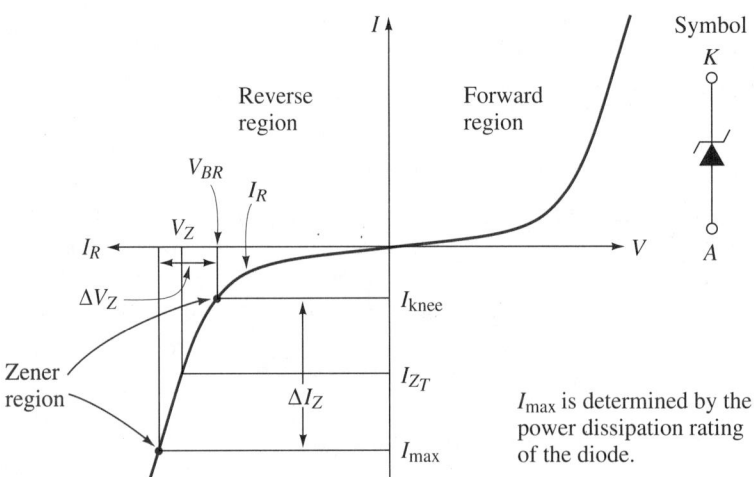

FIGURE 4.11 Zener diode *V-I* curves.

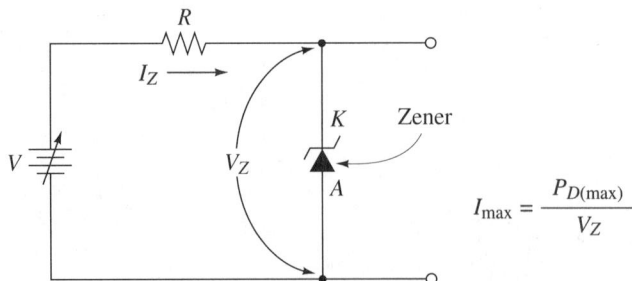

FIGURE 4.12 Zener diode test circuit.

4.10 TROUBLESHOOTING ZENER DIODES

Let's analyze the possible trouble spots in Figure 4.13. V_1 (V_{DC}) should be about 0.7 V because the zener is forward-biased.

V_2 (V_{BC}) should indicate the zener voltage (V_Z) or 8 V. A zener should have no more than its rated voltage (V_Z) across its terminals.

Assume that the source voltage is acceptable in every situation and the measured voltages are as shown for each. The typical problems are listed in Table 4.1 for each set of voltages.

Troubleshooting zener diodes

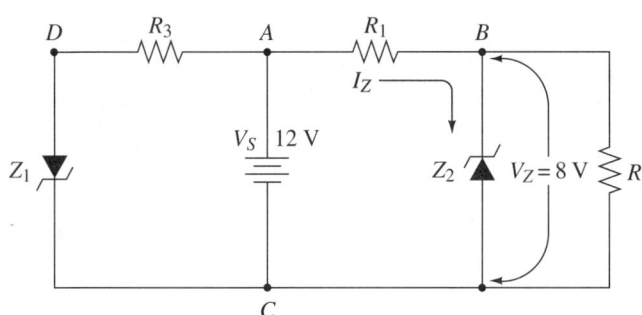

FIGURE 4.13 Troubleshooting zener diodes.

TABLE 4.1 Typical Problems for Zener Diodes

V_{AC}	V_{AB}	V_{BC}	V_{AD}	V_{DC}	Condition
12	4	8	11.3	0.7	Normal
12	4	8	0	12	Z_1 open
12	4	8	12	0	Z_1 short
12	4	8	12	0	R_3 open
12	0	8	11.3	0.7	R_1 short
$I_{Z(max)}$ may be exceeded.					
12	12	0	11.3	0.7	R_1 open
12	12	0	11.3	0.7	Z_2 short
12	12	0	11.3	0.7	R_2 short
12	$V_S\left(\frac{R_1}{R_1+R_2}\right)$	$V_S\left(\frac{R_2}{R_1+R_2}\right)$	11.3	0.7	Z_2 open
12	4	8	11.3	0.7	R_2 open
Z_2 may open due to excess current.					
Z_1 very hot			0		R_3 short

4.11 CRITICAL ZENER DIODE PARAMETERS

When replacing a zener diode, the important values to check are the

1. V_Z rating
2. $I_{Z(max)}$ rating
3. $P_{D(max)}$ rating

4.12   TROUBLESHOOTING LIGHT-EMITTING DIODES

Working with light-emitting diodes

Light-emitting diodes (LEDs) are designed so that the energy released when electrons recombine across the *pn* junction (Figure 4.14) is in the frequency (wavelength) of visible light. Various colors are available, depending on the material used in the diode. The symbol, shown in Figure 4.15, is a diode with energy rays being emitted.

The three characteristics of the anode lead are shown in Figure 4.16.

Recombinations at the *pn* junction create photons of energy released as visible or IR light.

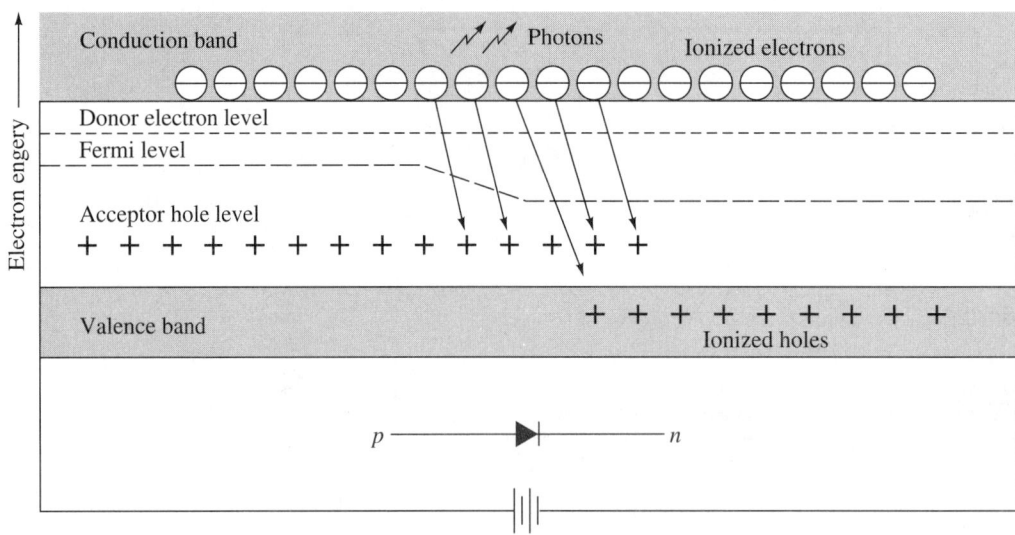

FIGURE 4.14 The LED.

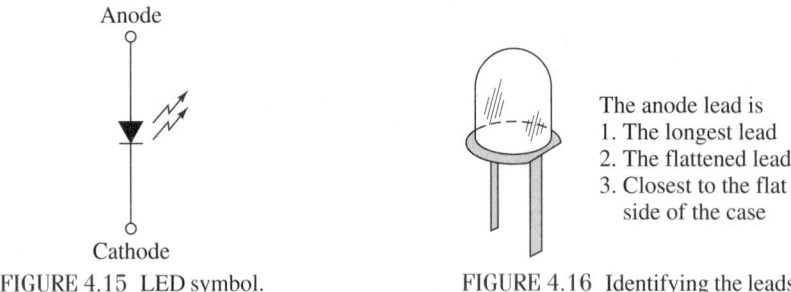

Anode

Cathode

FIGURE 4.15 LED symbol.

The anode lead is
1. The longest lead
2. The flattened lead
3. Closest to the flat side of the case

FIGURE 4.16 Identifying the leads.

4.13   LED CHARACTERISTIC CURVES

The characteristic curve for a typical diode is shown in Figure 4.17. From the curve you can see that the voltage dropped across the forward-biased LED ranges from 1 V to 3 V, and the LED breaks down with a reverse voltage of –3 V to –10 V.

Because of their low forward current rating (20 mA to 100 mA), a series resistor is always used to limit current to a safe level below the maximum rated current for the diode.

The wavelength of radiated energy can be in the visible light region or in the infrared region.

Caution — Never look into an infrared LED. You risk damaging the retina of your eye.

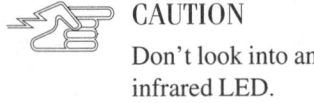

CAUTION

Don't look into an infrared LED.

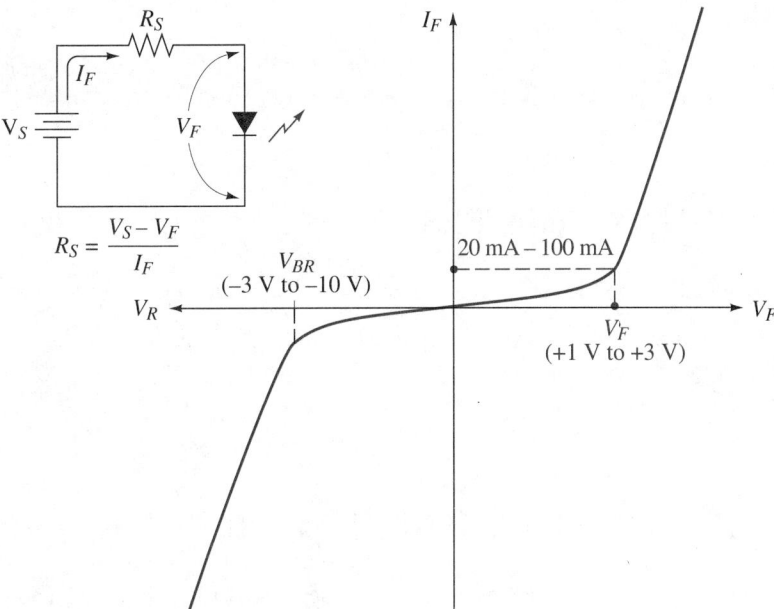

FIGURE 4.17 LED characteristic curve.

The intensity of the LED light is controlled by the level of forward current. For the brightest possible display, use the smallest value of series resistance possible without exceeding the maximum forward current rating, $I_{F(max)}$.

Therefore, in the equation $R_S = V_F/I_F$, I_F would be $I_{F(max)}$ and V_F would be $V_{F(min)}$ from the curve.

4.14 TESTING LEDS

Although LEDs can be tested like a normal *pn* junction diode, if voltage is available and the LED is out, you can be pretty certain that the LED is faulty.

Use the high ranges when performing ohmic tests.

Ohmic Testing

When performing ohmic tests, as shown in Figure 4.18, use the high ranges because higher voltages are required to forward-bias the LED. For example, red LEDs require 1.5 V whereas green and yellow LEDs require 2.0 V and infrared LEDs, about 1.0 V.

Therefore, DMMs usually are not used to test LEDs. DMMs have voltages on the ohms ranges that are too low and cannot forward-bias the LED. Table 4.2 shows test results.

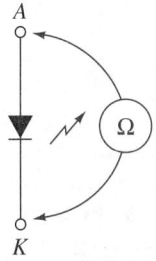

FIGURE 4.18 Ohmic testing.

TABLE 4.2 Ohmic Tests

Ohmmeter	LED Condition
Condition 1	
Low (<1000)	Normal
> 1000	Bad
Condition 2	
> 1000	Normal
< 1000	Bad

Test Condition

	1	2
A	+	−
K	−	+

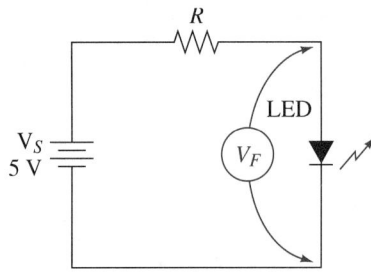

FIGURE 4.19 Testing the LED.

TABLE 4.3 Voltage Testing

V_F Is	LED Condition
0	Shorted if R and V_S are acceptable.
5	Open
1–3	Normal

Voltage Testing

Performing voltage measurements on a circuit like that in Figure 4.19 requires a few simple checks. Typical voltage tests and corresponding LED conditions are listed in Table 4.3.

4.15 CRITICAL PARAMETERS OF LEDS

When replacing an LED, the most important parameters to consider are the following:

1. The wavelength of energy emissions — visible light, infrared, etc.
2. I_F — the forward current rating
3. V_F — the forward voltage rating
4. V_{BR} — the reverse breakdown voltage rating

4.16 TESTING MULTISEGMENT DISPLAYS

The most common multisegment display is the seven-segment LED shown in Figure 4.20. By forward-biasing the LEDs, the numbers 0–9 can be displayed. Each segment (LED) can be tested with the ohmmeter in the same way that a normal LED is tested. Figure 4.20 shows the ohmmeter connected to test segment F of the common-anode display and segment C of the common-cathode display. Be sure to use the $R \times 10,000$ range or higher.

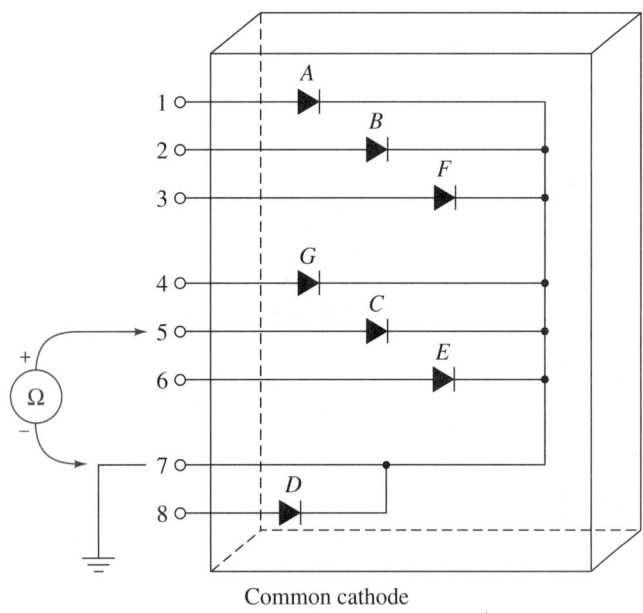

Common cathode

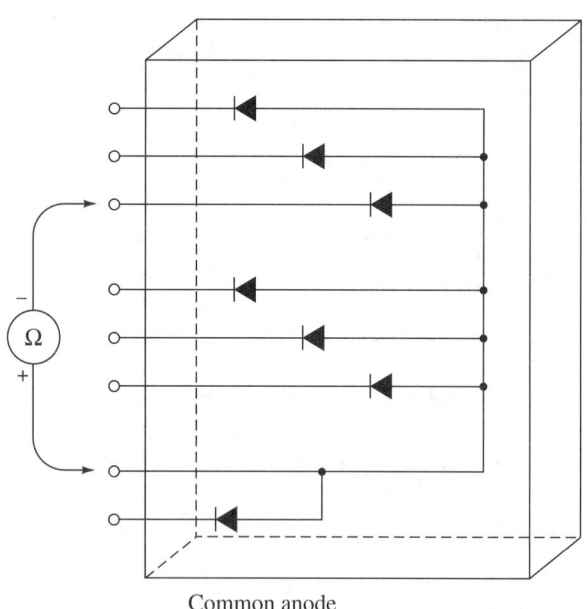

Common anode

FIGURE 4.20 Multisegment displays.

4.17 TESTING PHOTODIODES

The photodiode is operated in reverse-bias mode, and the value of reverse current (I_R) is determined by the intensity of light striking the *pn* junction. See Figure 4.21. The normal reverse current of the typical photodiode when not excited (known as the *dark* current, I_D) is only a few nanoamperes. However, when excited by an external light source, this reverse current increases to several microamperes. This is known as the *light* current

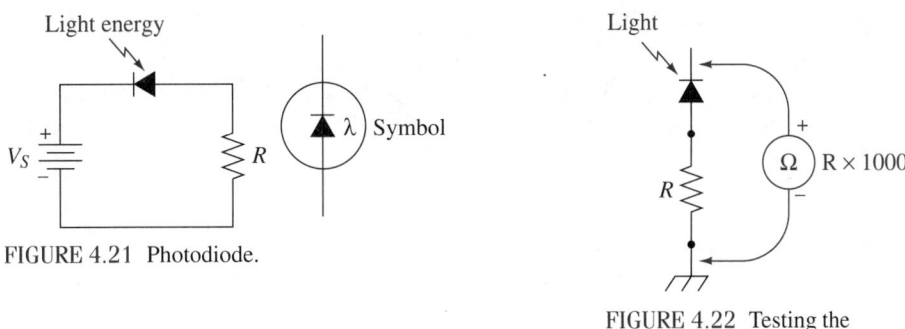

FIGURE 4.21 Photodiode.

FIGURE 4.22 Testing the photodiode.

(I_L). For example, the MRD500 photodiode has a dark current of 2 nA, which increases to 9 μA under the influence of light. Thus the dynamic resistance of the photodiode decreases with an increase in I_R. A particular photodiode is most sensitive to a given wavelength of light, as specified by the manufacturer's data sheet. The MRD500 is most sensitive to 800 nm, or 375 THz.

4.18   TESTING PHOTODIODES OUT OF THE CIRCUIT

A photodiode can be tested out of the circuit the same as any *pn* junction diode, except that the photodiode is reverse-biased. Figure 4.22 shows a simple test circuit for a photodiode. Photodiodes are commonly used in fiber-optic circuits, where they operate as receivers for the data signal being transmitted. A typical circuit is shown in Figure 4.23.

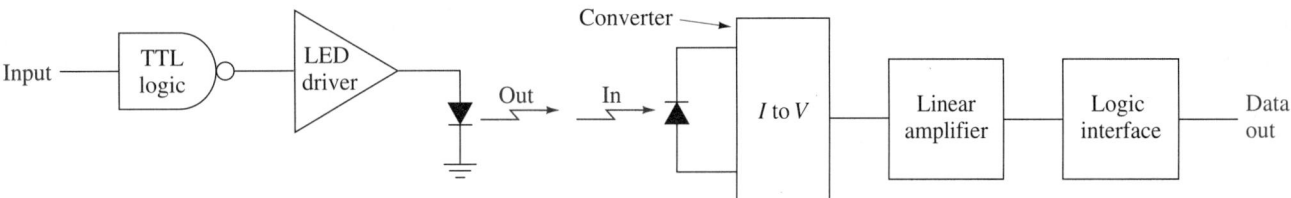

FIGURE 4.23 Typical circuit.

4.19   CRITICAL PARAMETERS OF PHOTODIODES

When selecting a photodiode, the most critical parameters to consider are

1. Dark current (I_D)
2. Light current (I_L)
3. Sensitivity (the increase in reverse current per specified amount of light intensity)
4. Spectral response (the frequency at which the diode response peaks)

4.20   VARACTOR DIODES

Here is another situation where the diode is operated in the reverse-bias mode. The varactor diode is designed to make use of the capacitance across the depletion layer (*pn* junction), which changes with the level of dc voltage applied. The capacitance change may be as much as 10 : 1. These diodes are commonly found in digital tuning circuits. The device symbol is shown in Figure 4.24.

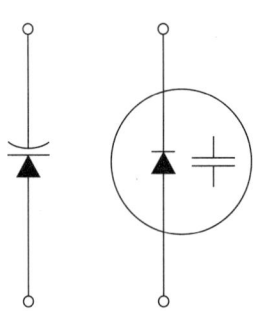

FIGURE 4.24 Varactor diode symbols.

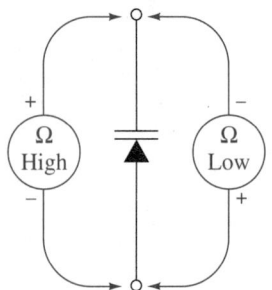

FIGURE 4.25 Testing the varactor diode.

4.21 TESTING VARACTOR DIODES

Figure 4.25 shows how the ohmmeter connections are made and the normal condition values in testing varacter diodes. Be sure to keep the ohmmeter set on the high ranges for this test.

4.22 TUNNEL DIODES

Tunnel diodes are very heavily doped with germanium or gallium arsenide to conduct with minimal forward-bias voltage. They have the strange V–I characteristic curve shown in Figure 4.26. For voltages greater than V_B, the tunnel diode performs like a normal diode.

4.23 TESTING TUNNEL DIODES

Tunnel diodes can be tested with the ohmmeter, as shown in Figure 4.27.

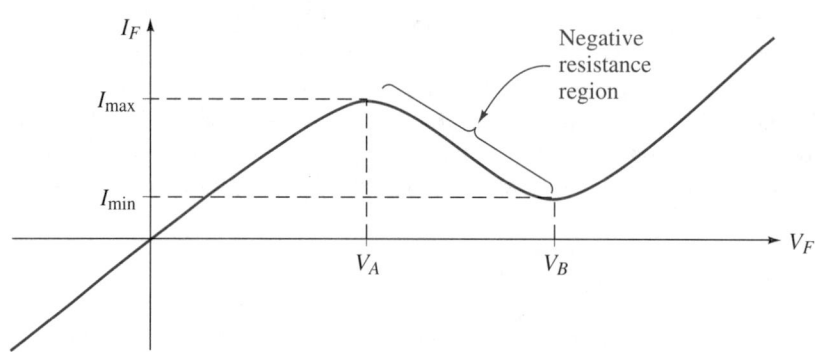

FIGURE 4.26 Tunnel diode characteristic curve.

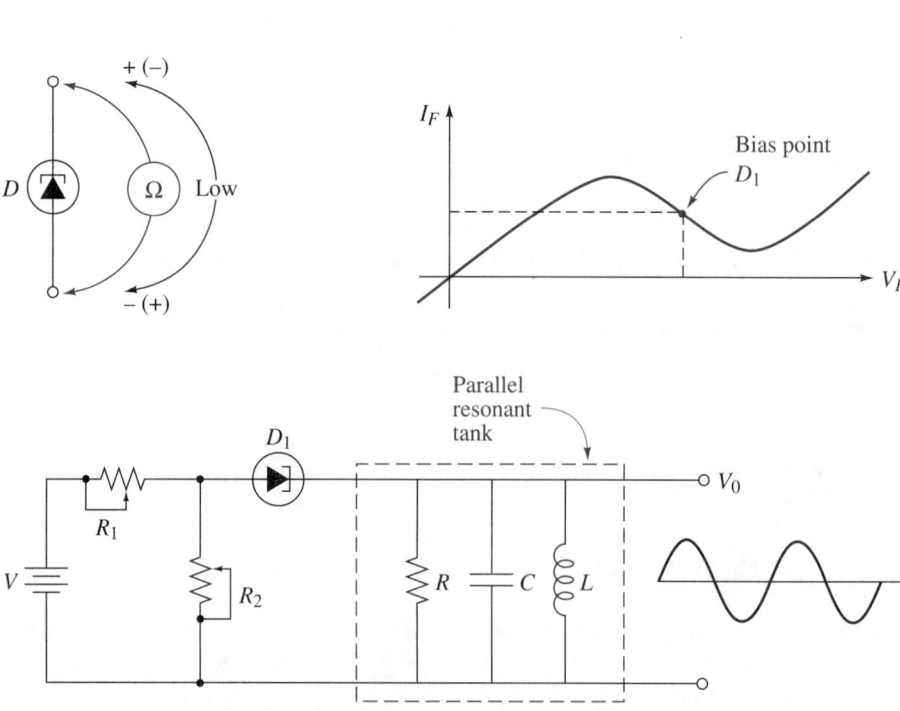

FIGURE 4.27 Testing the tunnel diode.

Adjust R_1 & R_2 to bias D_1 in the center of its negative resistance region. The output should be a sustained oscillation, as shown.

4.24 SHOTTKY DIODES (HOT CARRIER DIODES)

The Shottky diode is found in high-speed switching circuits. The junction for this diode is a metal-silicon junction, which creates a very large current when a forward-bias voltage is applied. Because there are so many conduction-band electrons available in the metal, the device turns on and off very fast. It is, therefore, most useful in very high-frequency applications. The symbol is shown in Figure 4.28.

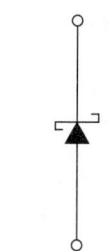

FIGURE 4.28 Shottky diode symbol.

4.25 TESTING SHOTTKY DIODES

Shottky diodes can be tested as shown in Figure 4.29. Results of the test are given in Table 4.4.

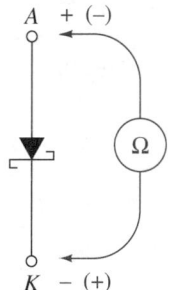

FIGURE 4.29 Testing the Shottky diode.

TABLE 4.4 Test Results

A	K	Reading	Condition
+	−	Low ohms	Acceptable
+	−	High ohms	Open
−	+	High ohms	Acceptable
+	−	Zero	Shorted
−	+	Low ohms	Shorted

4.26 PIN DIODES

A PIN diode is a *pn* device sandwiched around a layer of intrinsic silicon, as shown in Figure 4.30. It has the characteristic of a variable resistance when forward-biased and a fixed capacitance when reverse-biased. It is primarily used in high-frequency modulation circuits.

4.27 TESTING PIN DIODES

PIN diodes can be tested as shown in Figure 4.31. Test results are shown in Table 4.5.

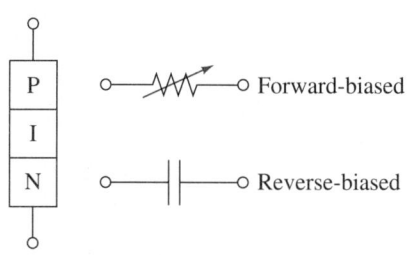

FIGURE 4.30 PIN diode.

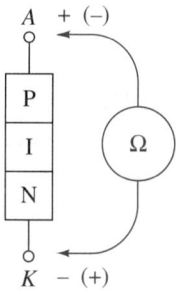

FIGURE 4.31 Testing the PIN diode.

TABLE 4.5 Test Results

A	K	Reading	Condition
+	−	Low	Acceptable
−	+	Very high	Acceptable
+	−	High	Open
−	+	Low	Shorted

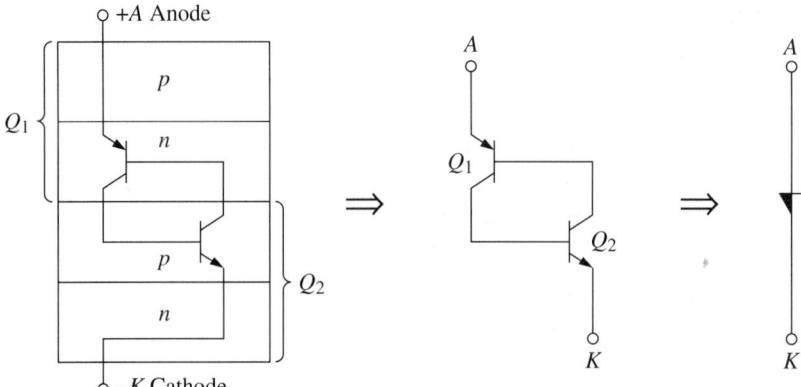

FIGURE 4.32 The silicon unilateral switch.

4.28 THE SILICON UNILATERAL SWITCH

The silicon unilateral switch (SUS, Figure 4.32) is a two-terminal, four-layer (*pnpn*) device that is triggered into conduction when a forward-bias voltage exceeding the threshold voltage (V_{BO}, forward breakover) is applied. Also known as a latch, Shockley diode, *pnpn* diode, or four-layer diode, the SUS conducts current in one direction only. Once turned on, the SUS remains on until the forward current reduces to a minimum level, known as the *holding current*.

When the SUS is in the on state, the forward resistance drops, and, therefore, the voltage across the device (V_{AK}) drops to a minimum value.

4.29 TESTING SILICON UNILATERAL SWITCHES

Dynamic

The characteristic curve is shown in Figure 4.33; it can be derived using the test circuit shown. When V_S is increased so that V_{AK} exceeds the breakover voltage (threshold) rating

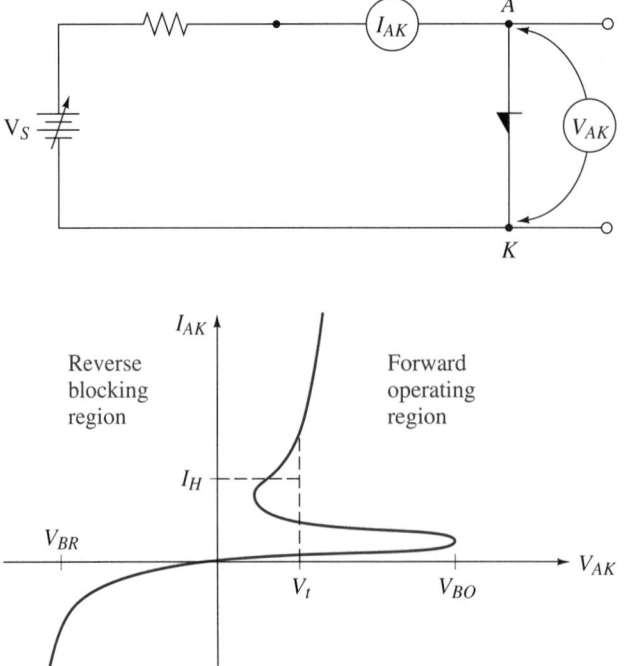

FIGURE 4.33 Characteristic curve.

of the SUS, the device turns on and conducts heavily, driving Q_1 and Q_2 into saturation. Then V_{AK} drops to a low value known as the average on-state voltage (V_t).

When the current drops below the holding current (I_H), the SUS turns off and remains off until the forward-blocking voltage is exceeded again.

Ohmmeter Test

The ohmmeter can be used to test the SUS, as shown in Figure 4.34. Table 4.6 shows the test results.

4.30 SILICON-CONTROLLED RECTIFIERS

Like the SUS, the silicon-controlled rectifier (SCR) is turned on when a forward threshold voltage is reached. Unlike the SUS, however, the SCR can be controlled to reduce the value of a breakover voltage required for turn-on. This is done by applying a voltage and current to the gate of the device. Figure 4.35 shows the physical construction, symbol, and characteristic curve for the SCR.

The SCR conducts in one direction only and is called a *unilateral device*. SCRs can switch currents up to 2000 A.

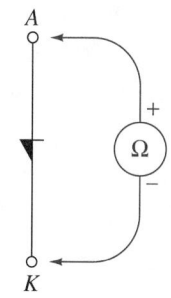

FIGURE 4.34 Testing the SUS.

TABLE 4.6 Test Results

If Ohmmeter Indicates	The SUS Is
∞	OK
Low value	Shorted

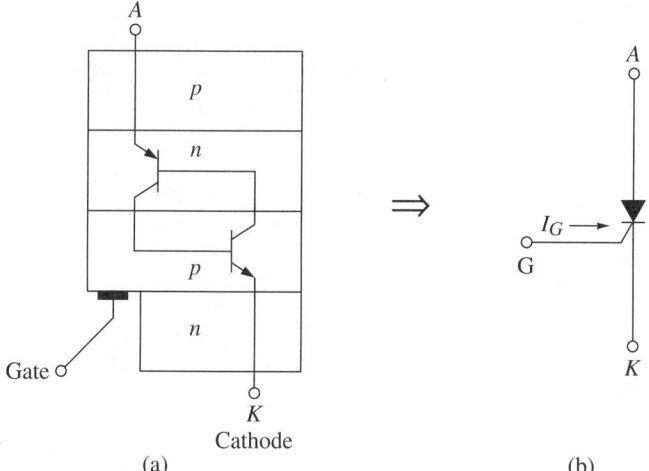

(a)   (b)

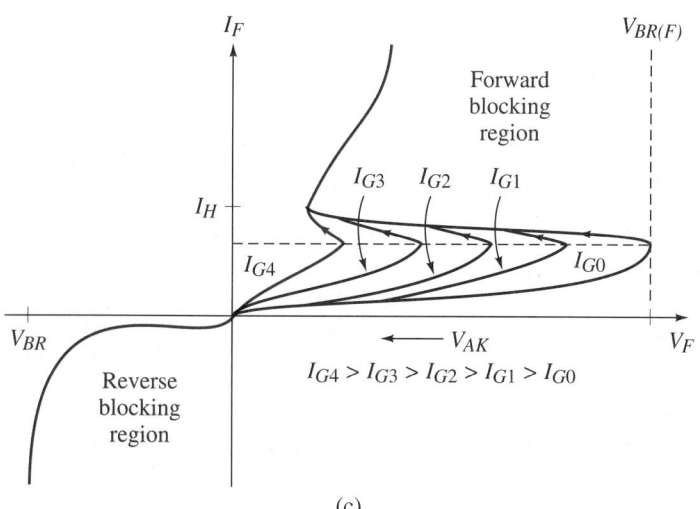

(c)   FIGURE 4.35 The SCR.

4.31 TESTING SILICON-CONTROLLED RECTIFIERS

Test circuit to determine the dynamic characteristics of the SCR.

Dynamic Tests

A test circuit that can be used to determine the dynamic characteristics of the SCR is shown in Figure 4.36. The gate current (I_{GT}) and gate-cathode voltage (V_{GK}) control the forward-bias voltage (V_{AK}) at which the SCR fires. When using this circuit, the following sequence is appropriate:

1. Set V_S below the forward blocking-voltage level.

2. Set $V_G = 0$ and close switch S_1.

3. The SCR should be off. Increase V_G until the SCR fires (turns on).

4. Read the value of I_{GT}. This value is the minimum value of gate-trigger current required to fire the SCR at this anode-cathode voltage setting.

5. Read the value of V_{AK}. It should be near zero if the SCR is conducting and equal to V_S if the SCR is not conducting.

6. If the SCR is off, increase the value of V_G. If it remains off, the SCR is open and must be replaced.

7. If the SCR is good and conducting, reduce the value of V_S to the turn-off point while observing I_A. This value of I_A at turn-off is the holding-current level for the SCR.

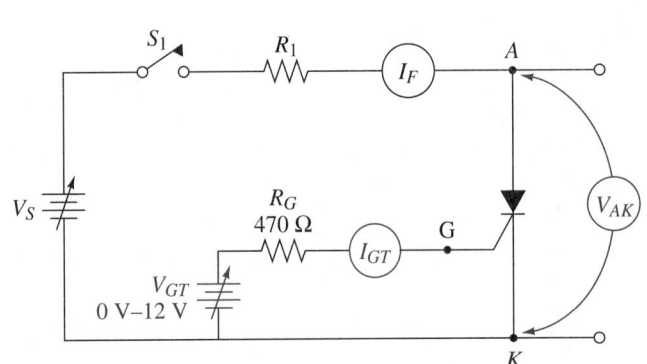

FIGURE 4.36 SCR test circuit.

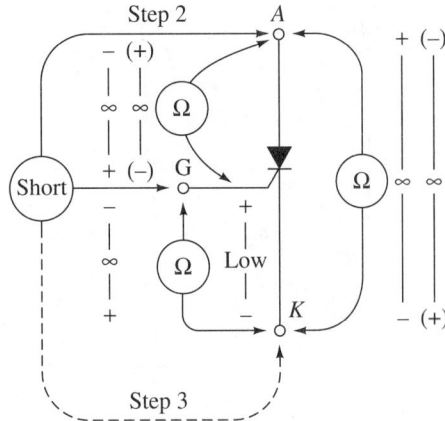

FIGURE 4.37 SCR tests.

Ohmmeter Tests

Refer to Figure 4.37.

Step 1. The anode-cathode can be tested by connecting the ohmmeter as shown. An infinite ohms value indicates a good device and a low ohms value indicates a shorted one.

Step 2. If a jumper wire is used to short the gate to anode, the SCR should switch on and indicate a low value of ohms.

Step 3. Remove the jumper wire and the SCR should stay on.

Step 4. Now use the jumper wire to short the anode to the cathode, and the SCR should switch off.

In-circuit testing of the SCR is preferred. If the SCR is open, forced triggering will do nothing. If it is shorted, it will conduct regardless of the polarity of V_{AK}. Voltage measurements at the gate, anode, and cathode should determine whether or not the SCR is faulty. If this is not practical, conduct the dynamic tests covered earlier in this section.

4.32 DIACS

The diac is a switching device that conducts in both directions. The terms *anode* and *cathode* do not apply, and the terminals are usually identified as MT1 and MT2. The diac acts like two paralleled silicon unilateral switches. The construction and symbol are shown in Figure 4.38. The diac has a threshold (breakover) voltage for each polarity. When on, the diac will drop about 1 V between MT2 and MT1. It turns off when a minimum current (the holding current) is reached. The characteristic curve is shown in Figure 4.39.

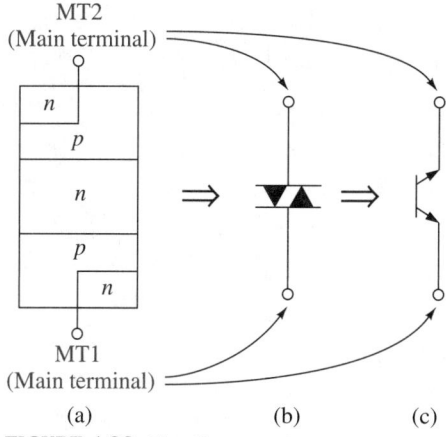

(a)   (b)   (c)

FIGURE 4.38 The diac.

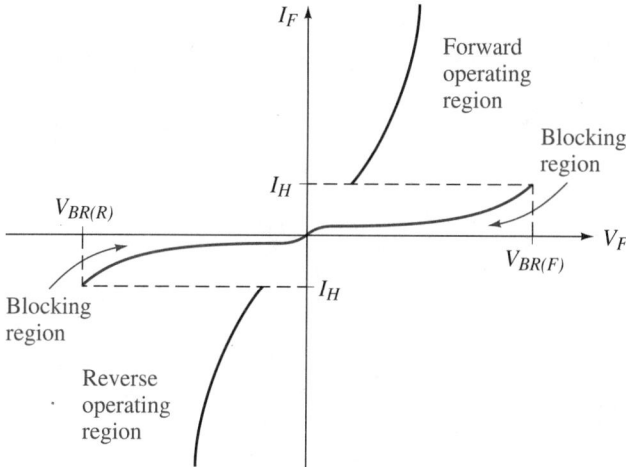

FIGURE 4.39 Diac characteristic curve.

4.33 TESTING DIACS

Ohmmeter Tests

The ohmic tests (Figure 4.40), should show an infinite value of resistance in both directions when the diac is not conducting. If a low resistance is shown, the diac is shorted. When conducting an in-circuit test for conduction, the voltage drop across the diac should be about 1 V.

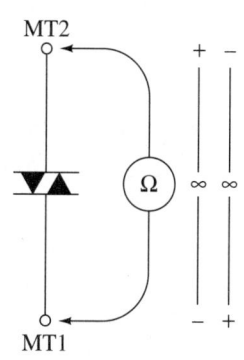

FIGURE 4.40 Diac tests.

4.34 SILICON BILATERAL SWITCHES

The silicon bilateral switch (SBS) shown in Figure 4.41, like the diac, is a bilateral device. However, the SBS is faster, breaks over at lower voltages, has a gate terminal, and is normally in an integrated circuit package. The SBS can handle current levels comparable to those of the diac.

4.35 TESTING SILICON BILATERAL SWITCHES

The SBS can be tested with the ohmmeter, as shown in Figure 4.42.

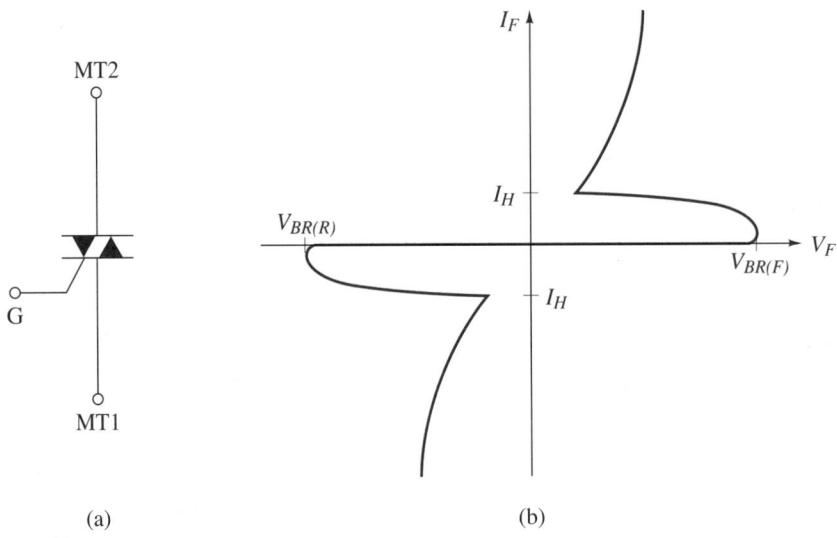

(a)   (b)

FIGURE 4.41 The SBS.

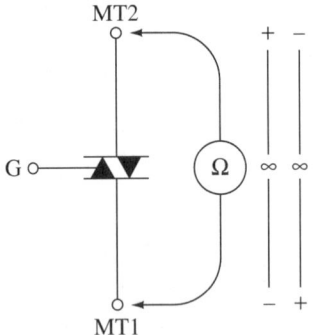

1. The ohmmeter should indicate infinite resistance when the SBS is off.

2. When turned on, the voltage drop from MT2 to MT1 will be about 1.0 V.

FIGURE 4.42 Testing the SBS.

4.36 TRIACS

The triac shown in Figure 4.43 is similar to two parallel SCRs. The device is bilateral. It cannot, however, switch rapidly and is best used at the low frequency of power line voltage (60 Hz).

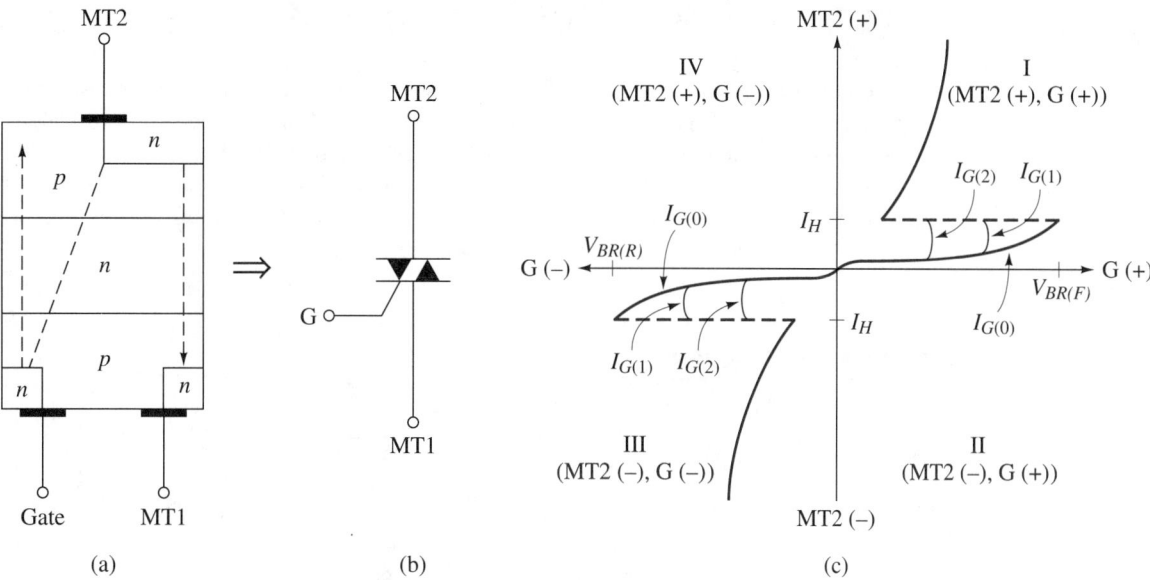

FIGURE 4.43 Triacs.

Figure 4.43(c) is the characteristic curve for the triac. Quadrant I operation has the same characteristic as the SCR. Quadrant I operation tells us the following:

1. Triacs are not normally turned on by using breakover voltages but rather by a gate-trigger voltage causing sufficient gate current (I_G) for triggering.

2. The triac is often falsely triggered by noise or even increases in temperature.

3. Once in conduction, the triac remains on unless the current drops below the holding current (I_H).

4. Electron current is from MT1 to MT2 through one-half of the device on the left side.

Quadrant II operation also gives current flow through the left side of the device.

Quadrant III and Quadrant IV operation provides current flow through the right half of the device. See Figure 4.44. The values of V_{GT} are found in the data sheet for the particular device.

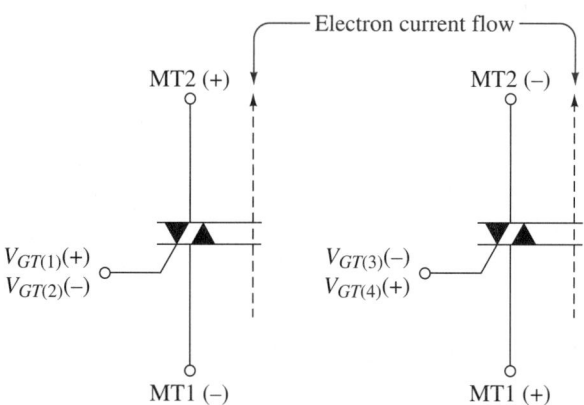

FIGURE 4.44 Quadrants III and IV.

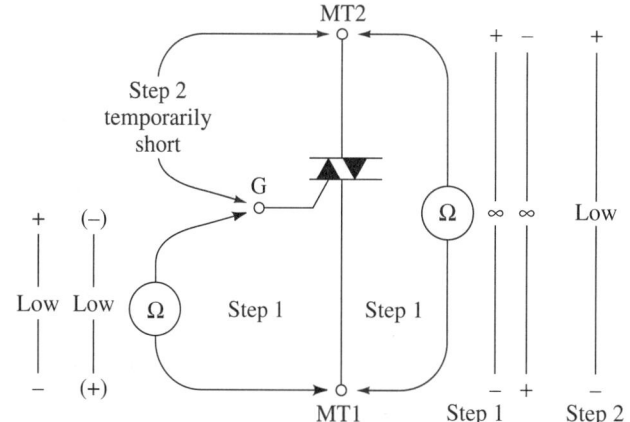

FIGURE 4.45 Testing triacs.

4.37 TESTING TRIACS WITH THE OHMMETER

The ohmmeter should indicate infinite resistance in either direction between MT2 and MT1, as shown in Figure 4.45. When you short the gate terminal to the MT2 terminal, the triac should switch on, giving a low resistance reading on your ohmmeter. Remove the short and the triac should remain on (Table 4.7). Since this will test only one-half of the triac, you need to reverse the ohmmeter leads and test the other half, as shown in Figure 4.46. There should be no continuity until you short the gate to MT2.

TABLE 4.7 Test Results

	G	MT2	MT1	Indication
Step 1:	+		−	Low
	−		+	Low
		+	−	∞
		−	+	∞
Step 2:	Short Gate to MT2.			
		+	−	Low
			+	

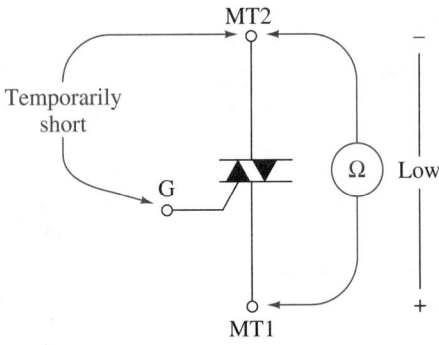

FIGURE 4.46 Testing the second half of the triac.

4.38 TESTING TRIACS IN CIRCUIT

When testing a triac in circuit, the voltmeter should show a very small voltage drop between MT2 and MT1 when the device has been triggered. If not, the triac is open and must be replaced.

4.39 TRANSISTORS

The bipolar transistor is a semiconductor device made of three layers of *p*- and *n*-type materials, as shown in Figure 4.47. The symbols for *npn* and *pnp* transistors are shown in Figure 4.48. As you can see from the physical construction, the transistor is actually

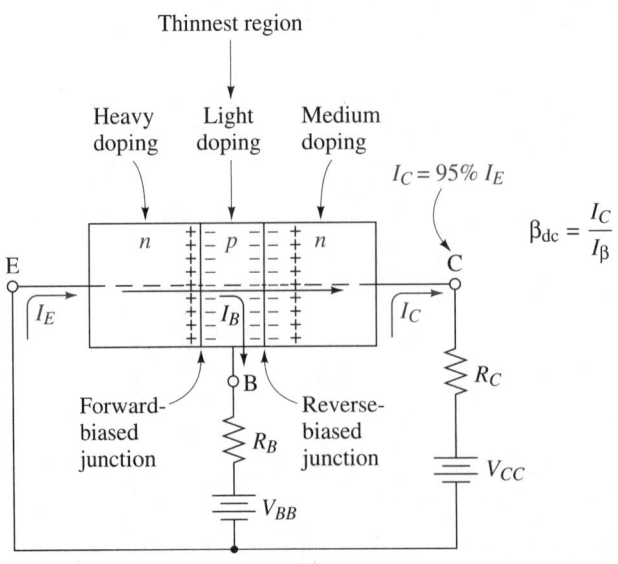

FIGURE 4.47 The transistor.

two *pn* junctions joined together. When properly biased by external voltages, the collector current will be about 99% of the emitter current. The remaining 1% will flow in the base circuit. The pertinent voltages are listed in Figure 4.49. How well the transistor operates as an amplifier is determined by its current gain, the dc beta, which is designated as h_{FE} or β.

$$\beta_{dc} = I_C/I_B \qquad (4.3)$$

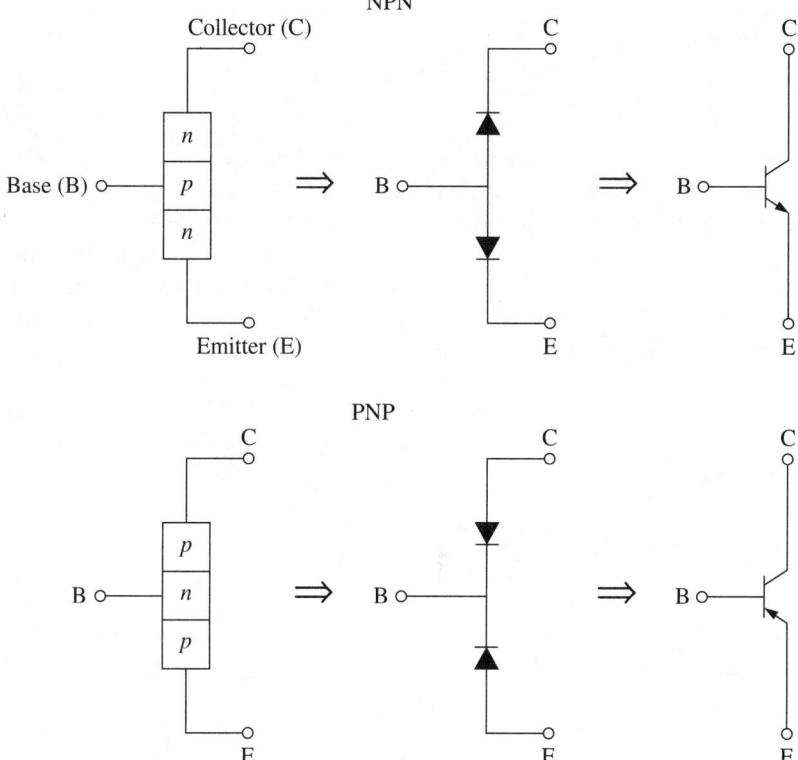

FIGURE 4.48 Transistor symbols.

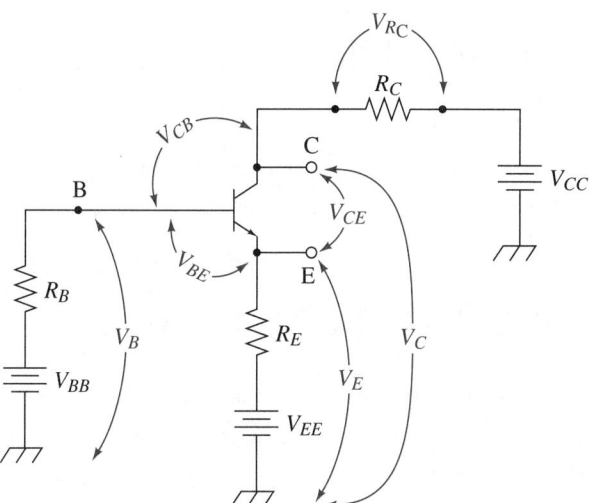

FIGURE 4.49 Transistor voltages.

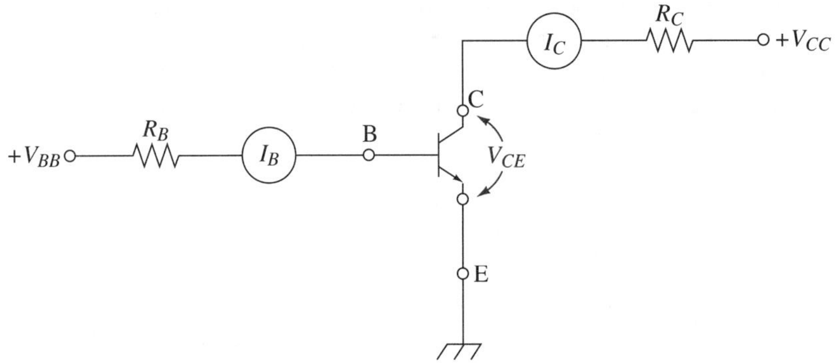

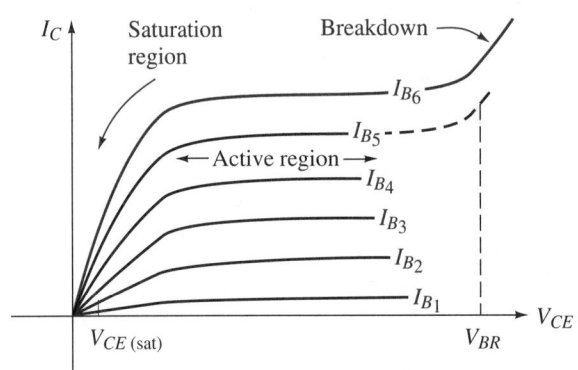

FIGURE 4.50 Test circuit.

β_{dc} can easily be measured by using the circuit in Figure 4.50.

The first step is to establish a fixed level of base current and then begin to increase V_{CC} from 0 to some maximum value not to exceed the manufacturer's listed breakdown voltage for the collector-emitter. By monitoring the level of collector current and plotting I_C versus I_B you can develop a characteristic curve for your transistor. If you vary the level of base current while making corresponding collector-current measurements, you will have a set of characteristic curves, as shown in Figure 4.51.

TABLE 4.8

Step	Ohmic Checks (npn)			
	B	E	C	Condition
1	−	✕	+	High
2	+	✕	−	Low
3	−	+	✕	High
4	+	−	✕	Low
5	✕	−	+	High
6	✕	+	−	High
(pnp)				
7	+	−	✕	High
8	−	+	✕	Low
9	+	✕	−	High
10	−	✕	+	Low
11	✕	+	−	High
12	✕	−	+	High

FIGURE 4.51 Characteristic curves.

4.40   TRANSISTOR FAILURE MODES

When the transistor is subjected to excessive levels of current,

1. The base-collector *pn* junction will short circuit first and eventually burn open.
2. The base-emitter junction burns open immediately.
3. Collector to emitter shorts are common.

4.41   TESTING TRANSISTORS OUT OF THE CIRCUIT

Leakage (Ohmic) Tests with the Ohmmeter

A few simple checks with the ohmmeter can determine the condition of your transistor (Figure 4.52). Table 4.8 indicates the pertinent measurements across the junctions for both *npn* and *pnp* transistors.

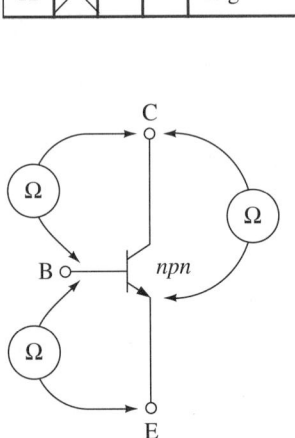

FIGURE 4.52 Transistor tests.

Be careful: Some VOMs have range settings with voltages high enough to cause excessive levels of current, which can damage the transistor. In general, you should avoid using the R × 1 range.

CAUTION

Watch out for high voltages!

Transistor Testers

The *conduction tester* in common use is rather like an ohmmeter with a built-in transistor socket. See Figure 4.53. This is primarily a go–no go test.

The *oscillator tester* basically places the transistor in a simple oscillator circuit and uses a meter to indicate the presence or absence of oscillation. A typical circuit is shown in Figure 4.54.

The *beta tester* uses a simple circuit to measure the collector current and the base current. In some testers the meter measuring the collector current is calibrated in beta units (Figure 4.55).

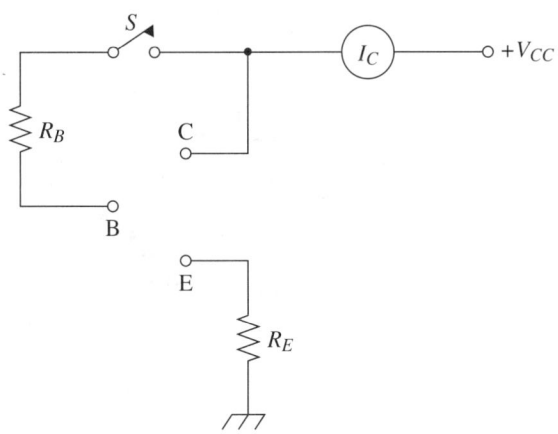

FIGURE 4.53 Go–no go test.

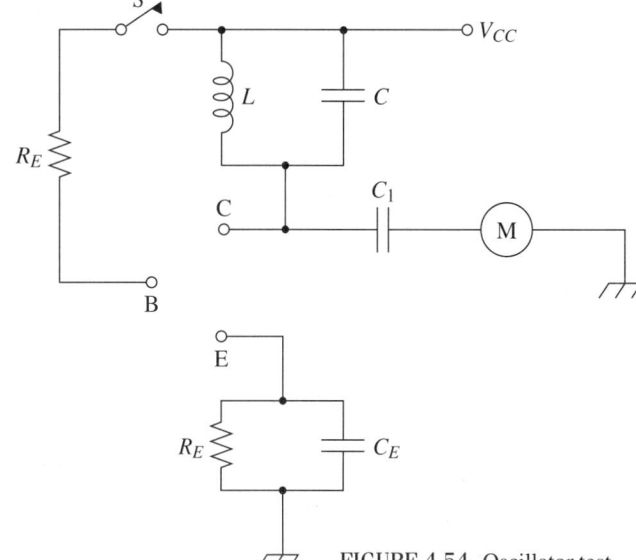

FIGURE 4.54 Oscillator test.

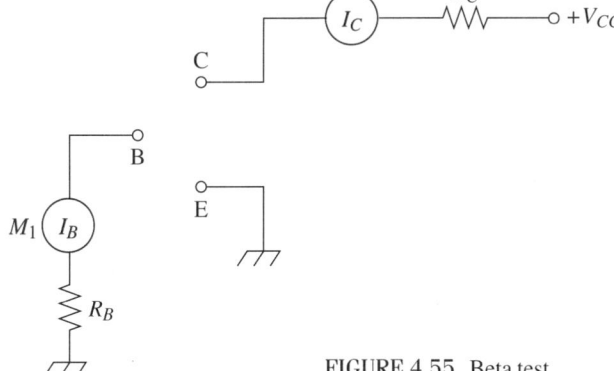

FIGURE 4.55 Beta test.

The *curve tracer* uses a built-in oscillator circuit to display the I_C-versus-V_{CE} curves. It is important to remember that most transistors housed in metal cases have the collector tied internally to the metal case.

When troubleshooting transistor circuits, the basic method used is to measure the dc voltage levels at the IC pins or the discreet component terminals. Any absence of voltage or abnormal readings will give you a place to start. You can learn a great deal by looking at the terminal voltages of the transistor. For normal amplifier action, the base-emitter junction must be forward-biased and the base-collector junction must be reverse-biased. For *npn* transistors, the base must be more positive than the emitter, and the collector is

Measure dc voltage levels.

positive. For *pnp* transistors, the base will be more negative than the emitter, and the collector has to be negative.

4.42   STATIC ANALYSIS OF TRANSISTOR CIRCUITS

The external dc voltage levels at the terminals of the transistor determine the transistor's quiescent (dc) operating point. The quiescent (Q) point sets the position of the transistor on the set of collector curves. By performing a dc analysis of the circuit, as shown in Figure 4.56, you can calculate the pertinent voltages, locate the Q point, and establish the dc load line for the circuit. Once the load line has been drawn, the operating current and collector-emitter voltage can be readily determined. If this Q point is not near the center of the dc load line, the circuit will probably go into saturation or cutoff and the ac output will be distorted.

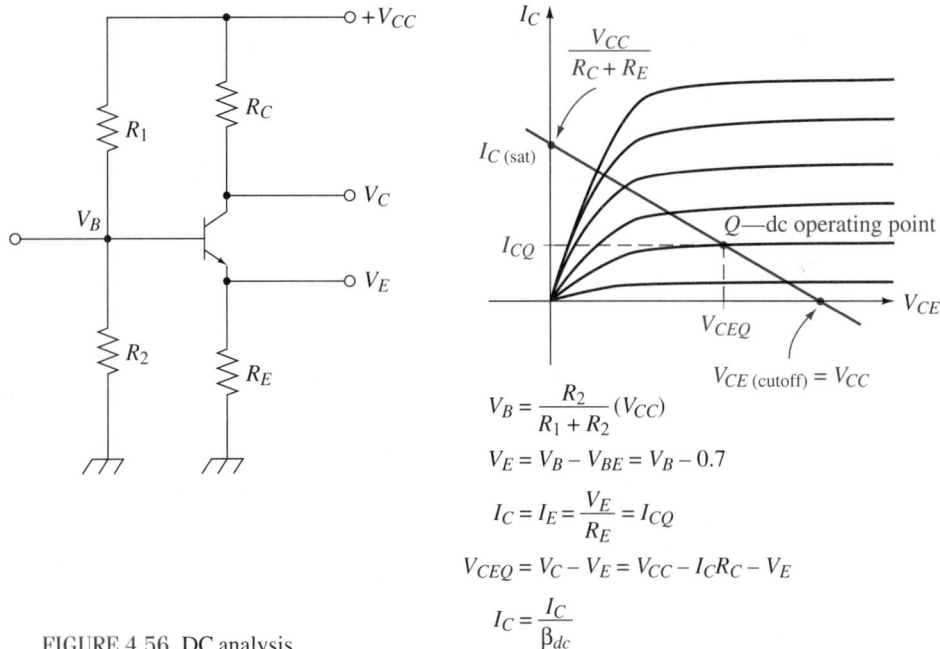

$$V_B = \frac{R_2}{R_1 + R_2}(V_{CC})$$

$$V_E = V_B - V_{BE} = V_B - 0.7$$

$$I_C = I_E = \frac{V_E}{R_E} = I_{CQ}$$

$$V_{CEQ} = V_C - V_E = V_{CC} - I_C R_C - V_E$$

$$I_C = \frac{I_C}{\beta_{dc}}$$

FIGURE 4.56 DC analysis.

4.43   TROUBLESHOOTING TRANSISTOR CIRCUITS

Analyze the dc trouble spots.

Let's analyze the possible dc trouble spots of a circuit like the one in Figure 4.57. First, we will redraw the circuit to use *pn* junctions in place of the transistor. You should verify that Figure 4.57 accomplishes this. Table 4.9 lists as many troubles as possible and also provides the measured voltages for the circuit. Take a few minutes to study these.

Let's now proceed to the next circuit (Figure 4.58). Here we have a voltage-divider-biased circuit. The normal voltages are given, and Table 4.10 provides a list of troubles. A closer look tells us the following:

- If R_1 *opens:*   The base current and base voltage (V_B) both go to zero. The transistor is biased off, and both the emitter voltage and current equal zero. Since $I_E = I_C$, $I_C R_C = 0$ and $V_C = V_{CC}$.
- If R_2 *opens:*   The current I_{R_1} ($I_B + I_{R_2}$) decreases to I_B. V_{R_1} decreases, V_B increases, and the transistor saturates. At this point $V_C = V_E$, $I_C = I_{C(sat)}$, $V_B = V_E + 0.7$, and $V_E = I_{C(sat)} R_E$.
- If R_C *opens:*   No collector current flows and $I_E = I_B$. The value of V_E will be very small, V_B will be $V_E + 0.7$, and V_C will equal V_E.

- If R_E opens: $I_E = I_B = I_C = 0$. The voltmeter across the emitter will complete the circuit and V_E will measure higher than normal. V_C will equal V_{CC}, V_B will be normal, V_E will be slightly higher, and V_C will equal V_{CC}.

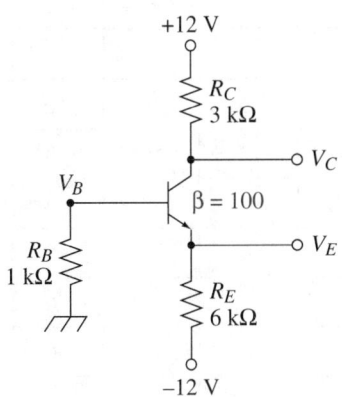

Redraw as:

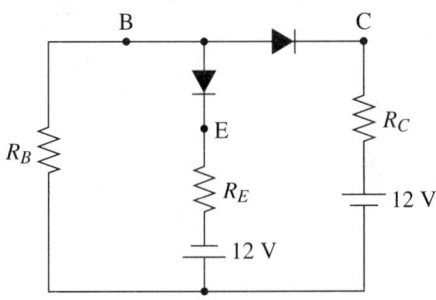

FIGURE 4.57 Troubleshooting.

TABLE 4.9 Possible Problems

Trouble	Measured Voltages		
	V_B	V_E	V_C
R_E short	−11.21	−12	−12
R_C short	−0.013	−0.644	+12
$R_C + R_E$ short	4.6	0.8	Transistor destroyed
B − E diode short	−1.67	−1.67	+12
C − B diode short	0.81	1.47	1.46
Open R_C	−2.29	−1.57	−2.27
Open R_B	−10.71	−12	+12
Open R_E	0	0	+12
All junctions bad	0	−12	+12

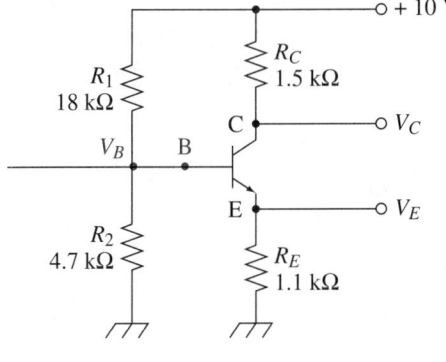

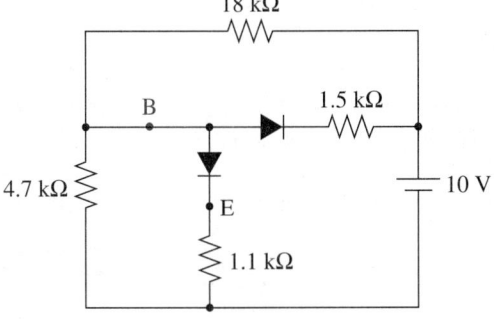

FIGURE 4.58 Troubleshooting.

TABLE 4.10 Possible Problems

Normal	2.03	1.36	8.17
Trouble	V_B	V_E	V_C
R_1 short	10	9.22	9.24
R_1 open	0	0	10
R_2 short	0	0	10
R_2 open	5.08	4.38	4.46
R_C short	2.03	1.36	10
R_C open	0.96	0.31	0.33
R_E short	0.72	0	0.09
R_E open	2.05	1.63	10
All junctions bad	2.05	0	10

When you get through with the *npn* circuit, go to the *pnp* circuit in Figure 4.59 and the corresponding troubles listed in Table 4.11.

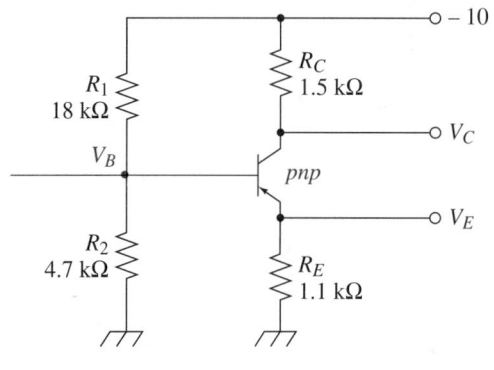

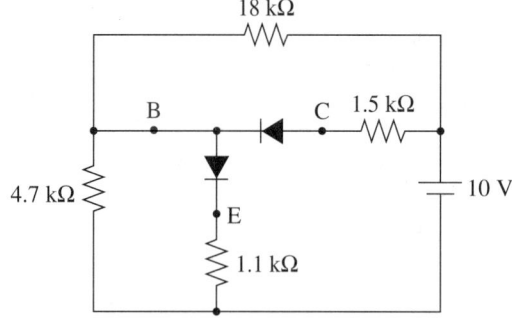

FIGURE 4.59 Troubleshooting.

TABLE 4.11 Possible Problems

Normal	−2.1	−1.42	−8.2
Trouble	V_B	V_E	V_C
R_1 open	0	0	−10
R_1 short	−10	−9.34	−9.3
R_2 open	−5.22	−4.49	−4.5
R_2 short	0	0	−10
R_C short	−2.1	−1.42	−10
R_C open	−1.02	−0.32	−0.3
R_E short	−.75	0	0
R_E open	−2.14	−1.73	−10
All junctions bad	−2.14	0	−10

4.44   THE "CLICK" TEST

One quick method to check if a transistor is functioning is to short-circuit the base-emitter junctions for a second and listen for a "click" sound in the speaker. Observe the collector voltage; if you see the V_C go up toward V_{CC} when removing the forward bias, the transistor is operating.

4.45   APPLYING A FORWARD BIAS

If you apply a forward bias to the base and the transistor is good, you should see an increase in the emitter voltage.

4.46   STATIC TROUBLESHOOTING
CLASS A AMPLIFIERS

How to troubleshoot class A amplifiers

A class A amplifier is one that operates over the complete 360° of the input cycle. The amplifier is operating *linearly,* with the output becoming an expanded version of the input signal. See Figure 4.60.

When doing the dc analysis of this amplifier, the transistor bias voltages are taken without any input signal being processed.

1. Measure V_{CE}. It should be approximately equal to $\frac{1}{2}V_{CC}$ as a typical value.
    If $V_C = V_{CC}$, the transistor is open or the bias voltages are wrong.
    If V_C is very low, the transistor is either saturated or shorted.

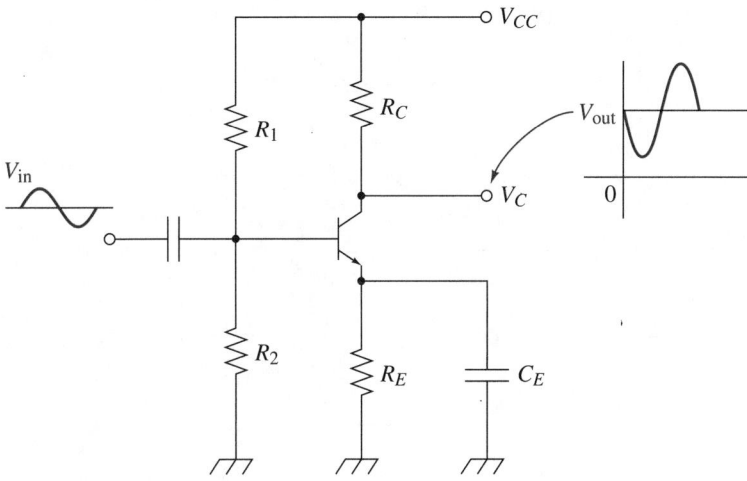

FIGURE 4.60 Class A amplifier stage.

2. Measure V_{BE}. If it is much lower than 0.7 V (silicon), the base-emitter diode may be shorted.

If V_{BE} is much greater than 0.7, the base-emitter diode is probably open.

Note: The polarity of the base voltage should be the same as that of the collector supply.

4.47 TROUBLESHOOTING CLASS B AMPLIFIER STAGES

The class B amplifier circuit operates for 180° of the input cycle and is used in the output power stage of amplifiers. It uses complementary transistor pairs such as the 2N3904 and the 2N3906 or two matching transistors coupled to transformer secondaries (also known as a *push-pull* stage). See Figures 4.61 and 4.62.

How to troubleshoot class B amplifiers

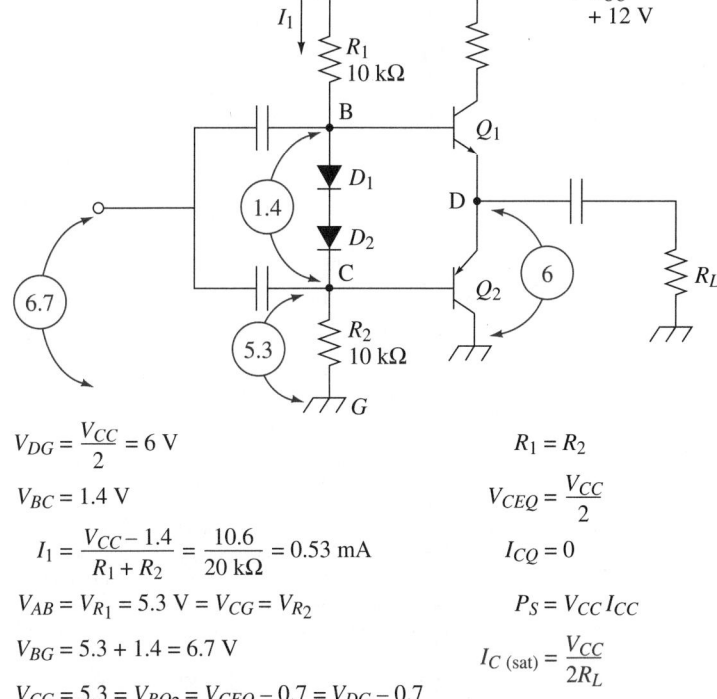

$$V_{DG} = \frac{V_{CC}}{2} = 6 \text{ V} \qquad\qquad R_1 = R_2$$

$$V_{BC} = 1.4 \text{ V} \qquad\qquad V_{CEQ} = \frac{V_{CC}}{2}$$

$$I_1 = \frac{V_{CC} - 1.4}{R_1 + R_2} = \frac{10.6}{20 \text{ k}\Omega} = 0.53 \text{ mA} \qquad\qquad I_{CQ} = 0$$

$$V_{AB} = V_{R_1} = 5.3 \text{ V} = V_{CG} = V_{R_2} \qquad\qquad P_S = V_{CC} I_{CC}$$

$$V_{BG} = 5.3 + 1.4 = 6.7 \text{ V}$$

$$I_{C \text{ (sat)}} = \frac{V_{CC}}{2R_L}$$

$$V_{CG} = 5.3 = V_{BQ2} = V_{CEQ} - 0.7 = V_{DG} - 0.7$$

FIGURE 4.61 Complementary symmetry.

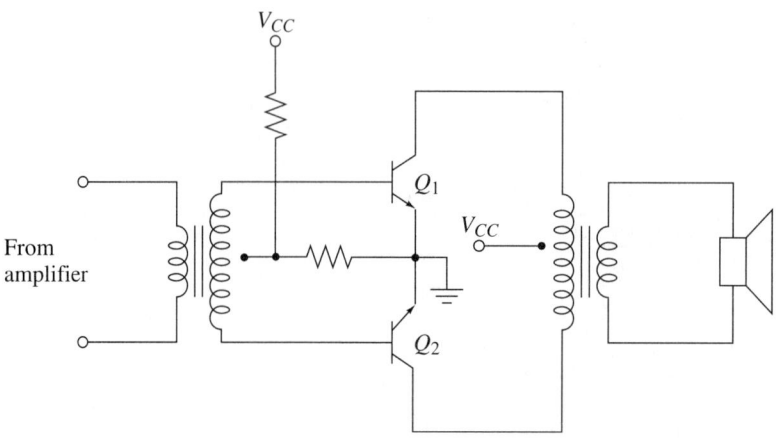

FIGURE 4.62 Push-pull amplifier.

The circuit can be easily checked by measuring the bias voltages. Each transistor should be biased slightly "on" to avoid the *crossover distortion* shown in Figure 4.63. Thus, with no input signal, V_{BE} should be about 0.7 V.

The *push-pull* stage in Figure 4.62 can be quickly checked by measuring the collector and emitter voltages. The complementary symmetry amplifier stage of Figure 4.61 is the more popular and least expensive. The characteristic curves are shown in Figure 4.64.

Table 4.12 lists the normal voltage levels for the circuit along with a selection of troubles.

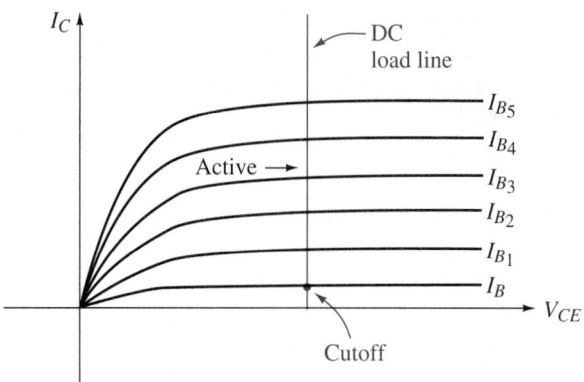

FIGURE 4.63 Crossover distortion.

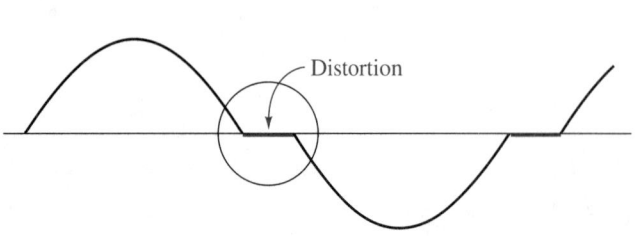

FIGURE 4.64 Characteristic curve.

TABLE 4.12 Normal Voltage Levels and Possible Problems

V_{DG}	V_{CG}	V_{BC}	Condition
6	5.3	1.4	Normal
>6	>5.3	>6.7	R_2 open
<6	≈0	≈0	R_1 open
Low	Low	Low	D_1 acceptable; D_2 open
Low	5.3	5.3	Q_1 open
High	5.3	5.3	Q_2 open

4.48 TROUBLESHOOTING CLASS C AMPLIFIER STAGES

The class C amplifier circuit, commonly used in radio frequency amplifiers, operates (conducts) for a very short period of time, much less than 180° of the input signal. The basic circuit is shown in Figure 4.65. The negative voltage on the base biases the transistor into cutoff. Therefore, $I_C = 0$, $V_{CE} = V_{CC}$, and the dc load line is vertical, as in Figure 4.66.

With no ac input signal, $V_{CEQ} = V_{CC}$ and $I_{CQ} = 0$. Also check V_B, V_{BE}, and V_E. If $V_E = 0$,

- The base emitter diode may be opened.
- The collector emitter may be shorted.
- The collector may be opened.

If V_E is too high, the transistor is shorted.

How to troubleshoot class C amplifiers

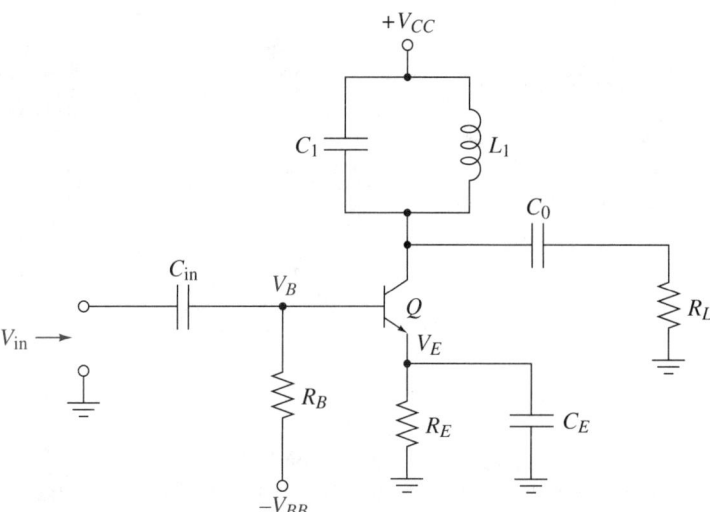

FIGURE 4.65 Class C amplifier.

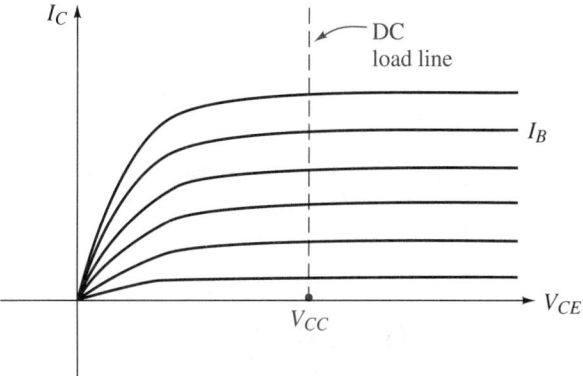

FIGURE 4.66 DC load line.

4.49 TROUBLESHOOTING UNIJUNCTION TRANSISTORS

The unijunction transistor (Figure 4.67) is a triggered device. The emitter voltage (V_E) is increased until a threshold voltage (V_P) is reached, at which time the E-B_1 junction becomes forward-biased ($V_E = V_k + 0.7$ V) and R_{B_1} decreases as current begins to flow.

How to troubleshoot unijunction transistors

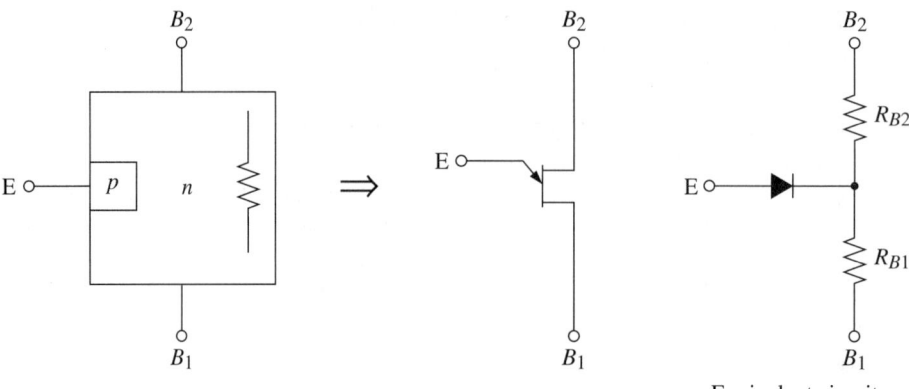

Equivalent circuit

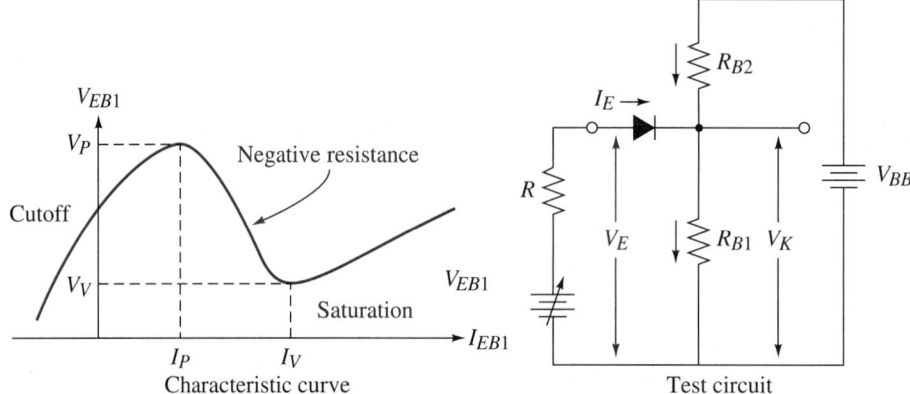

FIGURE 4.67 Unijunction transistor.

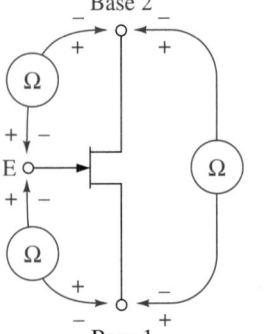

FIGURE 4.68 Testing the UJT.

TABLE 4.13 Ohmic Tests

E	B_1	B_2	Condition
+	−	✕	Low
−	+	✕	High
+	✕	−	Low
−	✕	+	High
✕	+	−	Medium
✕	−	+	to high

The decreasing R_{B_1} allows more current to flow and the bulk resistance, R_{B_1}, decreases even further until the minimum (valley point) is reached. The value of V_k is found by calculating the voltage across the lower base resistance using the voltage-divider equation, as follows:

$$V_k = \frac{R_{B_1} V_{BB}}{R_{B_1} + R_{B_2}}$$

and

$$V_k = nV_{BB} \tag{4.4}$$

n is known as the *intrinsic standoff ratio* and is found on the data sheet for the particular device. So,

$$V_E = nV_{BB} + 0.7 \text{ V} \tag{4.5}$$

4.50   TESTING UNIJUNCTION TRANSISTORS

The unijunction transistor (UJT) is normally tested in the circuit under power. However, you can perform ohmic tests with your ohmmeter as shown in Figure 4.68 and Table 4.13.

4.51   THE PROGRAMMABLE UJT (PUT)

The PUT (programmable unijunction transistor) is a four-layer device much like the SUS/SCR. It can be programmed to trigger at different voltage levels (Figure 4.69).

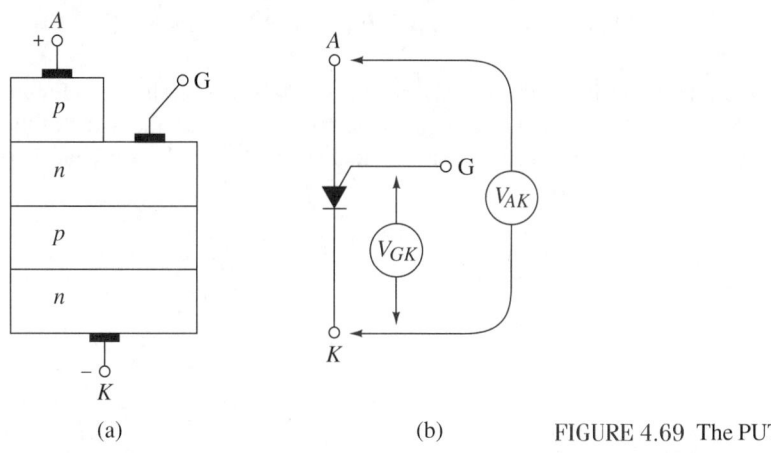

(a)   (b)   FIGURE 4.69 The PUT.

4.52 TESTING THE PUT

Dynamic

A simple test circuit is shown in Figure 4.70. The device turns on when $V_{AK} = V_{GK} = V_P$. Once turned on, the device conducts until the current level drops to I_p. The intrinsic standoff ratio is provided by $R_1 - R_2$ and is derived from the equation

$$n = R_2/(R_1 + R_2) \qquad (4.6)$$

It can, therefore, be varied by changing the values of R_1 and R_2.

Ohmic

The ohmmeter can be used to test the PUT, as shown in Figure 4.71. Table 4.14 shows test results.

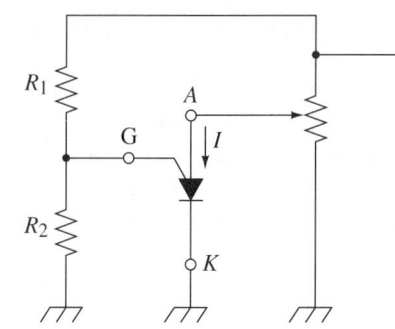

$$V_G = \frac{R_2}{R_1 + R_2}(V_S)$$

$$n = \frac{R_2}{R_1 + R_2}$$

$$V_P = V_G + 0.7 \text{ V}$$

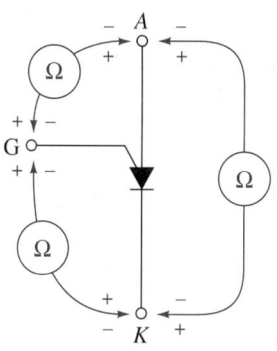

FIGURE 4.71 Testing the PUT.

TABLE 4.14 Test Results

A	K	G	Condition
+	−	⊠	High
−	+	⊠	High
+	⊠	−	Low
−	⊠	+	High
⊠	+	−	High
⊠	−	+	High

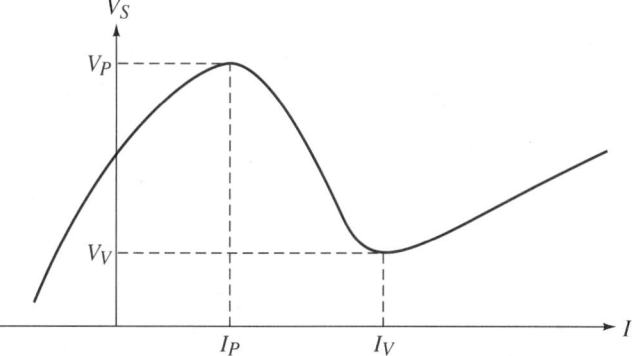

FIGURE 4.70 PUT test circuit.

4.53   THE JUNCTION FIELD-EFFECT TRANSISTOR

The junction field-effect transistor (JFET) is a three-terminal device, as shown in Figure 4.72. Its operation is based on the application of an electric field to vary the width of the channel and control the source-to-drain current. A wider channel offers less resistance to current flow.

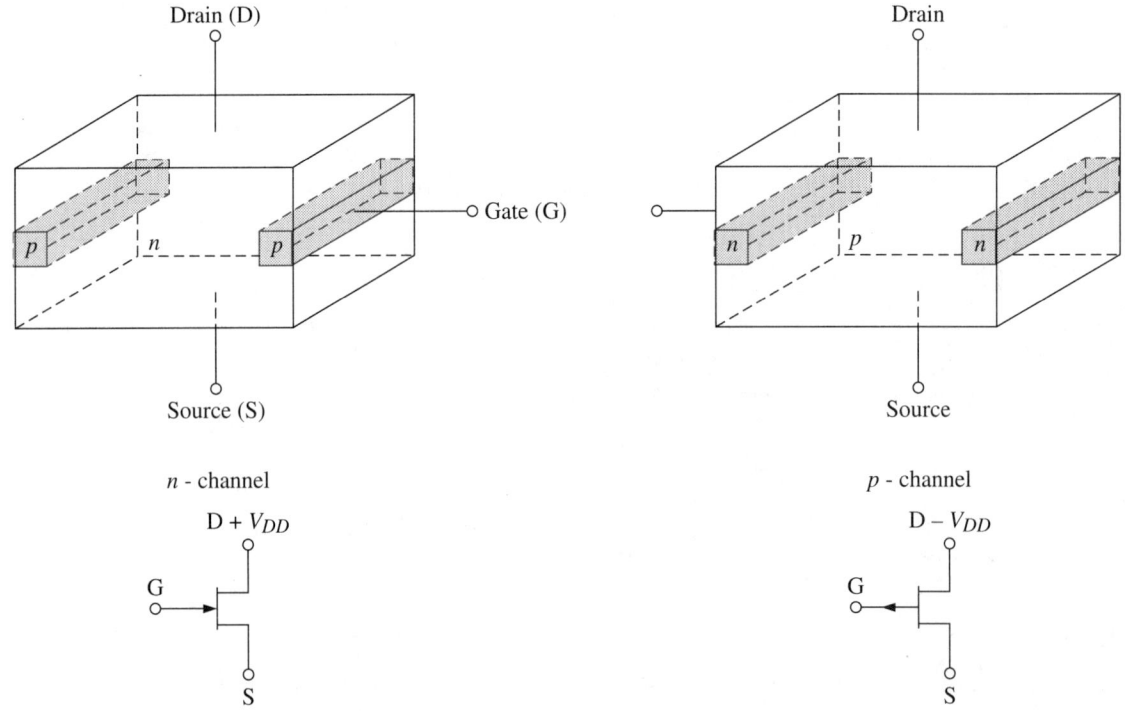

FIGURE 4.72   The JFET.

It is possible to decrease the gate width by applying a reverse bias from the gate-source junction, which creates a depletion region. The depletion region causes less current to flow.

Let's take a closer look at this device: We will start with the circuit in Figure 4.73 and close switch S_2 while keeping S_1 open. Now, let's slowly increase V_{DS} while observing I_D. The curve trace is the characteristic drain curve for a JFET. At a value of V_{DS} called the pinch-off voltage (V_P), the drain current levels off and remains constant until the breakdown voltage is reached. This value of current is the maximum for the device and is called the I_{DSS}. See Figure 4.74.

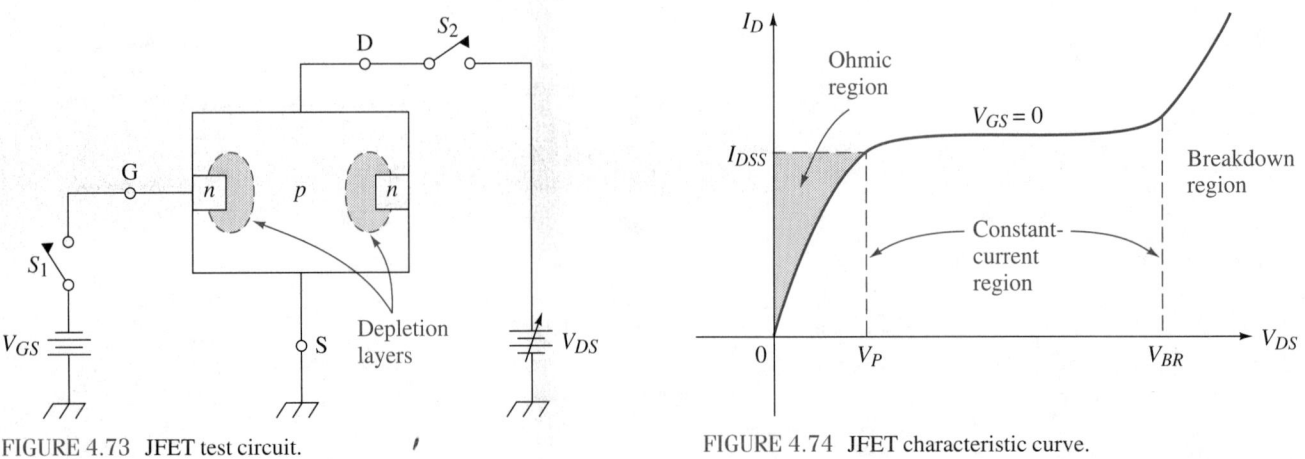

FIGURE 4.73   JFET test circuit.

FIGURE 4.74   JFET characteristic curve.

Next, we will open S_2, close S_1, and vary the level of V_{GS} to -1 V; a depletion layer will again form, as shown in Figure 4.75. Now close S_2 and increase V_{DS}. A new drain curve will be traced. By increasing the value of V_{GS}, a complete set of drain curves (Figure 4.76) is created.

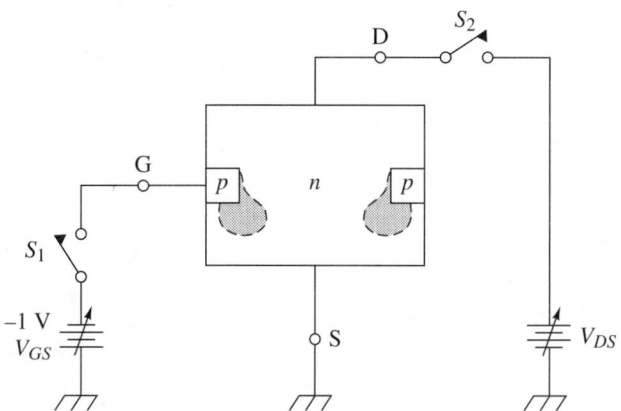

FIGURE 4.75 JFET test circuit.

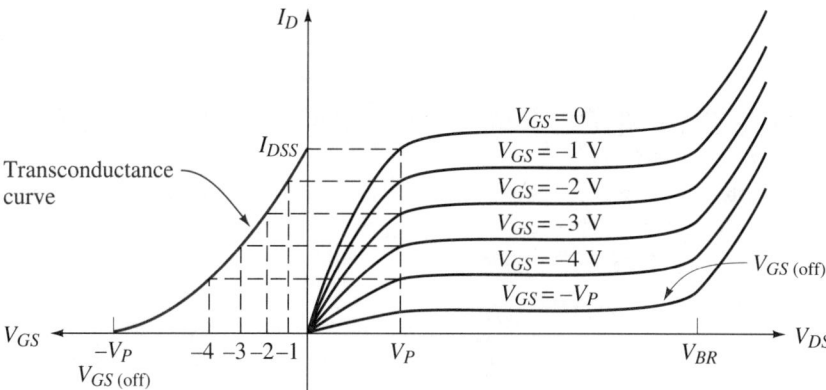

FIGURE 4.76 JFET drain curves.

As V_{GS} is made more negative,

1. The JFET pinches off at lower and lower values of V_P.
2. The constant current value of I_D decreases.
3. I_D drops to zero for values of $V_{GS} = V_P$. This value of V_{GS} is known as $V_{GS(off)}$.

By plotting the drain current against the corresponding values of V_{GS}, we get a characteristic known as the *device transconductance:*

$$g_m = dI_D/dV_{GS} \qquad (4.7)$$

The transconductance curve is described by

$$I_D = I_{DSS}(1 - V_{GS}/V_{GS(off)})^2 \qquad (4.8)$$

g_m is the slope of the curve and can be found at any point by

$$g_m = g_{mo}(1 - V_{GS}/V_{GS(off)}) \qquad (4.9)$$

where g_{mo} = maximum value of g_m at $V_{GS} = 0$. We can approximate the value of g_{mo} from

$$g_{mo} = \frac{2I_{DSS}}{V_{GS(off)}} \qquad (4.10)$$

Figure 4.77 shows a JFET circuit with voltage-divider biasing. The transconductance curve and dc load line are also shown in Figure 4.77(b).

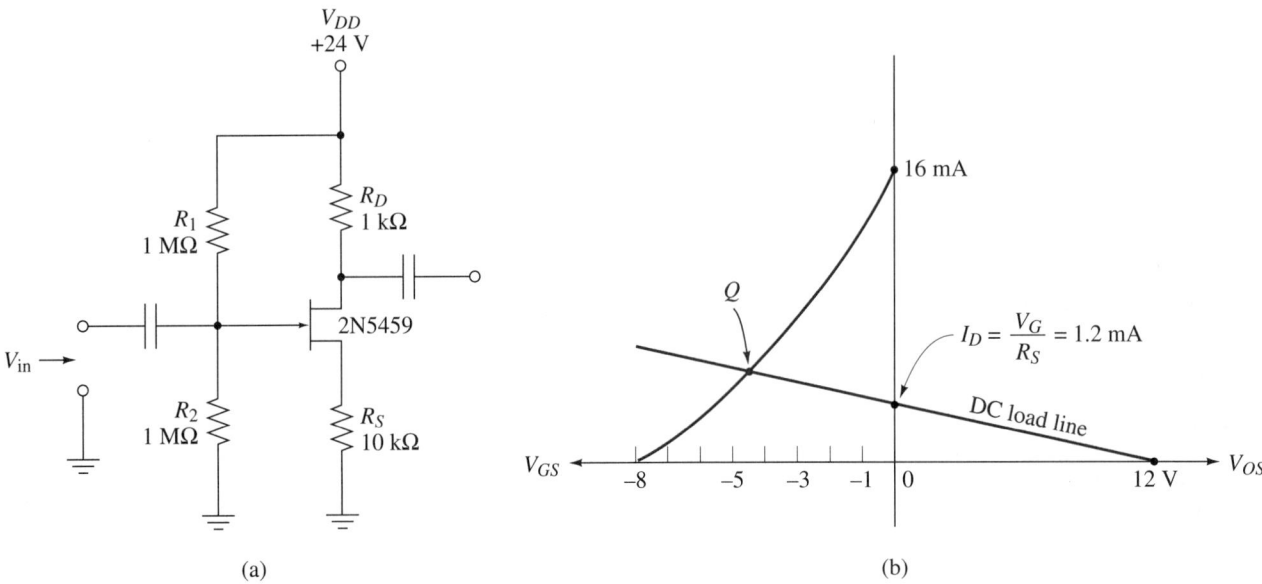

(a)   (b)

FIGURE 4.77 JFET circuit with voltage-divider bias.

4.54   TROUBLESHOOTING JFET CIRCUITS

JFETs can be tested in the circuit. Since there is only one junction that can become open or short, troubleshooting is fairly easy.

Consider the standard voltage-divider biased circuit in Figure 4.78.

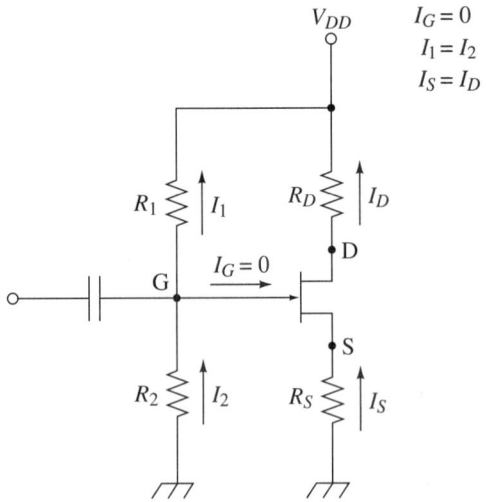

FIGURE 4.78 Standard VDB circuit.

Shorted Junction

If the gate-source junction becomes shorted, V_{GS} will be equal to zero and I_D will increase to the value of I_{DSS}. I_S will no longer equal I_D, and I_1 will no longer equal I_2. Note, also, that $I_G > 0$.

Open Junction

If the gate source junction becomes open, $V_{GS} = 0$, $I_D = I_{DSS}$, and $I_G = 0$.
To check a JFET circuit:

1. Check the output voltage.
2. Check the input voltage.
3. Check V_{DD}.
4. Check the ground connections.

Table 4.15 lists some of the common faults for the circuit and the expected results.

TABLE 4.15 Common Faults

Common Fault	Results
R_1 open	V_{GS} increases. $I_2 = 0$ V_D increases. Circuit operates like a self-biased circuit.
R_2 open	JFET is destroyed. $V_2 = V_{DD}$ $I_{R_1} = 0$ $V_{R_1} = 0$; $V_G = V_{DD}$
R_D open	$V_D = 0$ All currents go to zero.
R_S open	$V_S = V_{DD}$ $V_{RD} = 0$

4.55   TESTING A JFET WITH AN OHMMETER

Since there is a *pn* junction between the gate and the channel, a ohmmeter can be used to test the JFET, as shown in Figure 4.79. Table 4.16 shows test results.

Testing JFETs in the circuit

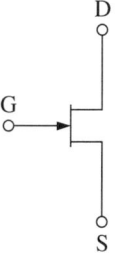

FIGURE 4.79   JFET ohmic tests.

TABLE 4.16

G	D	S	Condition
✕	+	−	Low ($R_{DS(on)}$)
✕	−	+	Low ($R_{DS(on)}$)
+	−	✕	Low
−	+	✕	High
+	✕	−	Low
−	✕	+	High

4.56   MOSFETS

Unlike the JFET, the gate of the metal oxide semiconductor FET, or MOSFET, is insulated by a layer of SiO_2 from the channel. Thus MOSFETs are also known as IGFETs (insulated gate FETs). Since there is no *pn* junction formed, the input resistance is extremely high. The JFET is operated only with a negative voltage on the gate (reverse-biased), creating a large depletion layer in the *pn* junction. This is known as the *depletion* mode of operation. Because the MOSFET's gate is insulated, it can be operated with either a positive or negative voltage on the gate. A positive voltage on the gate will attract electrons toward the gate (or enhance the conduction) and is called the *enhancement* mode.

Caution! The insulated layer is sensitive to static voltages and can be easily destroyed, *Handle it with care.*

CAUTION

Handle with care.

1. Avoid touching the pins.
2. Ship MOSFETs with shorting rings for protection.
3. Keep your body at the common potential when working with MOSFET circuits.
4. Wear a wrist ground strap.
5. Be careful: Do not ground your body when working with voltages greater than 30 V$_{ac}$ or 60 V$_{dc}$.

4.57 DEPLETION-TYPE MOSFETS

The depletion-type MOSFET, or D-MOSFET (Figure 4.80), can operate in either the depletion or the enhancement mode. If V_{GS} is negative (V_{GS} reverse-biased), the channel becomes void of carriers (depleted). Thus the channel resistance increases.

Positive values of V_G (V_{GS} forward-biased) will attract electrons to the channel and enhance current flow. The channel resistance effectively decreases.

The characteristic transconductance curve is shown in Figure 4.81(a) and the set of drain curves is shown in Figure 4.81(b).

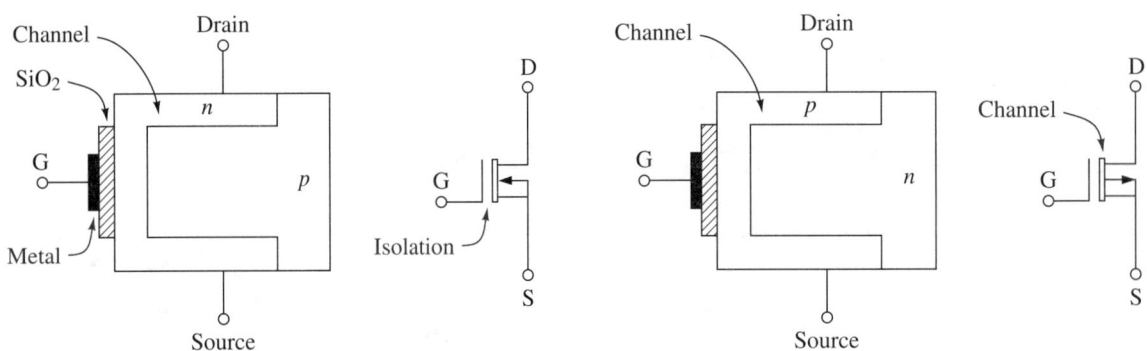

FIGURE 4.80 The D-MOSFET.

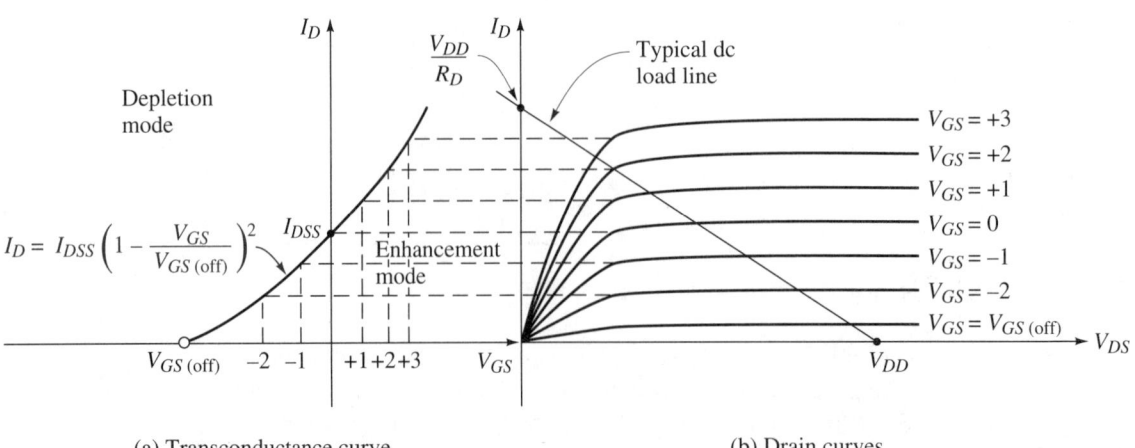

(a) Transconductance curve

(b) Drain curves

FIGURE 4.81 Characteristic transconductance curve.

4.58 ENHANCEMENT MOSFETS

The enhancement MOSFET (E-MOSFET) operates only in the enhancement mode and V_G must be greater than zero. Figure 4.82 shows the construction and symbols for p-channel and n-channel devices. The transconductance curve is shown in Figure 4.83.

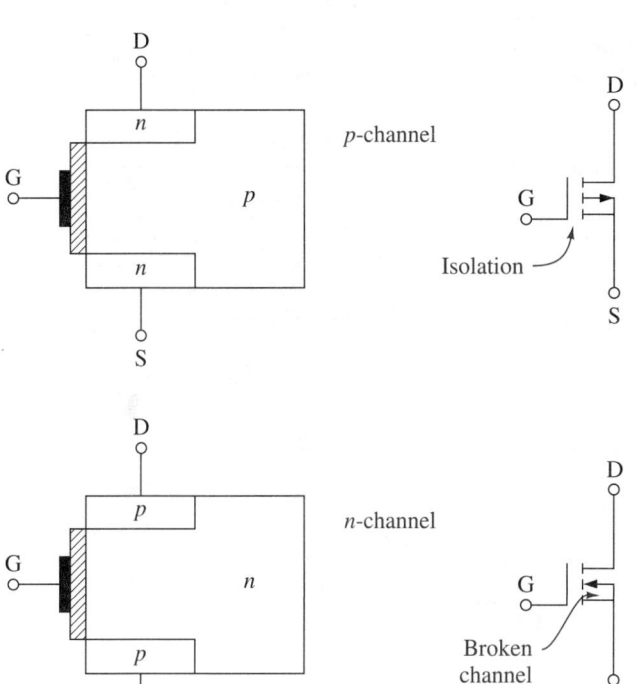

FIGURE 4.82 The E-MOSFET.

$$I_D = \frac{I_{D\,(on)}\,[V_{GS} - V_{GS\,(th)}]^2}{[V_{GS\,(on)} - V_{GS\,(th)}]^2}$$

FIGURE 4.83 Transconductance curve.

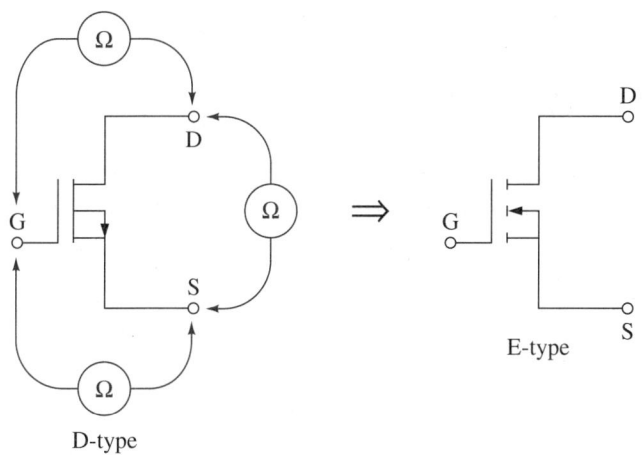

FIGURE 4.84 Ohmic tests.

4.59 TESTING MOSFETS WITH AN OHMMETER

D- and E-type MOSFETs can be tested with an ohmmeter, as shown in Figure 4.84. Table 4.17 shows test results.

TABLE 4.17 Test Results

G	D	S	Condition
D-type			
⊠	+	−	Low
⊠	−	+	Low
+	⊠	−	High
−	⊠	+	High
+	−	⊠	High
−	+	⊠	High
E-type			
⊠	+	−	High
⊠	−	+	High
+	⊠	−	High
−	⊠	+	High
+	−	⊠	High
−	+	⊠	High

4.60   COMPLEMENTARY MOSFETS

CAUTION:

Handle MOSFETs carefully.

Complementary metal oxide semiconductor (CMOS) logic circuits have contributed tremendously to digital circuitry. One of the MOSFET's primary applications is the CMOS inverter in Figure 4.85. Table 4.18 shows the corresponding truth table. Let's look closer at this circuit. The inverter converts one logic level to another.

The advantages of this circuit over transistor circuits are as follows:

- Dissipate very small amounts of power
- Consume very small levels of current and, therefore, drive many other circuits

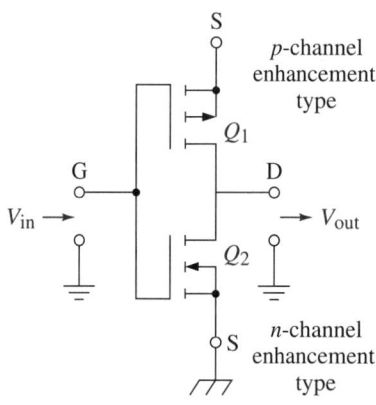

FIGURE 4.85 CMOS.

TABLE 4.18 Truth Table

V_{in}	Q_1	Q_2	V_{out}
0	On	Off	+ 12
+ 12	Off	On	0

4.61   HANDLING MOSFETS

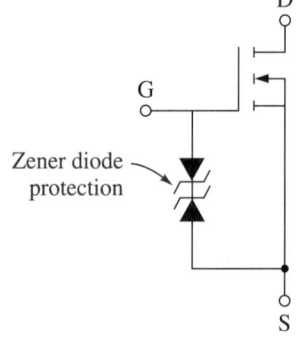

FIGURE 4.86 MOSFET with protection.

The thin layer of SiO_2 is easily destroyed; even touching the leads of the device can generate enough static electricity to destroy it. Most devices include built-in protection in the form of zener diodes, as in Figure 4.86. You should observe the following precautions:

1. Store MOSFETs in conductive foam (not styrofoam).
2. Handle them as little as possible.
3. Always turn off the power before removing or inserting MOSFET devices.

4.62   PHOTORESISTORS

A photoresistor is a light-sensitive resistance device usually made of cadmium sulphide (CdS) or cadmium selenide (CdSe). Figure 4.87 shows the symbol and typical charac-

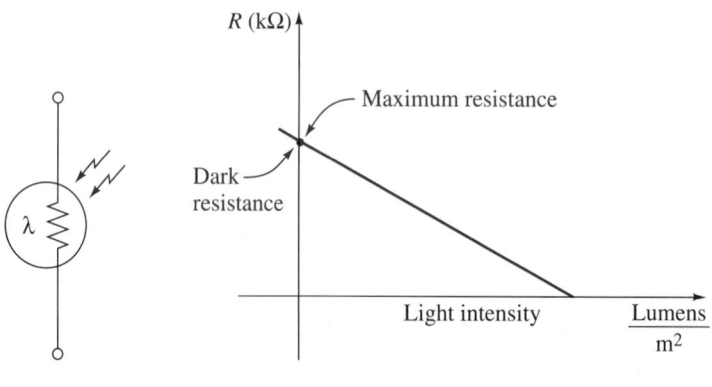

(a) Symbol
(b) Characteristic curve

FIGURE 4.87 Photoresistor.

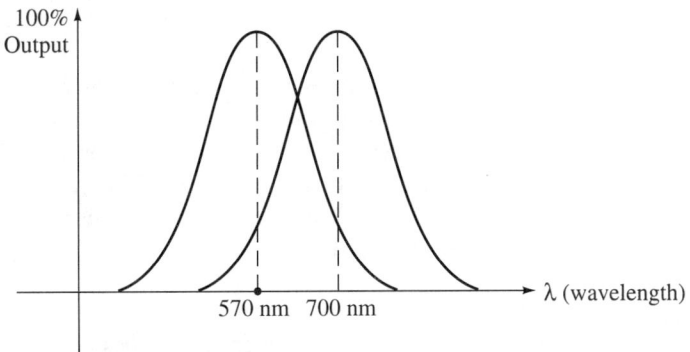

FIGURE 4.88 Spectral response.

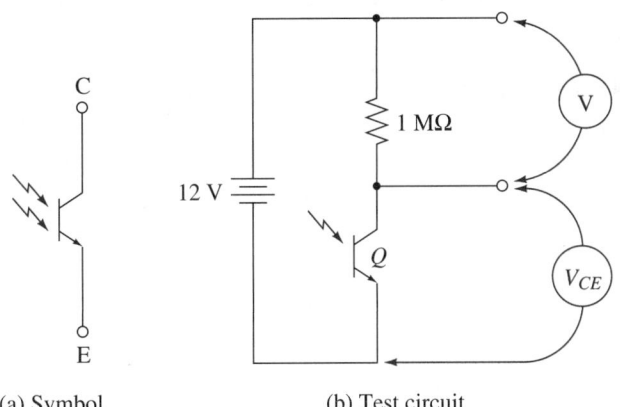

(a) Symbol   (b) Test circuit

FIGURE 4.89 Phototransistor.

TABLE 4.19 Truth Table

Q	V_{CE}
Off	12 V
On	Low

teristic curve of resistance versus light intensity. The maximum resistance (called the *dark resistance*) equals the value of resistance of the device in total darkness.

The CdS cell is most sensitive to wavelengths of 570 nm and the CdSe, to 700 nm, as shown by the spectral response curves in Figure 4.88.

4.63 TESTING PHOTORESISTORS

An ohmmeter will very quickly tell you whether the resistance of the device changes with exposure to light.

4.64 PHOTOTRANSISTORS

A three-terminal photodetector is one whose collector current is controlled by the intensity of light impinging on the base. Figure 4.89 shows the symbol and typical dynamic test circuit. Table 4.19 shows the corresponding truth table.

4.65 TESTING PHOTOTRANSISTORS WITH AN OHMMETER

Figure 4.90 shows how an ohmmeter is used to test a phototransistor.

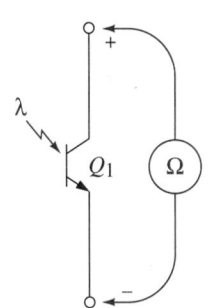

The ohmmeter should indicate a low value when Q_1 turns on.

FIGURE 4.90 Ohmic tests.

4.66   OPTOISOLATORS

An optoisolator uses light to couple the input signal from a photoemitter to the output through a photodetector. Figure 4.91 shows a typical design of an optoisolator. As shown, the outputs can come in the form of transistors, triacs, SCRs, and Darlington transistors.

The main use of an optoisolator is to isolate electrically low-voltage circuits from high-voltage circuits. It is possible with these devices to protect equipment and circuits that are highly sensitive to voltage transients. They are also highly effective at reducing ground-loop problems.

They can be used to control hundreds of volts in the output circuits by very low voltage circuits (5 V, for example) in the input circuit. It is common to find these devices being used to gate SCRs, triacs, etc.

Be careful— there may be dangerously high voltages present. Grounded equipment, such as your oscilloscope, must be used carefully. The typical voltage drop across the input LED is about 1.0 V.

Table 4.20 shows the pertinent parameters of the 4N38.

CAUTION

Watch for dangerously high voltages.

4.67   TESTING OPTOISOLATORS

The LED is tested like a normal diode and the output is tested like a normal transistor.

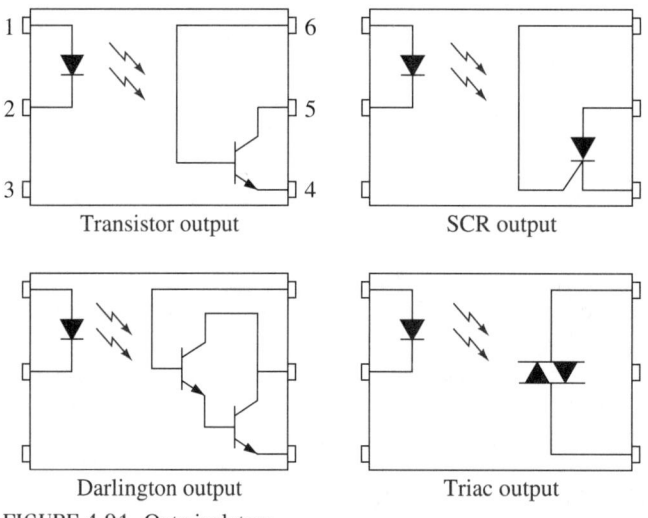

FIGURE 4.91   Opto isolators.

TABLE 4.20   4N38
Transistor Output

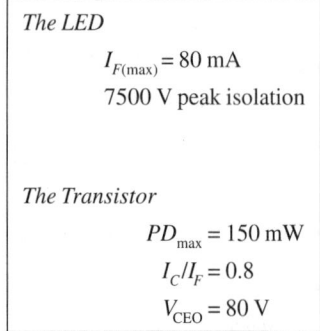

The LED

$I_{F(max)} = 80$ mA

7500 V peak isolation

The Transistor

$PD_{max} = 150$ mW

$I_C/I_F = 0.8$

$V_{CEO} = 80$ V

4.68 THE OPTOINTERRUPTERS

The optointerrupter (Figure 4.92) uses a photoemitter and photodetector to detect the presence of an object. Normally, light reaches the detector unless something passes through the gap and interrupts the beam.

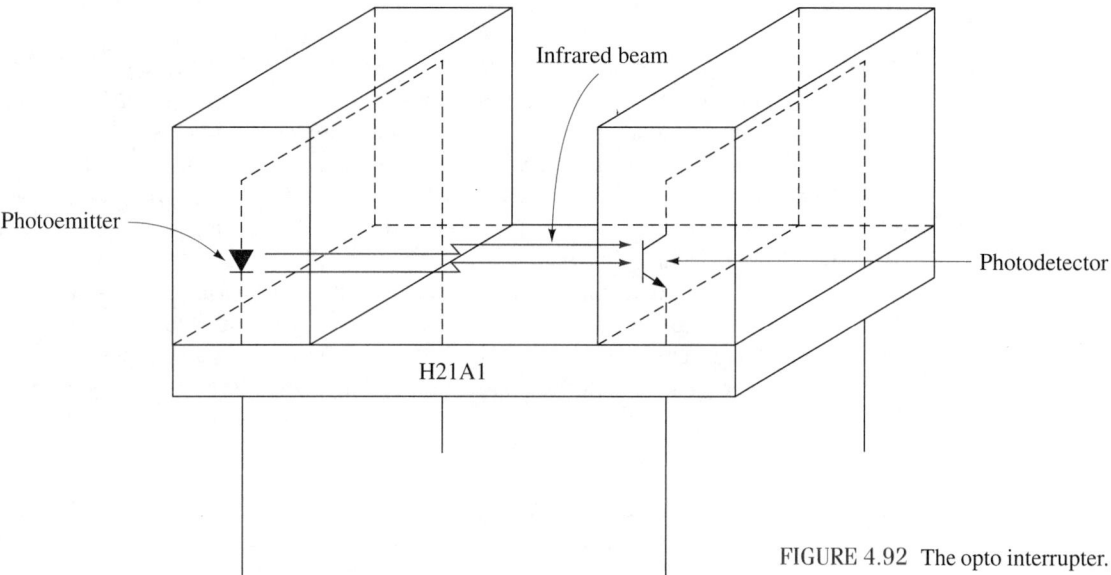

FIGURE 4.92 The opto interrupter.

4.69 TESTING OPTOINTERRUPTERS

A photoemitter can be tested with an ohmmeter like a normal diode, and the photodetector can be tested in the same way as a transistor.

4.70 SOLID-STATE RELAYS

The solid-state relay is a terrific isolation device and uses a dc input-control voltage to turn a larger ac voltage on or off in the output circuit, as shown in Figure 4.93. When V_{in} is positive on terminal 1 and negative on terminal 2, the photoemitter is forward-biased, and the light turns on the triac, coupling output terminal 3 to terminal 4. The triac can handle large values of current for the load.

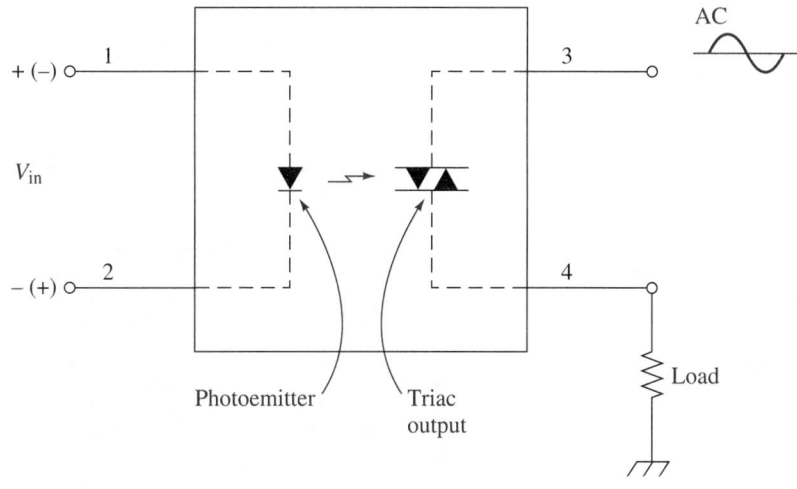

FIGURE 4.93 Solid-state relay.

4.71 TESTING SOLID-STATE RELAYS

The photoemitter can be tested like a normal diode, and the output device like a normal triac.

4.72 LED DISPLAYS

Typical common-anode-multiplexed seven-segment displays (LED displays) are wired with corresponding segments on a common bus. They are linked to one driver circuit, which provides the ground for each segment. A second switching circuit selects the place value being displayed. In Figure 4.94 for example, the decimal number 16.08 is displayed, and the digital circuitry grounds all pins (A, B, C, D, E, F, G, DP) and supplies positive V_{dc} voltage to pins 1, 2 and 4. However, this is not done all at once. It's done in steps, which are listed in Table 4.21.

The trick here is to display each step, one after the other, fast enough so that the human eye will see 16.08. This process is called *scanning*. As with a television picture, which is presented frame by frame at a frequency of 60 frames per second, the eye sees a continuous picture. The digital-clock pulse train accomplishes this by switching the control circuitry.

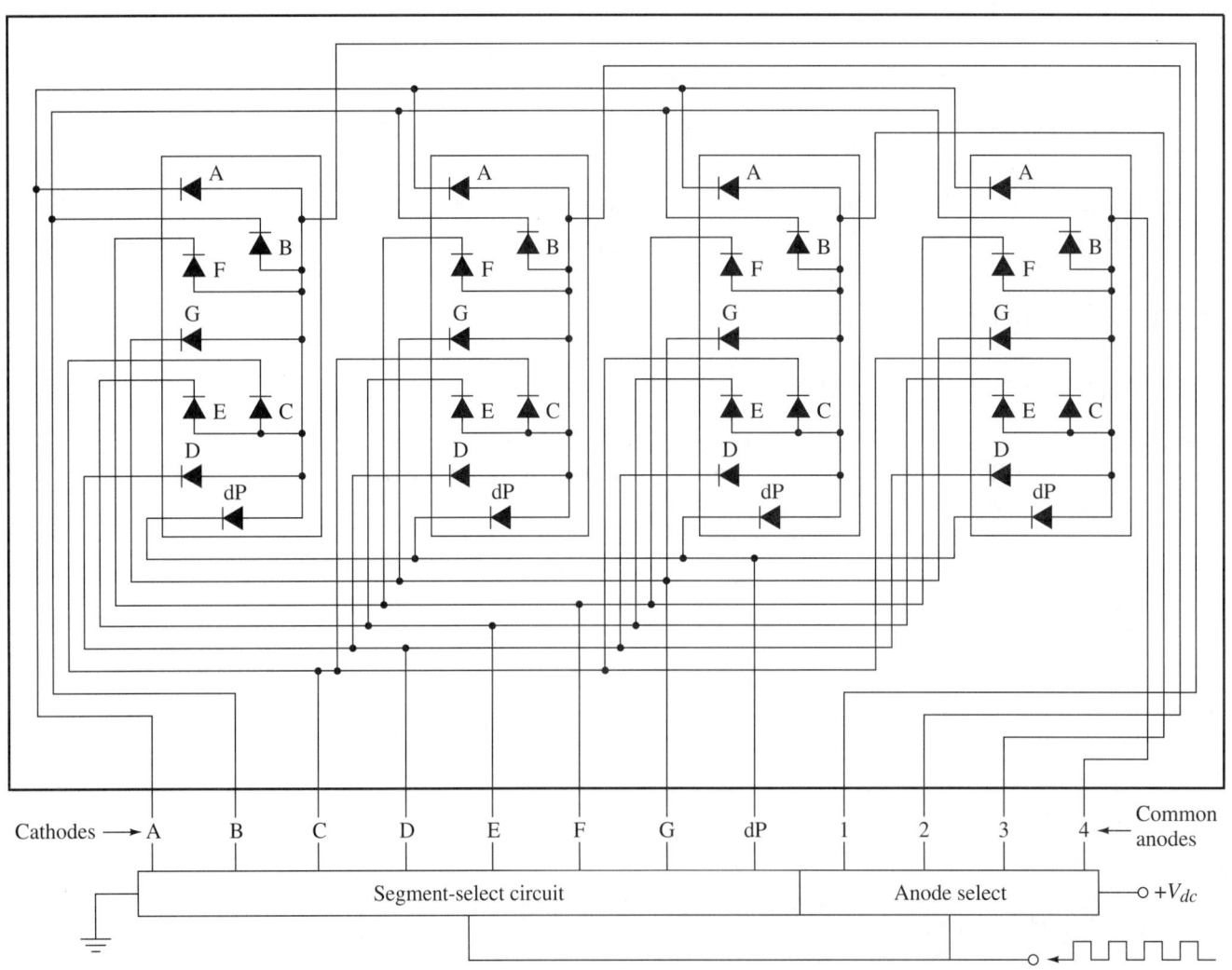

FIGURE 4.94 LED displays.

TABLE 4.21 Steps in LED Displays

Step 1	Ground B, C.
	Apply $+V_{dc}$ to 1.
	Display 1 0 0 0.
Step 2	Open B, C, and 1.
	Ground C, D, E, F, G.
	Apply $+V_{dc}$ to 2.
	Display 0 6 0 0.
Step 3	Open C, D, E, F, G, and 2.
	Ground DP.
	Apply $+V_{dc}$ to 2.
	Display 0 0.0 0.
Step 4	Open DP and 2.
	All switches are open.
	Display 0 0 0 0.
Step 5	Ground A, B, C, D, E, F, and G.
	Apply $+V_{dc}$ to 4.
	Display 0 0 0 8

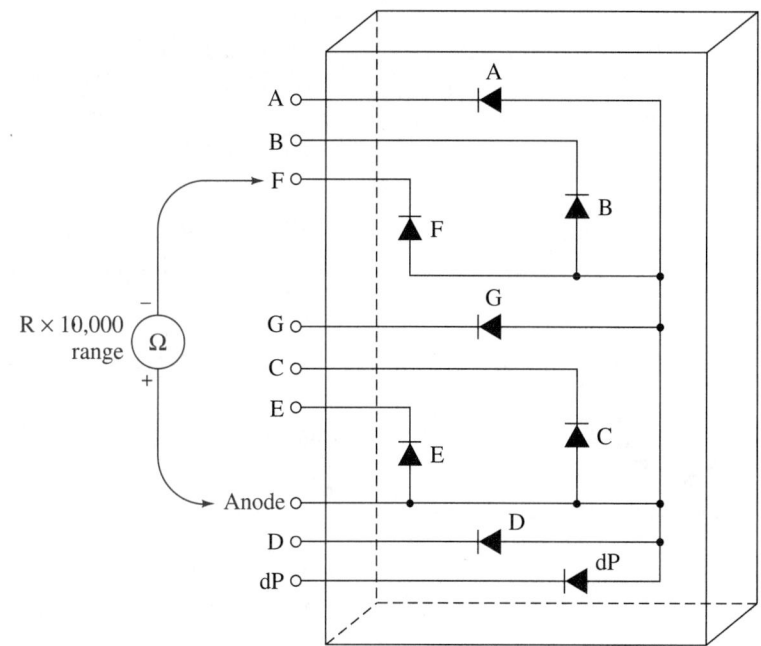

FIGURE 4.95 Testing LEDs.

4.73 TESTING LED DISPLAYS

Each individual LED segment of the display can be tested with an ohmmeter. You will have to use the highest range on your meter in order to forward-bias the diode. See the example in Figure 4.95.

4.74 TROUBLESHOOTING PRINTED CIRCUIT BOARDS

When a printed circuit board, or PCB, has a short circuit, it is either a solder bridge, a shorted capacitor, or an internally shorted integrated circuit. Tracing short circuits on PCBs is extremely time consuming if you don't want to damage the board. If you are lucky enough to have another board that is functioning normally, a series of comparison measurements will lead you to the faulty component.

Tracing short circuits on PCBs

You can check ICs chip by chip by spraying them with "freeze spray" and looking for erratic operations. This needs to be done with some care because temporary shorts can be introduced if the PCB is dirty, and static charges due to the spray can destroy field-effect transistors.

Using a logic pulser to inject a signal at a point and following the signal with a tracer is another method.

Visual Inspection

You should conduct a visual inspection of the PCB and look for the following:

1. Solder bridges and breaks on the foil
2. Loose connections

3. Missing components
4. Errors in the PCB layout
5. Diodes and polarized capacitors mounted backward
6. Burned or leaking components
7. Burned traces or discoloration

Ohmic Checks

You can use an ohmmeter to check the continuity of traces on the PCB.

4.75   TROUBLESHOOTING INTEGRATED CIRCUITS

How to check integrated circuits

Testing integrated circuits (ICs) is most often done in circuit under power as follows:

1. Test $V_{CC.}$
2. Check the ground connections.
3. Check the reset voltage and clock signals if it is a digital circuit.
4. Check the input and output voltages.
5. Make power-source measurements — some ICs require both $+$ and $-$ voltages.
6. Determine which pin is connected to the common ground.
7. If the IC has a conducting case, check for any internal connections to the ground pin. (Be sure to check your data sheets. Both ground pins are intentionally not internally connected on some chips, like the 8088.)

As shown in Figure 4.96,

- Digital ICs usually have $+V_{CC}$ on pin 14 and ground on pin 7.
- Linear ICs usually have $+V_{CC}$ on pin 11 and ground on pin 6.

    Before testing, always refer to the data sheets for correct pin assignments.

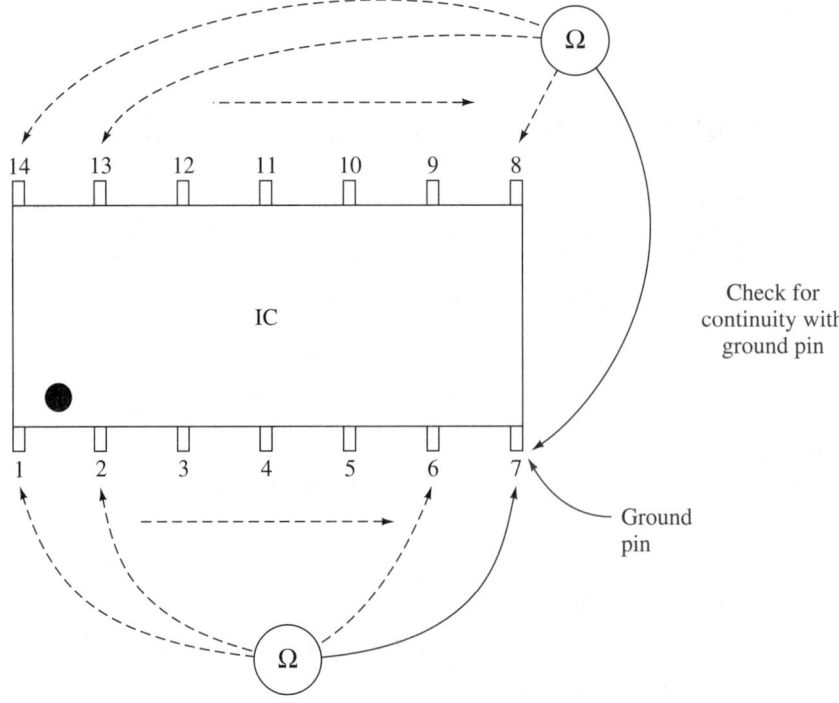

FIGURE 4.96   Checking ICs.

4.76 COMMON FAILURES

Most troubles with integrated circuits occur in four places:

- Capacitors
- Bad transistors
- Other ICs
- Diodes

4.77 ELECTROSTATICALLY SENSITIVE DEVICES

Electrostatically sensitive devices must be handled with care at all times. These devices include IGFETs and MOSFETs. Testing such devices in circuit is no problem; however, out-of-circuit tests can be troublesome. Be sure to follow this procedure:

1. Turn the power off.
2. Keep your body at the same potential as the device by using a wrist strap.
3. Short the leads of the device together when moving.
4. Connect the test leads to the device before removing the shorting rings.
5. Keep all soldering iron tips at ground potential.

4.78 PROBLEMS

1. Briefly explain why the barrier potential exists at a *pn* junction.

2. Is the diode in Figure 4.97 forward-biased or reverse-biased?

3. Examine the diode curve in Figure 4.98. What is the value of the breakdown voltage? Is this a silicon or germanium diode?

4. Does the ohmmeter in Figure 4.99 indicate a good diode?

5. What is the voltage across the zener diode in Figure 4.100?

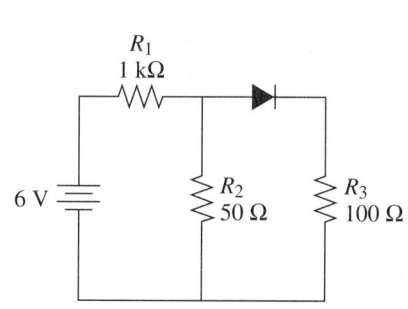

FIGURE 4.97 Problem 2.

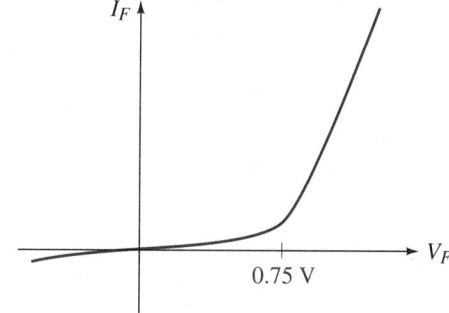

FIGURE 4.98 Problem 3.

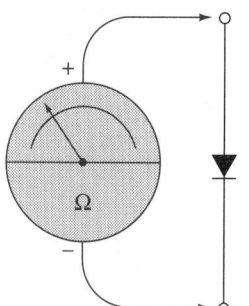

FIGURE 4.99 Problem 4.

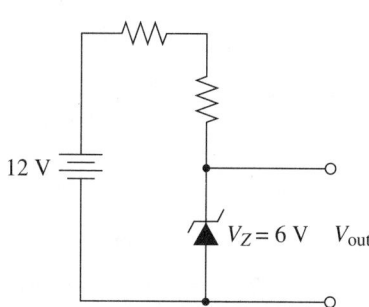

FIGURE 4.100 Problem 5.

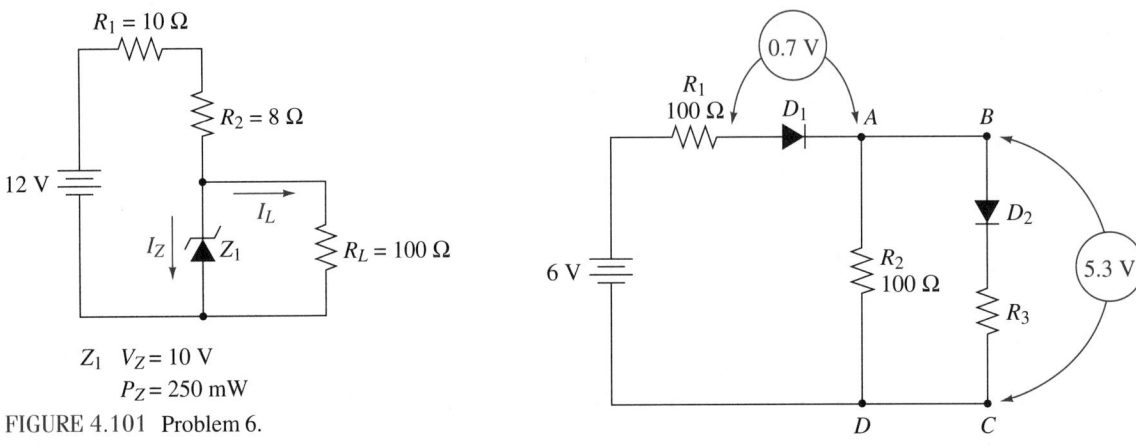

FIGURE 4.101 Problem 6.

FIGURE 4.102 Problem 7.

6. Calculate the maximum value of zener current in Figure 4.101.

7. Is diode D_2 in Figure 4.102 good?

8. Identify all possible troubles in Figure 4.102.

9. The LED in Figure 4.103 is not on. How do you explain this?

10. The numbers 4, 5, 6, 8, 9, and 0 cannot be displayed on the seven-segment display in Figure 4.104. What is wrong, and how can you test your answer?

11. Does the ohmmeter in Figure 4.105 indicate a good SCR?

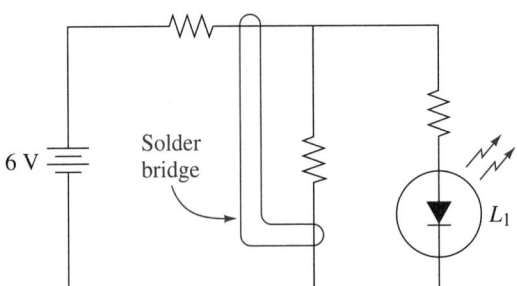

FIGURE 4.103 Problem 9.

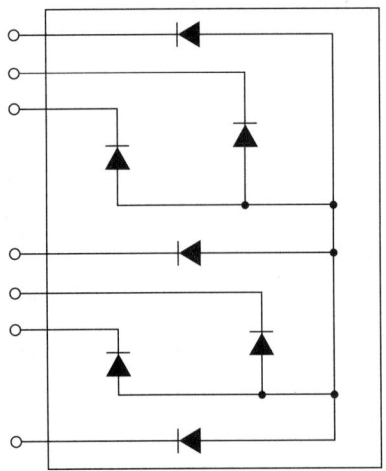

FIGURE 4.104 Problem 10.

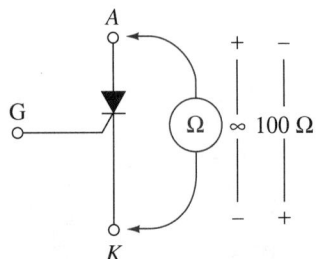

FIGURE 4.105 Problem 11.

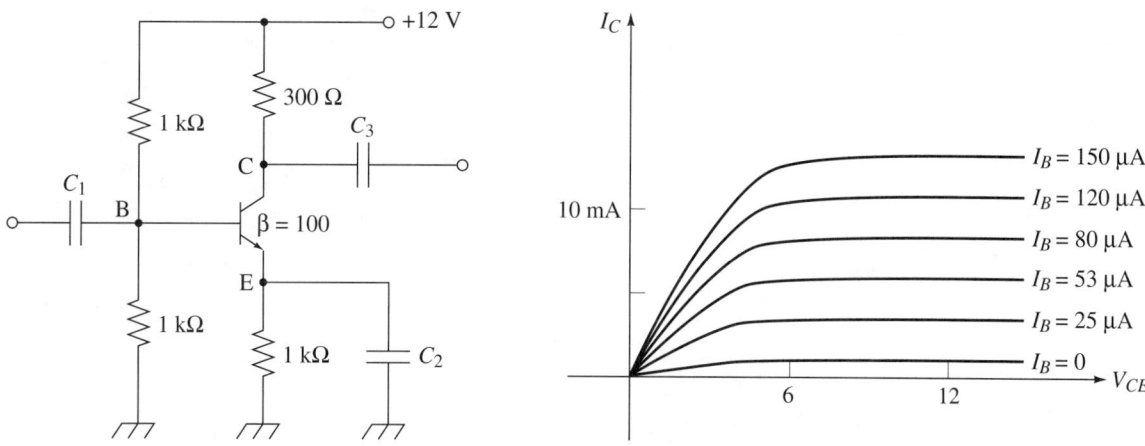

FIGURE 4.106 Problem 13.

12. Explain how a triac can be tested with an ohmmeter.

13. Perform a complete dc analysis of the transistor circuit in Figure 4.106. Draw the dc load line on the collector curves.

TABLE 4.22 Problems 14 and 15

Condition A	Condition B
$V_B = 6.0$ V	$V_B = 6.0$ V
$V_{BE} = 0.7$ V	$V_{BE} = 6.0$ V
$V_{CE} = 6.7$ V	$V_E = 0$ V
$V_E = 5.3$ V	$V_C = 12$ V
$V_{RC} = 0$ V	$V_{CE} = 12$ V
$V_C = 12$ V	$V_{RC} = 0$ V

14. Given the measurements listed in Table 4.22, condition A, what is the most likely problem?

15. If condition B exists (Table 4.22), what is the most likely problem?

16. Identify the problem with the class B circuit in Figure 4.107.

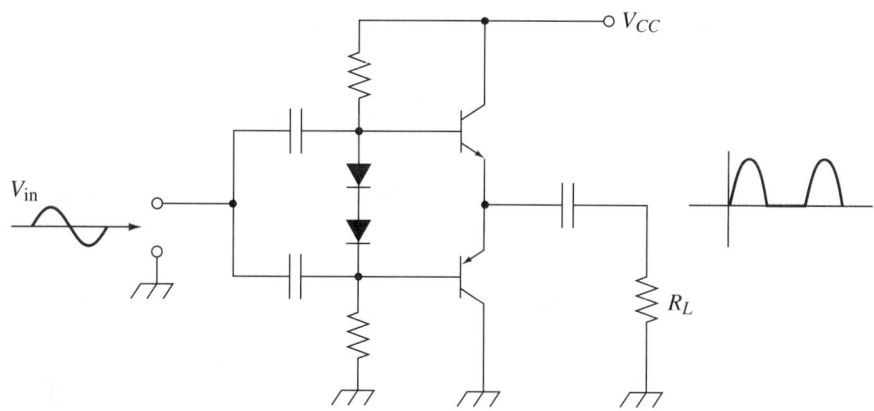

FIGURE 4.107 Problem 16.

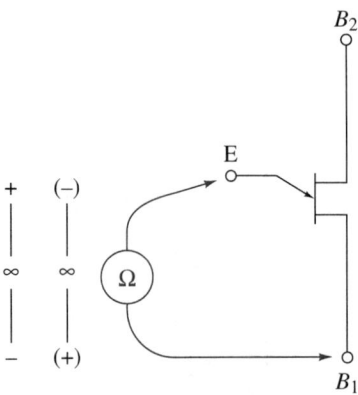

FIGURE 4.108 Problem 17.

17. Is the UJT in Figure 4.108 bad?

18. You determined that the JFET in Figure 4.109 has been destroyed. What are some possible reasons? Describe, step by step, how you will troubleshoot the circuit.

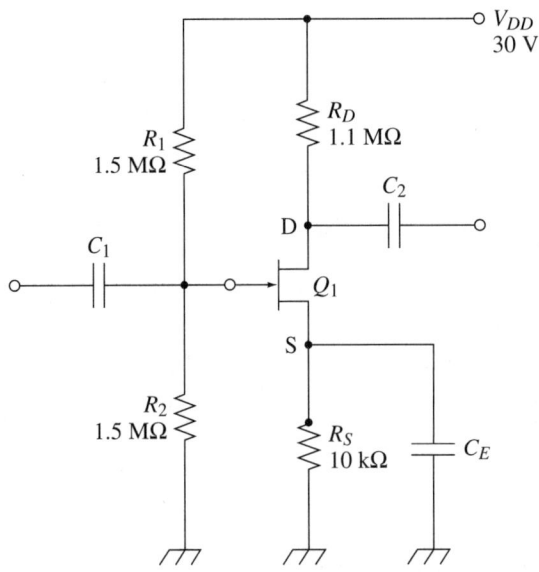

FIGURE 4.109 Problem 18.

19. List some of the cautions that should be observed when handling MOSFETs.

20. If the load in Figure 4.110 becomes shorted, how will the input to the solid-state relay be affected?

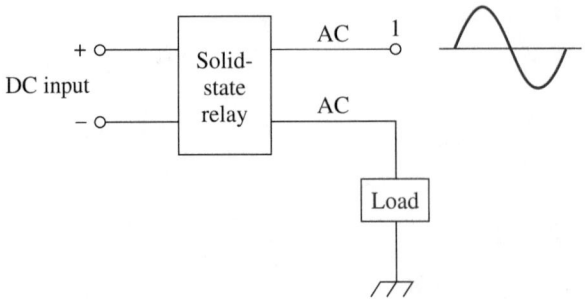

FIGURE 4.110 Problem 20.

4.79 EXERCISE 1: TROUBLESHOOTING THE CE CIRCUIT

The purpose of this exercise is to analyze the effects of various troubles on the common emitter (CE) circuit.

Components:

1 2N3904 *npn* transistor
1 10-kΩ resistor
1 2.2-kΩ resistor
1 3.6-kΩ resistor
1 1.0-kΩ resistor
2 1-μF capacitors
1 10-μF capacitor

Analyze the effects of problems with the common emitter circuit.

Construct the circuit as shown in Figure 4.111. Calculate the Q point parameters and draw the dc load line on the characteristic curves in Figure 4.112. Then inject, one by one, the faults listed in Table 4.23. After injecting the fault, make voltage measurements and complete the table as shown.

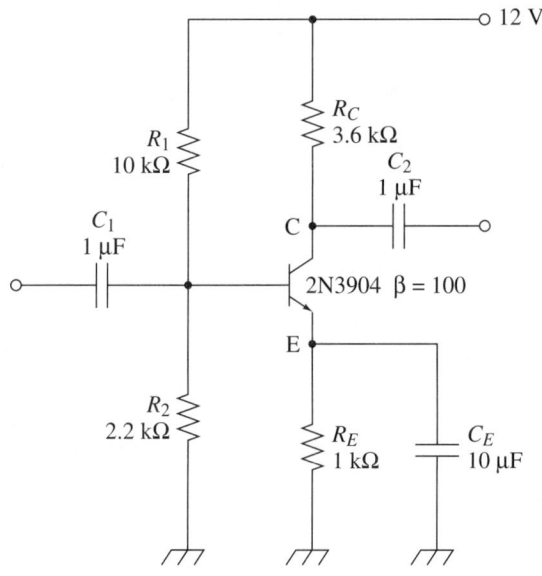

FIGURE 4.111 Exercise 1.

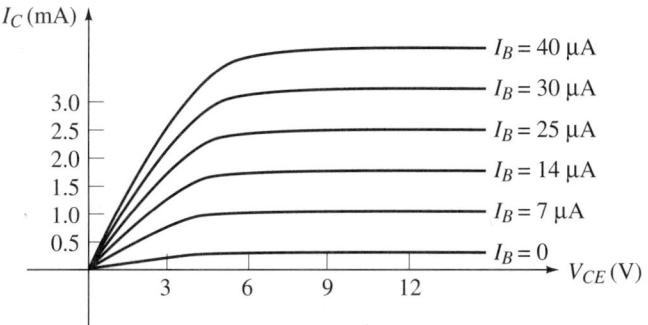

FIGURE 4.112 Exercise 1.

TABLE 4.23 Exercise 1

	Measurements				
Fault	V_B	V_E	V_{CE}	V_{RC}	V_C
R_1 open					
R_2 short					
R_C open					
R_C short					
R_E open					
R_E short					
C_E short					

5

TROUBLESHOOTING THE POWER SUPPLY

5.1 THE BASIC POWER SUPPLY

You won't find another system of circuits that has been used as often as a basic power supply. Every electronic system has to have a source of dc voltage to operate the various components. The power supply must convert the ac power-line voltage to a dc voltage of appropriate value. Since the power supply is common to all elements within a system, any problem in it will affect the entire system. Power supplies can be low voltage to provide the voltages required to operate transistor stages and integrated circuits or high voltage to provide the high levels of voltage required by television picture tubes, for example. Figure 5.1 shows a block diagram for each type.

Let's take a closer look at the basic low-voltage power supply. From Figure 5.1, you can see that it consists of four blocks:

1. Transformer
2. Rectifier
3. Filter
4. Regulator

Some power supplies are more complicated than others; however, they all break down into these four building blocks. We are going to investigate each part and then put them together in a complete system.

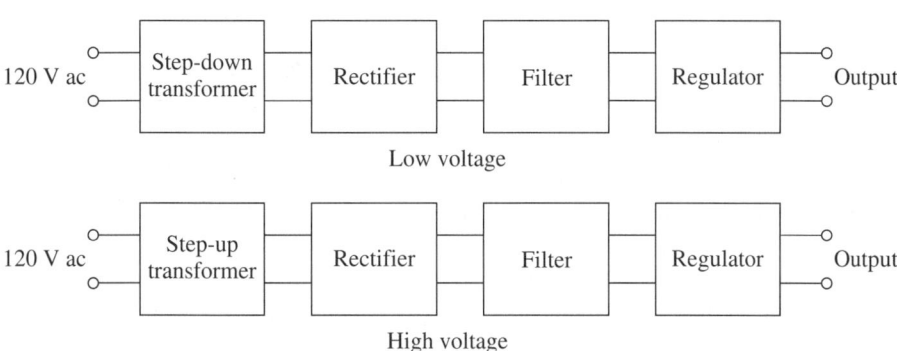

FIGURE 5.1 Block diagram.

5.2 TRANSFORMERS

The transformer is a magnetic circuit. The currents flowing in the primary winding induce a changing magnetic field that becomes concentrated in the iron core. This changing magnetic field induces currents in the secondary windings. Faraday's law tells us that a potential difference will be created across the coil terminals directly proportional to the rate of change of the magnetic flux and to the number of turns in the coil. Thus if the secondary coil has more turns than the primary, the output voltage will be greater than the input voltage. This type of a transformer is known as a *step-up transformer*. A *step-down transformer* is one whose secondary coil has fewer turns than the primary coil. With the step-down transformer, the secondary voltage is less than the primary. See Figure 5.2.

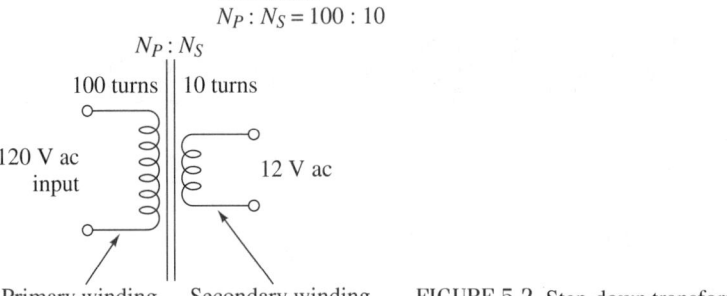

FIGURE 5.2 Step-down transformer.

The fundamental equations that apply to transformers are

$$N_P/N_S = V_P/V_S = I_S/I_P \tag{5.1}$$

and

$$P_P = P_S \tag{5.2}$$

where

N_P is the number of turns in the primary coil
N_S is the number of turns in the secondary coil
V_P is the primary voltage
V_S is the secondary voltage
I_S is the secondary current
I_P is the primary current
P_P is the primary power
P_S is the secondary power

Equation (5.1) shows that a transformer that steps up the voltage will, at the same time, step down the current so that the input power and output power will be equal.

5.3 TRANSFORMER FAULTS

Transformers rarely fail by themselves; instead, they usually fail as a result of excessive currents caused by some other shorted component. However, the following are typical faults:

1. Shorted primary or secondary turns
   Symptoms: Fuse blows and/or the transformer gets hot
             Low dc output voltage
2. Open primary or secondary turns
   Symptom: Output voltage equals zero
3. Shorts between windings and transformer core
   Symptom: Fuse blows

5.4   TESTING TRANSFORMERS

An open coil can be easily tested with your ohmmeter. Since the meter reads the dc resistance of the coil windings, a very high resistance reading will verify your suspicions.

If you suspect a short to the transformer core, an ohmic check will quickly verify or disprove this. Of course, you don't want to see any continuity here.

If any of these tests are positive, the only cure is to replace the transformer.

5.5   THE RECTIFIER STAGE

The purpose of the rectifier stage is to convert the incoming ac waveform to a pulsating dc waveform. There are three basic types in general use: the half-wave, the full-wave, and the bridge circuit. You will see the bridge rectifier circuit most often, followed by the full-wave circuit. These circuits can be either discreet components or integrated circuits. IC rectifiers are less expensive and easier to troubleshoot.

5.6   HALF-WAVE RECTIFIERS

In Figure 5.3, the diode becomes forward-biased when the input waveform is positive and exceeds the diode's barrier potential. Then the voltage across the load is equal to the input minus the drop across the diode. This drop is typically 0.7 V to 1.0 V. When the input drops below the diode's barrier potential and swings negative, the diode becomes reverse-biased, and the load voltage drops to zero. A dc voltmeter across the load will indicate the average voltage, or $0.318V_{2pk}$.

If the diode is reversed in the circuit, the output voltage will be negative.

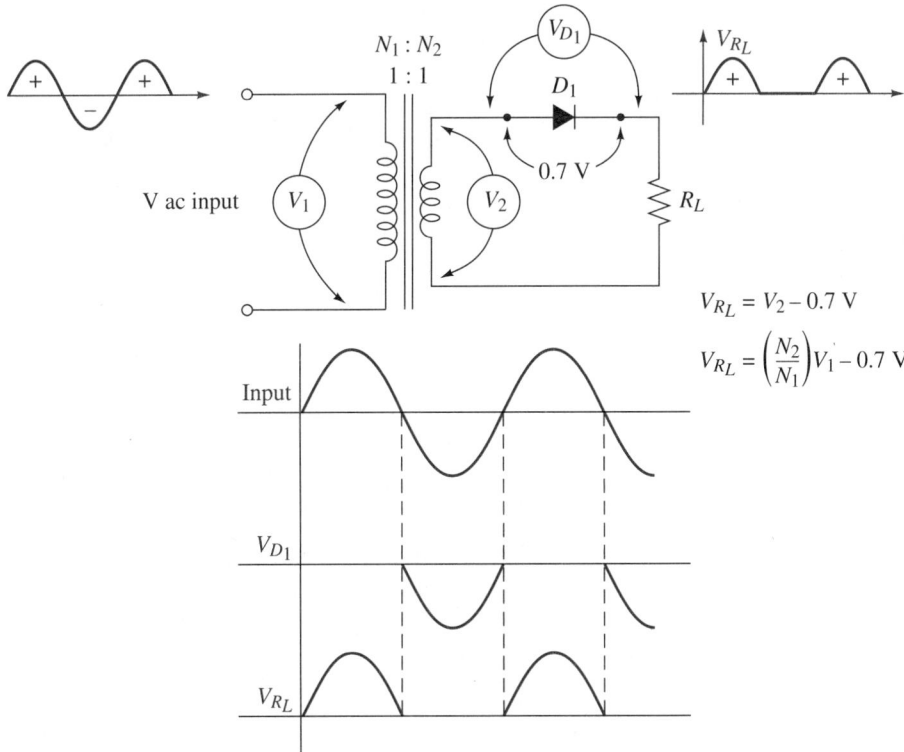

$$V_{RL} = V_2 - 0.7 \text{ V}$$

$$V_{RL} = \left(\frac{N_2}{N_1}\right)V_1 - 0.7 \text{ V}$$

FIGURE 5.3   The half-wave rectifier.

5.7 TYPICAL HALF-WAVE RECTIFIER FAULTS

The half-wave rectifier is a very simple circuit and can be easily analyzed. If the diode is shorted, the output voltage will be a sine wave whose value is equal to that given by the general transformer equation. The voltage drop across the diode will be zero.

If the diode is open, there will be no output voltage across the load, and the entire secondary circuit voltage will be dropped across the diode terminals. These conditions are summarized in Figure 5.4.

5.8 COMPONENT SUBSTITUTION

The important parameters to watch when replacing the diode in a half-wave rectifier are the following:

1. The average forward current (I_0) rating must exceed the maximum dc forward current in the circuit.
2. The minimum value of reverse working voltage (V_{RRM}) must exceed the peak inverse voltage, or PIV, which is equal to V_2, the peak voltage of the secondary circuit.

5.9 FULL-WAVE RECTIFIERS

The full-wave rectifier is easy to recognize because this circuit has two diodes, as shown in Figure 5.5. This circuit is like two half-wave circuits, one for the positive alternations of the input cycle and one for the negative alternations of the input cycle. The result is more pulses of output voltage and a higher level of average (dc) voltage. This value can be found by the equation

$$V_{avg} = 0.636V_{out(pk)} \tag{5.3}$$

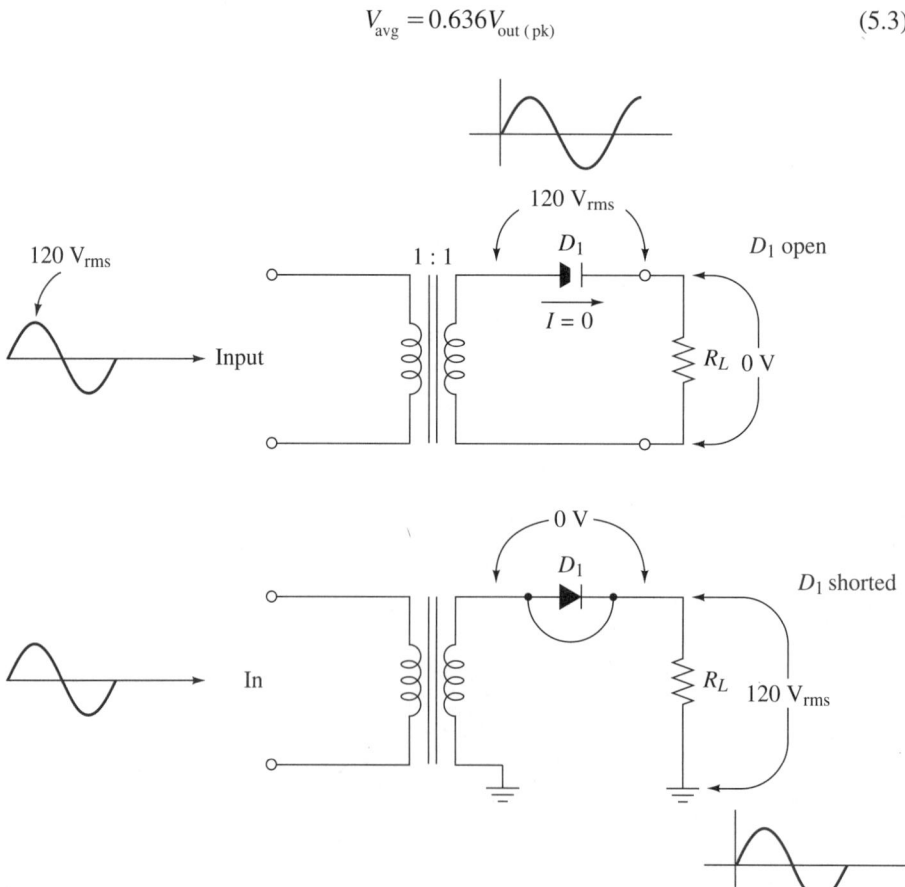

FIGURE 5.4 Half-wave rectifier faults.

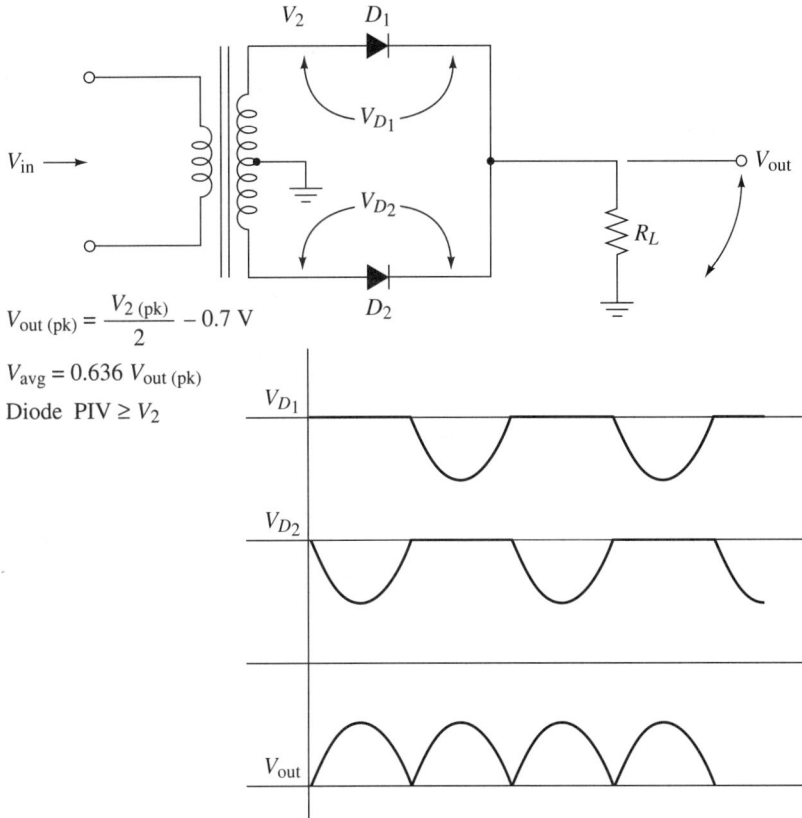

$$V_{out\,(pk)} = \frac{V_{2\,(pk)}}{2} - 0.7\ \text{V}$$

$$V_{avg} = 0.636\ V_{out\,(pk)}$$

Diode $\text{PIV} \geq V_2$

FIGURE 5.5 Full-wave rectifier.

5.10 TYPICAL FULL-WAVE RECTIFIER FAULTS

If a diode in the full-wave rectifier is shorted, the load will draw excessive current and most likely blow the main fuse.

If the diode is open, the output voltage across the load will have a waveform similar to that of a half-wave rectifier. These conditions are summarized in Figure 5.6.

5.11 COMPONENT SUBSTITUTION

By studying Figure 5.6 you can see that the PIV of any one diode is equal to the complete secondary voltage, or $\text{PIV} = V_2$.

5.12 FULL-WAVE BRIDGE CIRCUITS

It is easy to see why the bridge circuit is the most popular. It does not require a center-tapped transformer, and the output voltage is higher than that of other circuits. Four diodes are used in this circuit as shown in Figure 5.7. Only two diodes are conducting at a time. The direction of load current remains the same during both positive and negative alternations of the input waveform.

5.13 TYPICAL BRIDGE RECTIFIER FAULTS

The same faults listed for the half-wave rectifier also apply to the bridge rectifier circuit.

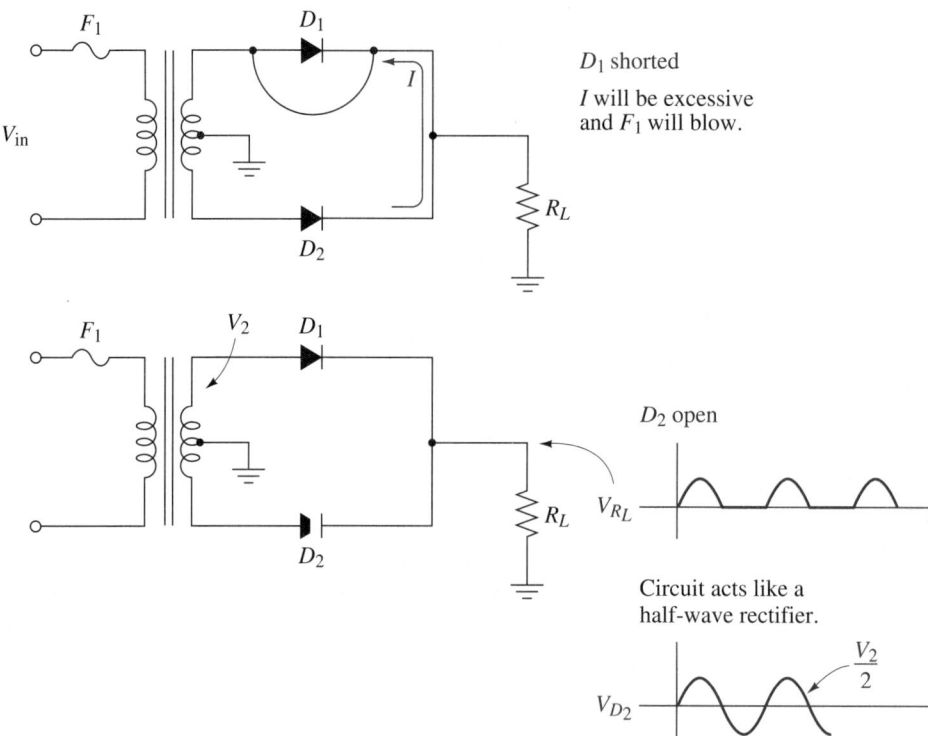

FIGURE 5.6 Full-wave rectifier faults.

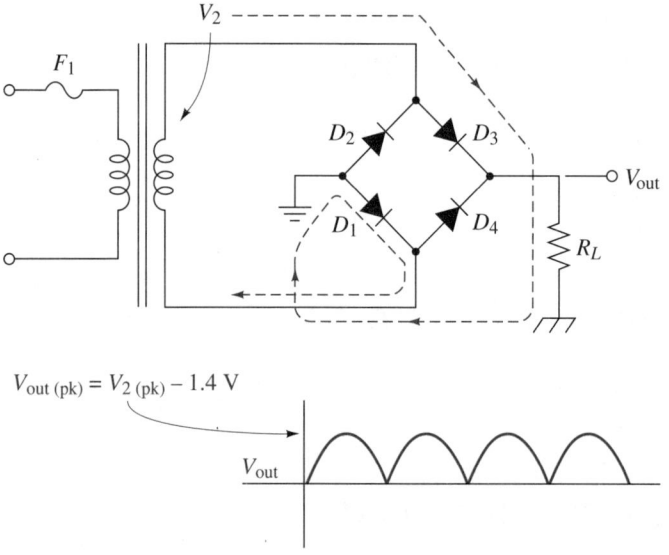

$$V_{out\,(pk)} = V_{2\,(pk)} - 1.4\ V$$

FIGURE 5.7 Bridge rectifier.

5.14 COMPONENT SUBSTITUTION

The PIV rating should be greater than the peak secondary voltage. The two diodes that are conducting share the dc load current. Therefore, the dc forward current rating should be greater than $0.5I_{dc}$.

5.15   THE FILTER STAGE

Don't forget that our ultimate goal with the power supply is to convert the incoming ac voltage waveform to a constant value of dc voltage. Thus the pulsating output from the rectifier must be filtered, or smoothed out, to get rid of the variations. Whatever variations remain after the filter stage are referred to as the ripple voltage, V_r.

5.16   CAPACITOR FILTERS

The capacitor filter is the simplest and most common filter. The capacitor filter in Figure 5.8 has a capacitor in shunt with the load resistance. At the instant of turn-on, the filter capacitor has no charge and acts like a short circuit. Therefore, a large current flows through the diode and is limited only by the resistance of the transformer secondary winding. The charge-time constant is very fast. During the negative alternation of the input waveform, the diode is not conducting and the capacitor begins to discharge through the load resistor. If the discharge-time constant is much longer than the period of the input waveform, the output voltage will drop very little before the next positive alternation recharges the capacitor. Thus the output voltage tends to be more stable.

Table 5.1 lists the pertinent parameters for the three types of rectifier circuits with the capacitor input filter.

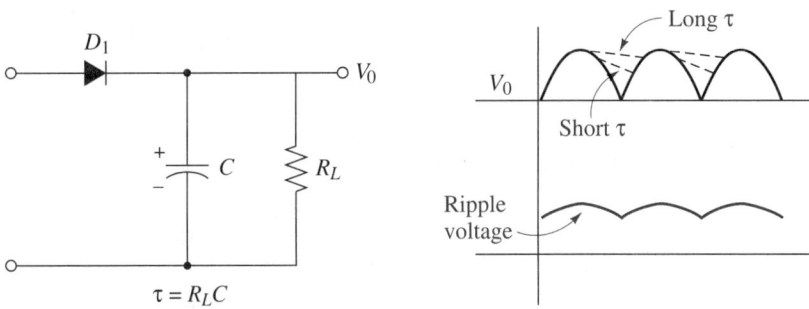

$$\tau = R_L C$$

FIGURE 5.8   The capacitor filter.

TABLE 5.1   Power Supply Rectifier with Capacitor Filter

	Half-wave	Full-wave	Bridge
V_{out}	$V_{2(pk)}$	$V_{2(pk)}/2$	$V_{2(pk)}$
Diode Current	I	$I/2$	$I/2$
Ripple Frequency	Frequency of V_{in}	2(frequency of V_{in})	2(frequency of V_{in})
Diodes	1	2	4
Diode PIV	$V_{2(pk)}$	$V_{2(pk)}$	$V_{2(pk)}$

5.17   TYPICAL CAPACITOR INPUT FILTER FAULTS

If the capacitor shorts or becomes extremely leaky, the load will be shorted, and the only resistance will be the dc resistance of the coil windings and the bulk resistance of the diode(s). As a result, the fuse will blow.

CAUTION

Electrolytic capacitors may explode.

If the capacitor opens, the ripple voltage will become excessive and the output voltage level will drop.

Remember that electrolytic capacitors often fail by shorting. They may blow up due to the excessive internal leakage.

5.18 THE VOLTAGE-REGULATOR STAGE

Up to this point in our power supply, we have taken the input ac voltage, rectified it into a pulsating dc voltage, and filtered the pulsations (ripple) to provide a smooth dc voltage.

Now, we need to regulate the dc voltage to do the following:

1. Provide specific values required by the circuits.

2. Maintain constant values of dc output voltage, regardless of changes in the input voltage. (This will, therefore, result in additional filtering action.) The action is known as *line regulation.*

3. Maintain constant values of dc output voltage regardless of changes in the load current demands. This action is known as *load regulation.*

Many manufacturers use a single regulation rating to indicate the maximum change in output voltage for specified ranges of input voltage and load currents. This rating is usually given in terms of percentage or microvolts per volt. See Figure 5.9.

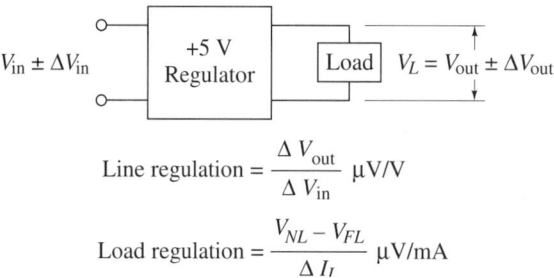

$$\text{Line regulation} = \frac{\Delta V_{\text{out}}}{\Delta V_{\text{in}}} \ \mu\text{V/V}$$

$$\text{Load regulation} = \frac{V_{NL} - V_{FL}}{\Delta I_l} \ \mu\text{V/mA}$$

FIGURE 5.9 Regulator ratings.

5.19 DISCRETE VOLTAGE REGULATORS

The most simple form of regulation is accomplished by using a zener diode. The circuit of Figure 5.10 has a zener diode placed in parallel with the load resistor. As you know, by keeping the zener current within the range of I_{ZK} to I_{ZM}, the voltage across the load will remain constant at a level equal to the V_Z of the diode. The drop across the series resistor, R_S, is equal to $V_S - V_Z$. The total circuit current is I_T and can be found by the equation

$$I_T = \frac{V_S - V_Z}{R_S} \tag{5.4}$$

The current through the load I_L can be found by dividing the zener voltage rating by the value of load resistance.

$$I_L = V_Z / R_S \tag{5.5}$$

The zener current is equal to the difference, or

$$I_Z = I_T - I_L \tag{5.6}$$

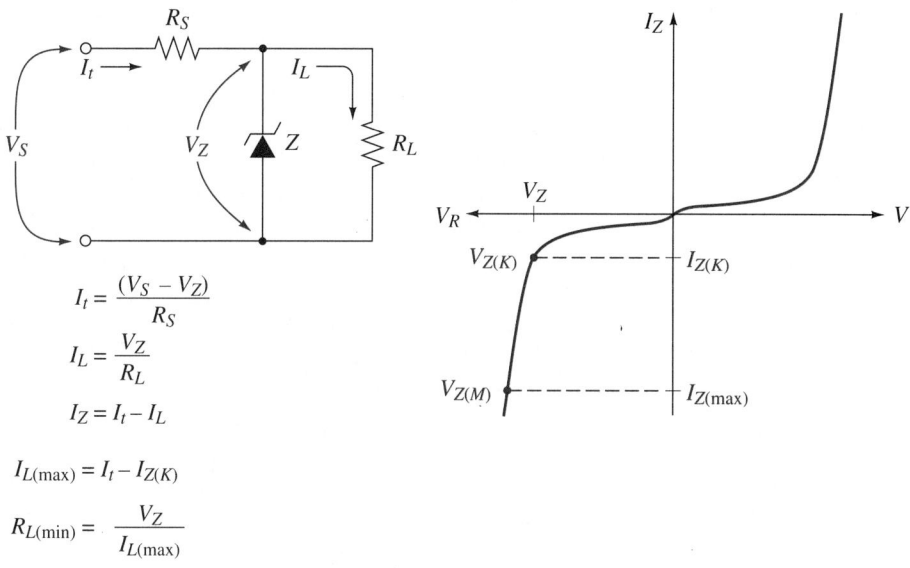

$$I_t = \frac{(V_S - V_Z)}{R_S}$$

$$I_L = \frac{V_Z}{R_L}$$

$$I_Z = I_t - I_L$$

$$I_{L(max)} = I_t - I_{Z(K)}$$

$$R_{L(min)} = \frac{V_Z}{I_{L(max)}}$$

FIGURE 5.10 Zener regulation.

5.20   TYPICAL ZENER REGULATOR FAULTS

If the zener diode shorts, the excessive current will probably cause the primary fuse to blow. The state of the diode can be tested with your ohmmeter.

If the zener diode opens, the regulation will be lost and the output voltage will increase. The ripple voltage will also increase.

The zener should be removed from the circuit and tested, as we have previously discussed.

5.21   DISCRETE SERIES FEEDBACK REGULATORS

The discrete series feedback regulator circuit uses a pass-through active device, as shown in Figure 5.11, to regulate the load current. It is important to remember that the entire load current passes through this device. This particular circuit uses an error-detection circuit to sample the output, compare this sample to a reference level, and generate an error voltage. The dc error voltage is then used to control the pass-through device and regulate the output. Because the pass-through device must handle the total load current, some form of current limiting must be used for protection. Various forms of current limiting exist, but the most popular are shown in Figures 5.12 and 5.13.

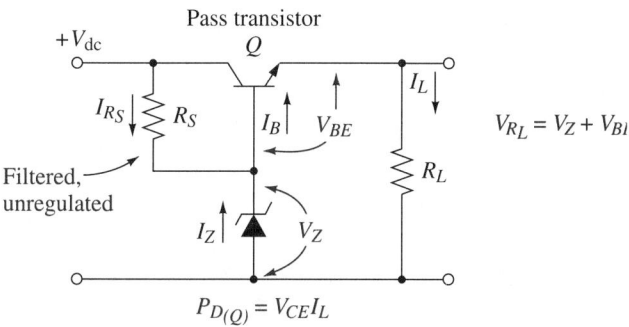

$$V_{RL} = V_Z + V_{BI}$$

$$P_{D(Q)} = V_{CE}I_L$$

FIGURE 5.11 Series regulator.

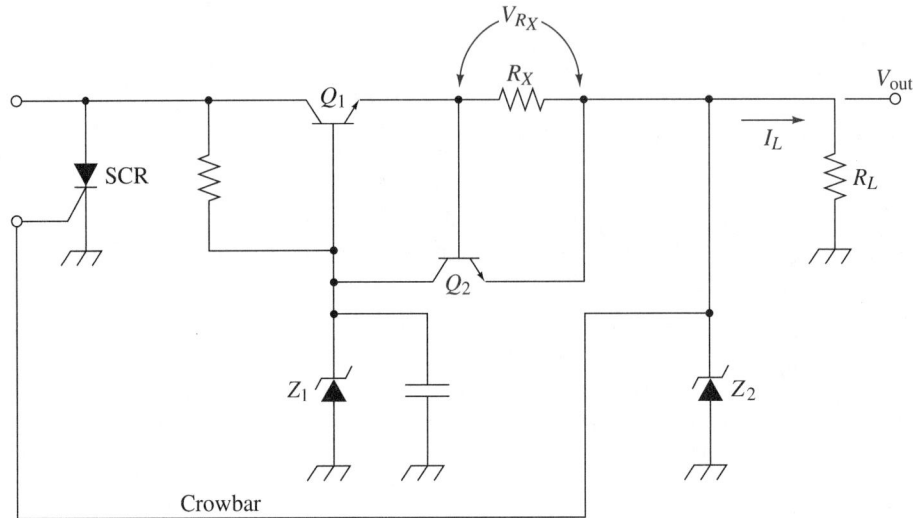

Q_2 is normally off.
If R_L shorts, I_L increases until $V_{RX} = 0.7$ V. Then Q_2 turns on and limits the value of I_L.
The crowbar monitors V_{out}. When V_{out} exceeds V_{Z2}, the SCR turns on and shorts V_{in} to ground.

FIGURE 5.12 Current limiting.

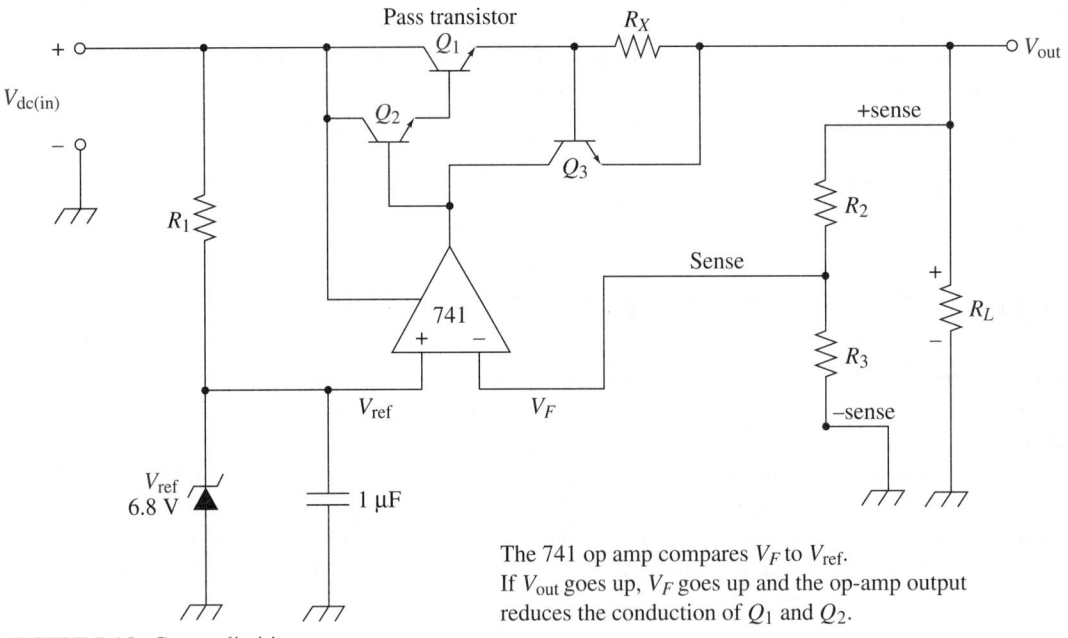

The 741 op amp compares V_F to V_{ref}.
If V_{out} goes up, V_F goes up and the op-amp output
reduces the conduction of Q_1 and Q_2.

FIGURE 5.13 Current limiting.

5.22   PROBLEMS IN SERIES PASS REGULATORS

The pass-through device must and does dissipate a great deal of power. It is the component that is most likely to fail. You can test the pass-through transistor the same way as any other transistor.

5.23   COMPONENT SELECTION

Obviously, the maximum power dissipation rating of the pass-through device is a critical parameter to observe when replacment is required.

5.24 DISCRETE SHUNT FEEDBACK REGULATORS

The shunt regulator is not as common as the series regulator just discussed. Review the typical circuit shown in Figure 5.14. Any changes in the load resistance will change the forward bias on the base of the shunt transistor and, therefore, the collector-to-emitter resistance. This increase or decrease in the shunt transistor's conduction will cause a corresponding change in the collector voltage, offsetting the initial change in the load voltage.

One very popular use of the shunt regulator is in the "crowbar" protection circuit shown in Figure 5.15. Here the shunt turns on for overvoltage conditions and reroutes the currents to ground.

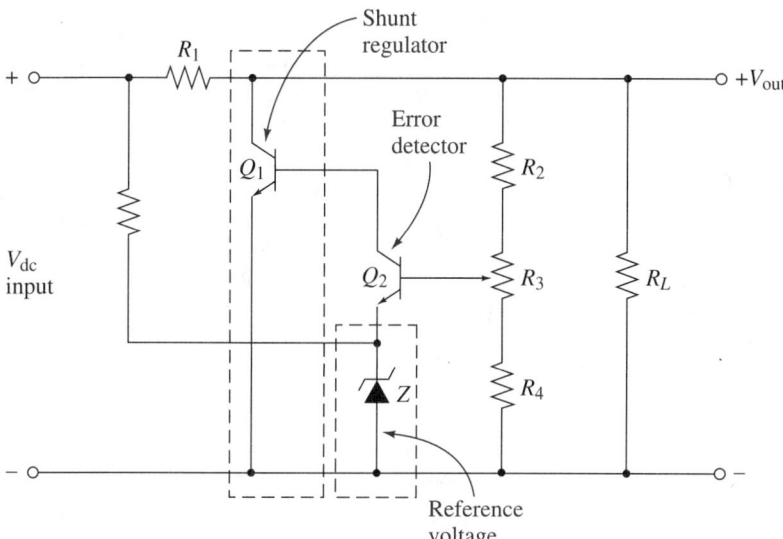

FIGURE 5.14 Typical circuit.

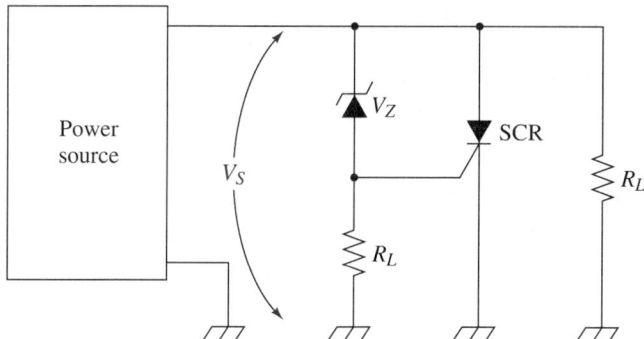

FIGURE 5.15 SCR crowbar.

5.25 IC VOLTAGE REGULATORS

Most IC regulators are three-terminal devices; however, some very popular devices like the LM723 have more than three terminals. IC voltage regulators have the following characteristics:

- Are found in medium and low power supplies
- Include precision references

- Include temperature compensation
- Include overload protection
- Include short-circuit protection

IC regulators are available in different versions, such as fixed positive, fixed negative, adjustable, and dual tracking. The fixed supplies provide specific values of output voltage and are available in a very wide range of positive and negative voltages. Figure 5.16 (a), (b), and (c) show typical circuits of each type. Inside the IC you will find a pass transistor and its associated circuitry, much like that of a series feedback regulator circuit. Therefore, IC regulators are really series regulators.

IC regulators like the LM723 are made of several building blocks, which include a voltage reference source, an error amplifier to compare the output to the reference, a pass-through device, and current-limiting circuitry. See Figure 5.17 for a typical circuit.

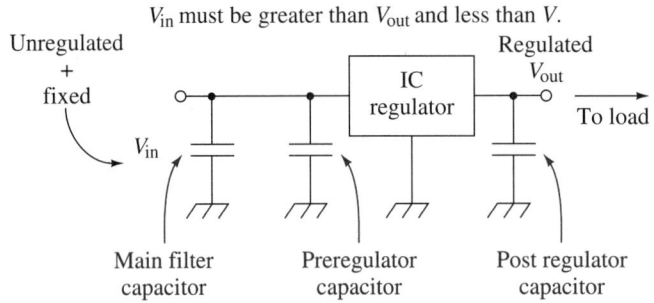

(a) Fixed regulator

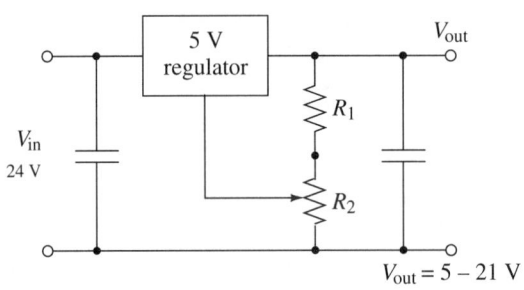

(b) Adjustable regulator

(c) Dual regulator

FIGURE 5.16 Typical circuits.

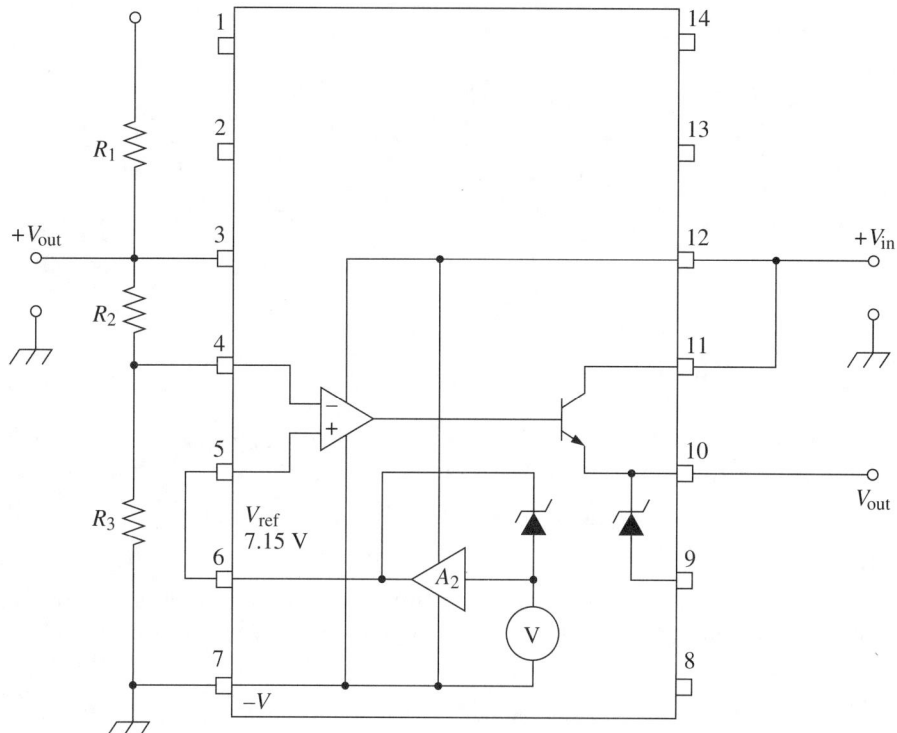

FIGURE 5.17 The LM 723.

5.26 TROUBLESHOOTING IC REGULATORS

It is very easy to troubleshoot an IC regulator circuit.

1. Check the input voltage and output voltage.

2. The input voltage must be at least 10% greater than the output for the device to regulate.

3. If the input voltage is acceptable, but the output is too high, you can assume that the IC is bad.

4. If the input voltage is acceptable, but the output is too low, you should first remove the load and retest. If the output voltage is normal with the load removed, there is a problem with the load. If the output is still too low with the load removed, the IC is bad.

Follow these steps to troubleshoot an IC regulator circuit.

5.27 TROUBLESHOOTING THE POWER SUPPLY

Excessive currents due to shorted loads most often take out the low-voltage power supply. If the fuse is blown, check to see if the failure was a "soft" or "catastrophic" one. A soft blow is possible due simply to the age of the fuse. However, short-circuit conditions will cause a catastrophic failure, and the fuse body will be black. When you replace the fuse, *never* use a new one with higher amperage ratings. Always use exact values or less.

A shorted power supply will have symptoms such as these:

Checks to make when a fuse is blown

- Blown fuse
- Hum in the output
- Hot components
- Smoke
- Crackling sounds

CAUTION

Be careful to never replace a fuse with one of a higher amp rating.

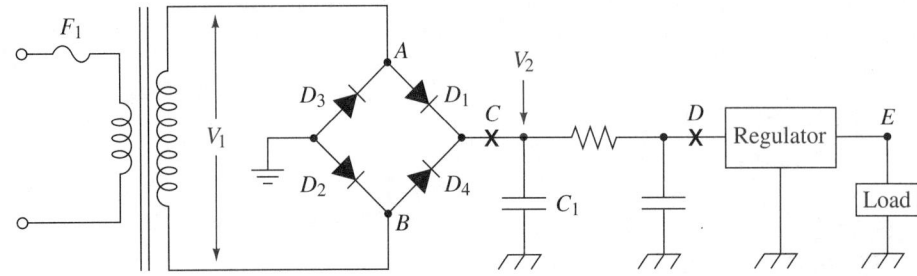

FIGURE 5.18 Complete power supply.

5.28 GENERAL PROCEDURE

Follow these procedures when troubleshooting a basic circuit.

In general, if you are troubleshooting the basic circuit shown in Figure 5.18, you should follow these procedures:

1. Remove the load. If the problem is gone, replace the load. If not, proceed to step 2.

2. Break the circuit at point *C,* about halfway. If the trouble disappears, reconnect point *C* and break at *D*. If the trouble again disappears, replace the regulator. If not, check the filter stage.

3. After breaking the circuit at point *C,* if the trouble did not disappear, proceed to points *A* and *B*. Break the circuit there. If the trouble is gone, check the diodes. If not, check the transformer.

Nearly every problem is accompanied by an increased level of ripple voltage on the output waveform. When a diode shorts, the transformer secondary becomes shorted with every half-cycle. This will blow the fuse. When electrolytics short, they leak, sweat, and even explode. Typical problems and their associated waveforms are displayed in Table 5.2.

When testing, be careful because the regulator's current-limiting circuitry may be operating due to the loading effects. Substitute a good load and check to see if the power

TABLE 5.2 Typical Problems and Their Associated Waveforms

Condition	V_1 Waveform	V_2 Waveform
Normal	⌒⌒	⋀⋀⋀⋀⋀
D_1 or D_2 shorted F_1 blows	⌒_⌒	⌣⌣
D_3 or D_4 shorted	⌐⌣	⌣⌣
D_1 or D_2 open	⌒⌒	⌣⌣
D_3 or D_4 open	⌒⌒	⌣⌣
C_1 shorted; Fuse blows or diodes burn	⌒⌒	——
C_1 open	⌒⌒	⌣⌣⌣

supply is acceptable. If it is, the load is at fault. If it is not, the regulator is at fault.

A regulator that is shorted to ground will act like a shorted capacitor.

5.29   THE SWITCHING POWER SUPPLY

The linear power supply has been the workhorse in industry for many years. As we have already seen, this power supply is characterized by use of a linear type of regulator, which is generally in series with the load. Thus the regulator is forced to handle the entire load current. Since there is, normally, a large voltage drop across the regulator, a great deal of power must be dissipated by the regulator's series pass transistor. This means that the linear power supply is naturally very inefficient. In fact, 50% efficiency is quite typical. To add to this, the fact that 60 Hz is the frequency being regulated throughout the system automatically means use of very large reactive components. The switching supply, on the other hand, derives a high frequency from the 60-Hz line frequency. This allows for a more efficient supply. In fact, 80% efficiency is most typical.

5.30   FUNDAMENTALS OF SWITCHING POWER SUPPLIES

By definition, a switching power supply accomplishes control and regulation of the output by use of a switching device. This switching device is controlled by a feedback network (loop) that continuously samples the output voltage, compares the sample to an established reference level, and provides an error signal to adjust the switching operation. Adjustment is achieved by varying the duty cycle or repetition rate of the switch. Since the switch is either on or off, the average power dissipation level is greatly reduced. This results in significant advantages. In fact, when compared with the typical linear power supply, a switching supply offers the following:

- Increased efficiency
- Lower operating temperatures
- Smaller size
- Less weight
- Operation over a wider range of input voltages
- Increased reliability
- Cost-effective performance
- Better line isolation (>60 dB)

More-advanced switching power supplies have eliminated the bulky power transformers, resulting in a smaller, lighter, and less-expensive device. A block diagram of this type of supply is shown in Figure 5.19. The incoming ac is rectified, filtered, and converted to a 20-kHz switching frequency before regulation. This negates the need for bulky 60-Hz components.

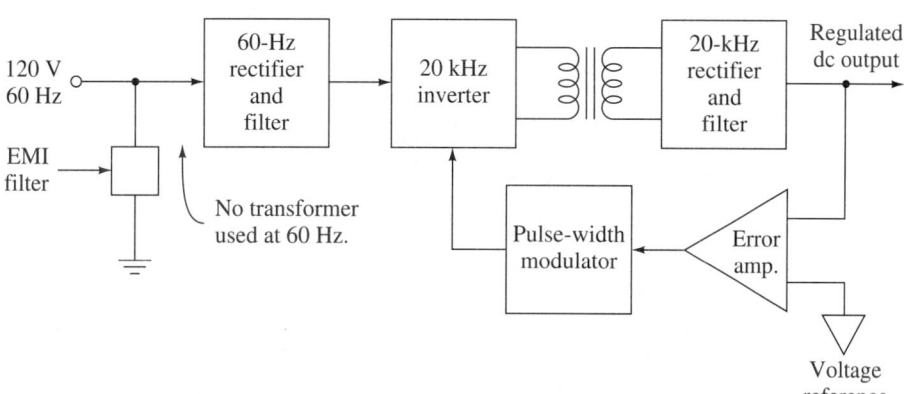

FIGURE 5.19 Block diagram of a switching power supply.

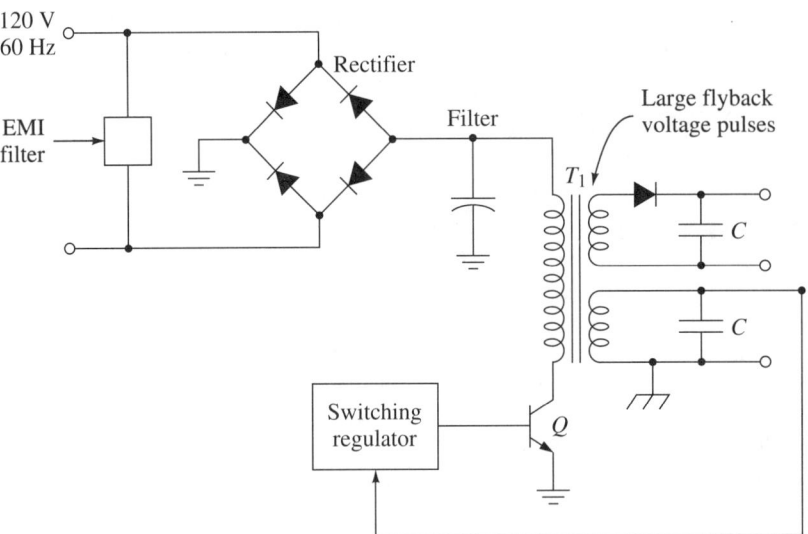

FIGURE 5.20 Switching power supply.

Figure 5.20 is a simplified schematic of a switching power supply. The 120-V, 60-Hz input voltage is full-wave rectified, filtered, and presented to the primary winding of transformer T1. Switching transistor Q is in series with the primary winding and, therefore, controls the supply. The switching transistor operates in two states only, fully saturated (on) or fully cut off (off). When saturated (on), a complete path allows current flow in the primary. When cut off (off), the path is opened. At the instant the switch opens, the large negative di/dt creates a very large flyback voltage pulse at the transformer secondary windings. The final voltage level available at the secondary terminal is determined by the turns ratio of the primary to secondary windings, the coupling factor, the switching frequency, and the duty cycle of the switching signal. Only the switching frequency and duty cycle are variable, however, and this fact is used to control the output voltage level. This task is performed by the switching regulator. Because the switching transistor operates in either the saturation or cutoff mode, its power dissipation level is greatly reduced.

If the power demand of the load is reduced, the switching regulator produces a high-frequency, low-duty cycle pulse. As demand increases, the regulator decreases the switching frequency and increases the duty cycle to deliver more power. The duration of the rectangular waveform is controlled by the logic circuitry of the pulse-width modulator inside the regulator. The pulse width will increase to boost the output voltage level and decrease to lower it. See Figure 5.21.

The switching rate can vary from 20 kHz to as much as 1 MHz. Today supplies use a single integrated circuit to accomplish the switching control. Because of the high frequencies used, all components are physically much smaller and lighter-weight than those processing 60-Hz signals.

5.31   SWITCHING REGULATORS

In order to reduce the energy loss during the switching process, an inductive element is used to provide a storage medium. In simple terms, the inductor stores energy in its magnetic field while the switch is on and releases this energy to the circuit during the off time of the switch. Control of the output voltage or current is then accomplished by varying the frequency (duty cycle) of the switching process. To analyze this energy-transformation process, let's consider the half-wave rectified load voltage in Figure 5.22. The diode can be considered to be a simple switch. The switching action is synchronized to the zero crossing of the input voltage. The addition of an inductor, as in Figure 5.23, helps to filter (smooth) the energy delivered to the load by releasing energy during the diode's (D_1) off time. Rectifier D_2 provides the current path for the inductor (L) during this period of time. A simple switching regulator is shown in Figure 5.24.

For this simple regulator, the switching rate is primarily determined by the inductor (L_1) and capacitor (C_1). The output voltage is a function of the on time of the signal to the base of Q_1. Essentially, the L_1C_1 filter produces a signal with a level equal to the mathe-

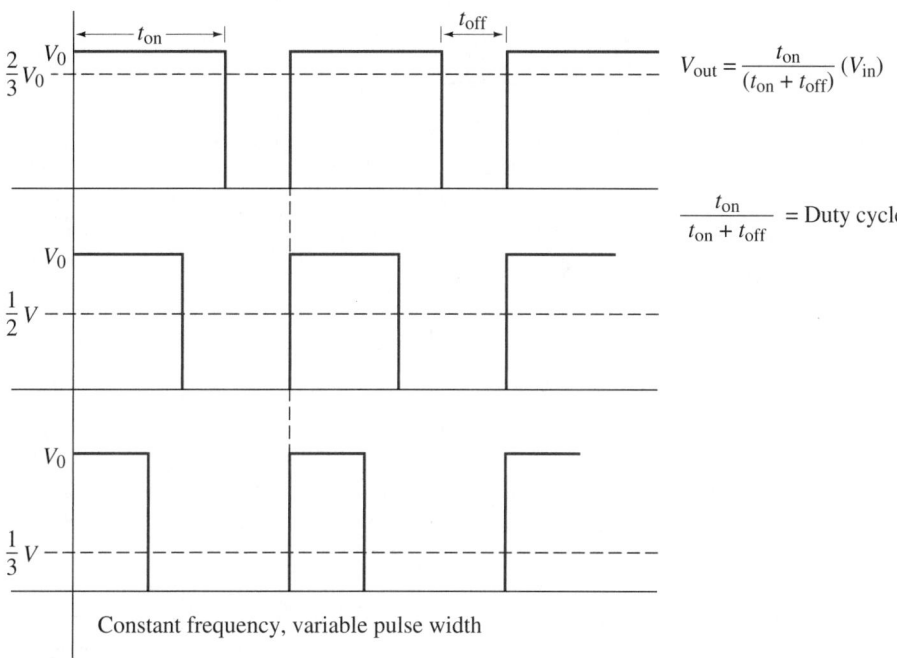

$$V_{out} = \frac{t_{on}}{(t_{on} + t_{off})}(V_{in})$$

$$\frac{t_{on}}{t_{on} + t_{off}} = \text{Duty cycle}$$

Constant frequency, variable pulse width

FIGURE 5.21 Pulse-width modulator output.

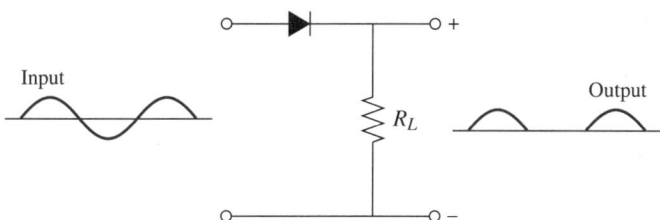

FIGURE 5.22 Half-wave rectifier.

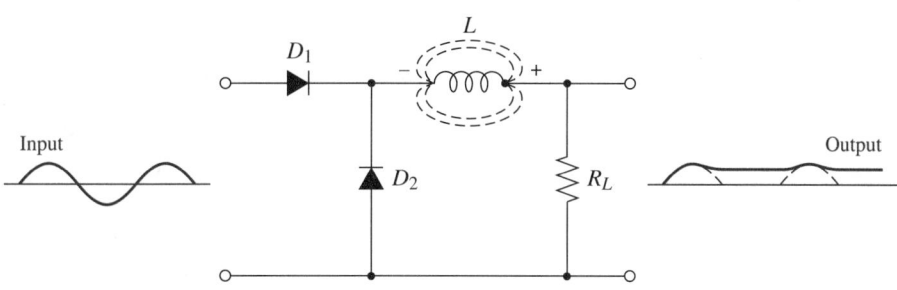

FIGURE 5.23 Filtered half-wave rectifier.

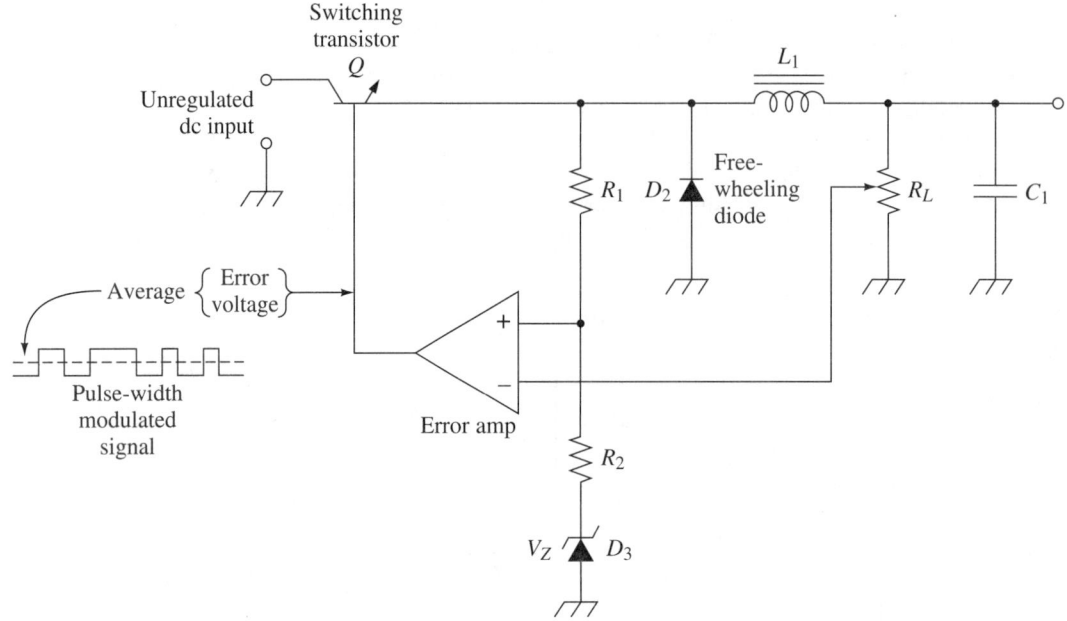

FIGURE 5.24 Switching regulator.

matical average of the switching pulses at the emitter of Q1. Note that the regulation is achieved by a combination of pulse width and pulse-repetition-rate modulation. Switching rates are typically 20 kHz and above.

In the self-excited switcher of Figure 5.24 the error amplifier compares the feedback (sampled output) voltage with the reference established by D_3. A drop in the output voltage causes V_{error} to go positive, turning Q_1 on. If V_{out} rises too high, V_{error} goes negative, turning Q_1 off.

The next step is to utilize a separate oscillator to produce the switching signal for the transistor. This easily results in a fixed-frequency, variable-pulse-width regulator, as in Figure 5.25.

Many monolithic integrated circuits are readily available today. A good example is the LM2574 produced by National Semiconductor. This switching regulator (Figure 5.26), is capable of driving a 0.5-A load with excellent line and load regulation. These units are simple to use, as seen in Figure 5.27.

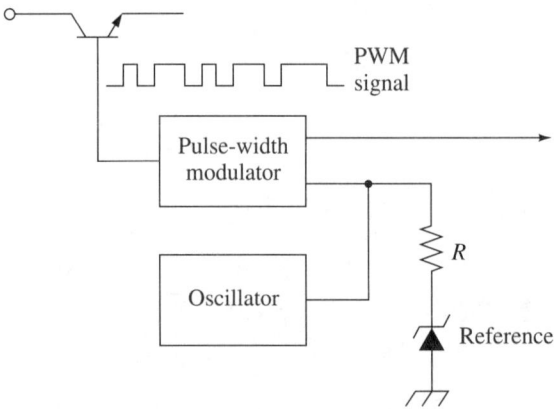

FIGURE 5.25 Fixed frequency switcher.

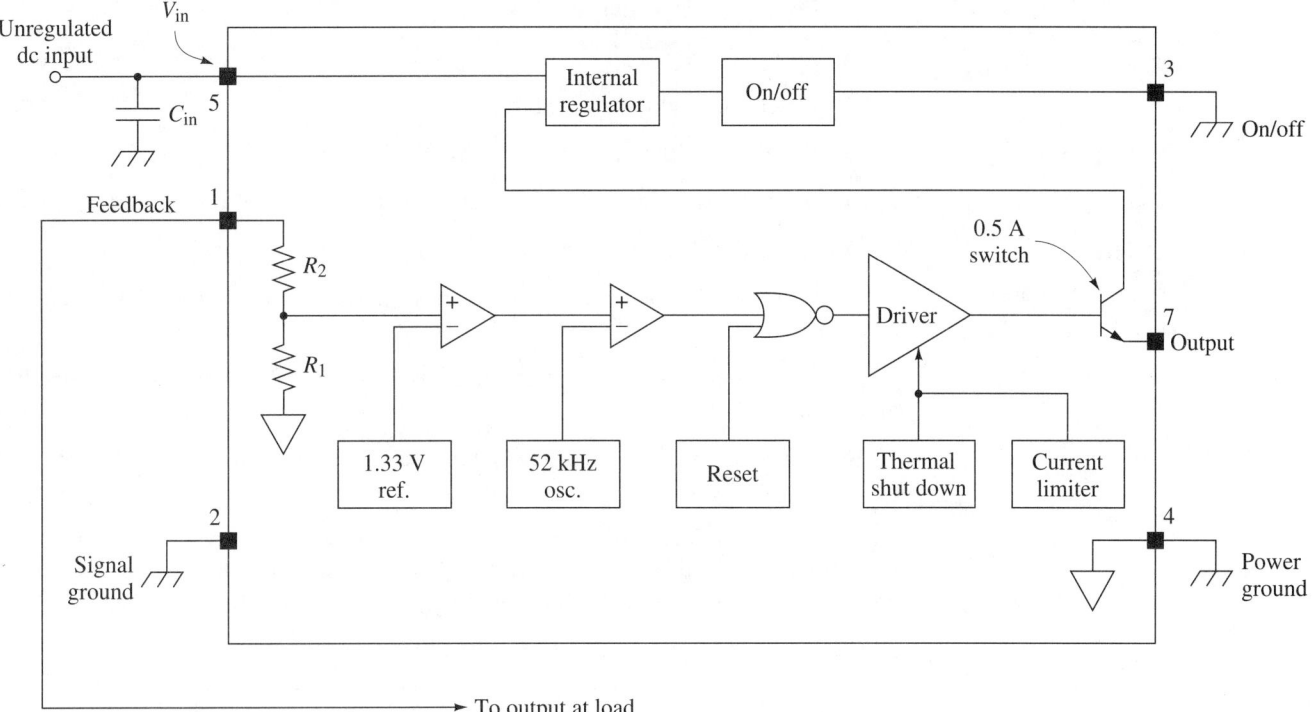

FIGURE 5.26 LM 2574.

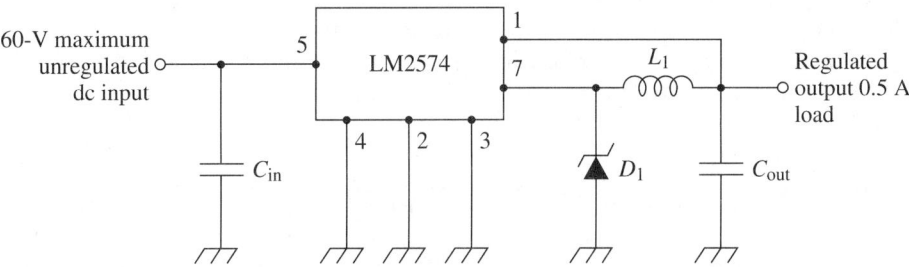

FIGURE 5.27 LM 2574 circuit.

5.32 TROUBLESHOOTING THE SWITCHING SUPPLY

Troubleshooting the switching power supply can be tricky. *Never operate a switching supply without a load.* This may create high-voltage transients, which can damage the switching transistor. *Do not use your oscilloscope to check the input section of the supply.* The input circuits are tied to earth ground and are very dangerous. Since your oscilloscope is also tied to earth ground, it is mandatory to use an isolation transformer for your protection. The best thing to do here is to avoid using the scope completely when making measurements on these circuits.

If the power supply is blowing fuses, check for a shorted switching transistor. Excessive current through the transformer primary is the most probable cause. *Be careful.* Voltages in excess of 300 V_{dc} are common in this section. In general, you should follow these steps when troubleshooting:

1. Remove all power.

2. Discharge the input filter capacitors.

 CAUTION
Don't forget the load.

 CAUTION
High voltages may occur.

Follow these steps when troubleshooting the switching power supply.

3. Remove the switching transistor.

4. Reapply power.

5. Check the condition of the fuse.

6. If the fuse is all right, replace the switching transistor.

7. If the fuse still blows after step 4, reinstall the switching transistor and continue.

8. Check the filter capacitors. *Be careful!* Make certain they are totally discharged.

9. If an input capacitor is found defective, check the condition of the rectifier diodes.

10. If all components test all right, check the EMI filter. Remove it from the supply and see if the fuse still blows.

11. Inspect the wiring and printed circuit boards for shorted lands and/or components.

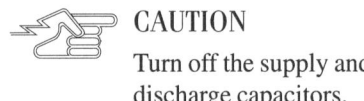

CAUTION

Be sure capacitors are discharged.

Overload conditions will normally put the switching supply in an idle state with minimum current flow. If this happens and the fuse does not blow, the short-circuit condition must occur after the transformer. If this is the case you should do the following:

1. Check the output circuits for shorted conditions.
2. Disconnect the output circuits one at a time and watch the condition of the fuse.
3. If the output circuits appear to be good, the next step is to check the feedback circuits.
4. Finally, check the transformer windings.

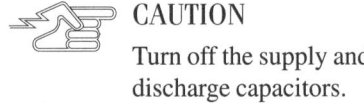

CAUTION

Turn off the supply and discharge capacitors.

The input voltage levels must be isolated from the output voltage levels. Therefore, different grounds are found on either side of the switching transformer. This situation is shown in Figure 5.20. It makes voltage isolation extremely necessary. Because of this, optical isolators or pulse transformers are typically used to achieve circuit isolation.

Always be sure to turn off the supply and discharge those capacitors.

5.33 TROUBLESHOOTING EXAMPLE 1

Troubleshooting circuit problems

Let's take a look at a realistic situation and analyze the circuit problems. Figure 5.18 shows the power-supply circuit. Initially, the output went to zero and then the supply started blowing fuses. Follow these steps to analyze and troubleshoot this problem (note that the regulator is a 5-V regulator):

Step 1. Remove load at point *E*.

Step 2. If the fuse does not blow, replace the load. If the fuse still blows, go to step 3.

Step 3. Break at point *C*.

Step 4. If the fuse still blows, check the diodes. If the fuse does not blow, go to step 5.

Step 5. Break at point *D*. If the fuse still blows, check the filter components. If the fuse does not blow, replace the regulator.

5.34 PROBLEMS

1. What are the basic building blocks of a power supply?

2. Identify the rectifier circuit in Figure 5.28.

3. Sketch the output waveform for the rectifier circuit in Figure 5.29.

4. If diode, D_1, in Figure 5.29 shorts, what will happen to the output voltage?

5. What is the purpose of the capacitor input filter?

6. If the capacitor in Figure 5.30 shorts, what will happen to the output voltage?

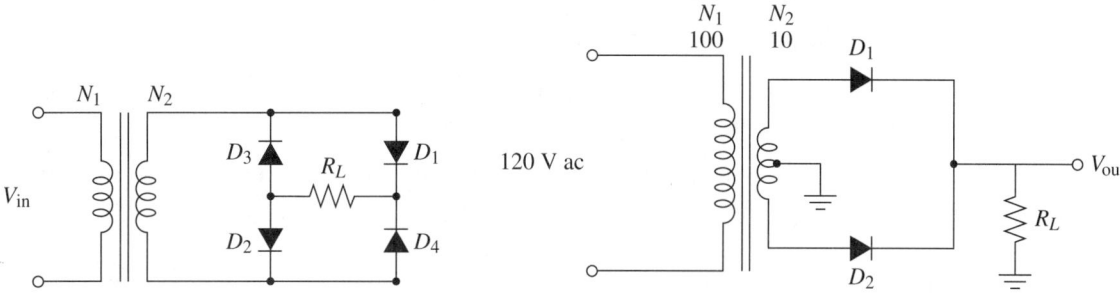

FIGURE 5.28 Problem 2.   FIGURE 5.29 Problem 3.

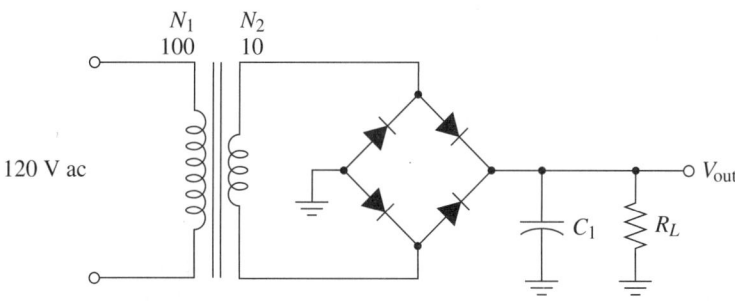

FIGURE 5.30 Problem 6.

7. How does the RLC time constant affect the ripple voltage?

8. What is the purpose of a regulator circuit?

9. You suspect the regulator in Figure 5.31 of being faulty. List the steps that you would use to troubleshoot the circuit.

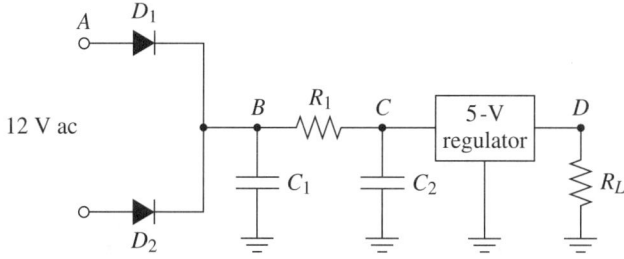

FIGURE 5.31 Problem 9.

10. Why is the output voltage in Figure 5.32 unregulated?

11. What component in the series regulator is most likely to fail?

12. Identify the trouble in Figure 5.33.

13. Describe the four basic types of IC regulators.

14. What should the regulated dc output voltage be in Figure 5.34?

15. Given the ripple voltage shown in Figure 5.34, is there a problem? What is it?

16. What are some of the most common transformer faults?

Problems 17–20 refer to Figure 5.35.

17. The output voltage for the power supply suddenly dropped to zero. What is your first troubleshooting step?

18. You examined the fuse and determined that a catastrophic failure occurred. What does this tell you?

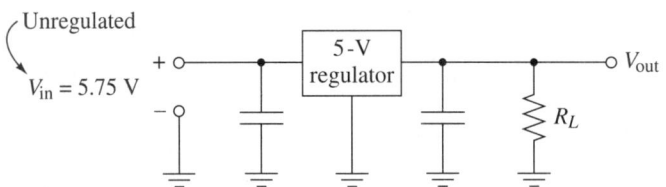

FIGURE 5.32 Problem 10.

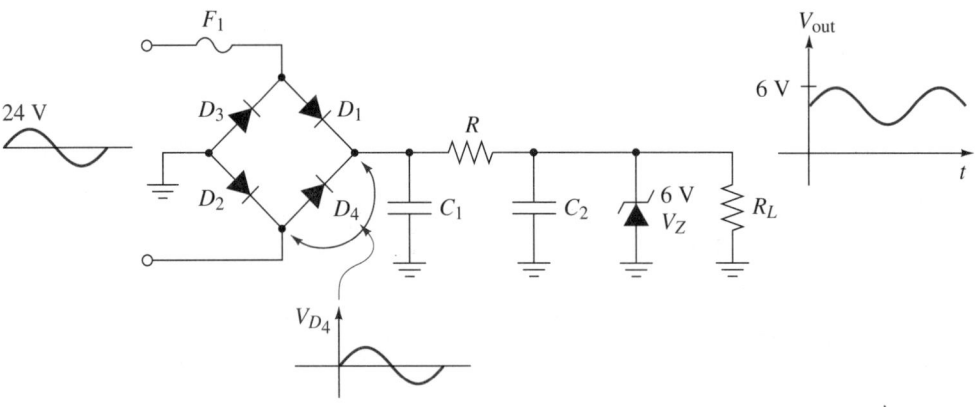

FIGURE 5.33 Problem 12.

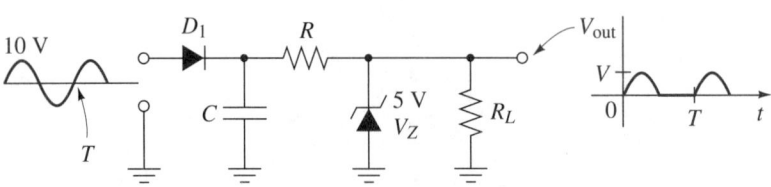

FIGURE 5.34 Problems 14 and 15.

19. You took various measurements and obtained the waveforms shown. What are your conclusions?

20. Complete Figure 5.36 by sketching the normal waveforms.

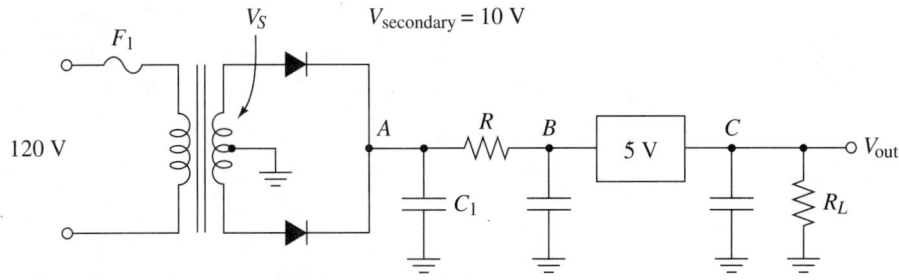

After breaking the circuit at point A,
the new fuse does not blow and waveform A is:

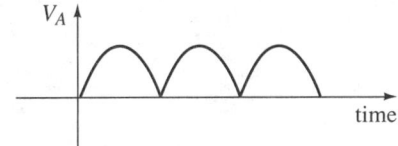

FIGURE 5.35 Problems 17–20.

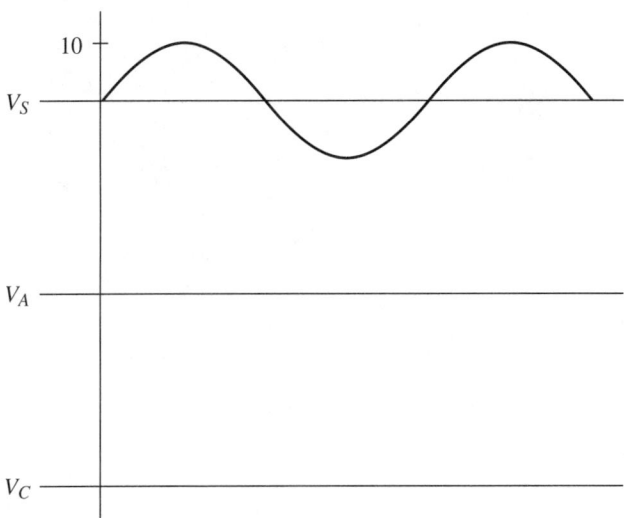

FIGURE 5.36 Problem 20.

6

AC TROUBLESHOOTING

6.1 INDUCTORS

When an inductor is subjected to alternating currents, a magnetic field is developed that expands and collapses with each alternation. The changing magnetic fields induce voltages across the coil due to self-induction. See Figure 6.1. The voltage can be determined by the equation

$$V = L \frac{di}{dt} \qquad (6.1)$$

Because of this effect, the inductor presents an opposition to alternating currents that is called its inductive reactance (X_L). The inductive reactance is found by

$$X_L = 2(3.14)fL \qquad (6.2)$$

As you can see, the reactance is directly proportional to both the coil inductance and the frequency of the input voltage. As the frequency increases, the reactance increases. Therefore, inductors are used to block ac and pass dc in tuning circuits, frequency-selective circuits, and oscillators.

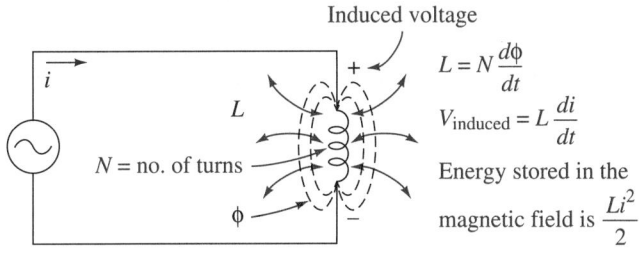

Induced voltage

$$L = N \frac{d\phi}{dt}$$

$$V_{induced} = L \frac{di}{dt}$$

Energy stored in the magnetic field is $\frac{Li^2}{2}$.

i

L

N = no. of turns

ϕ

FIGURE 6.1 Inductor in ac.

When inductors are added in series or parallel, the total value of inductance is found as shown in Figure 6.2.

All components have some inductance. A piece of wire has what is known as stray inductance. The values depend on the frequency involved. At radio frequencies and above component lead lengths, conductor paths and even printed circuit board traces can represent significant values of inductance.

These inductances can have significant impact on digital circuits operating at the current high speeds of 100–250 MHz.

It is important to remember that a coil resists any changes in the circuit current. Current does not change instantaneously in a coil. Therefore, the voltage across a coil will always lead the coil current by 90°. The waveforms are shown in Figure 6.3.

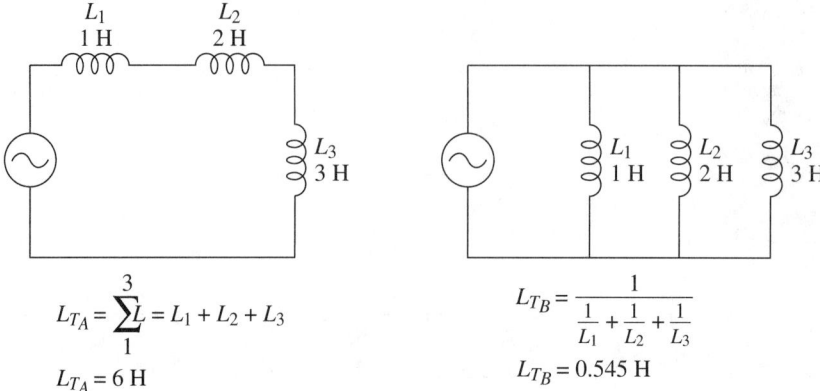

$$LT_A = \sum_1^3 L = L_1 + L_2 + L_3$$

$$LT_A = 6\ H$$

$$LT_B = \frac{1}{\frac{1}{L_1} + \frac{1}{L_2} + \frac{1}{L_3}}$$

$$LT_B = 0.545\ H$$

FIGURE 6.2 Series and parallel circuits.

Inductor voltage leads inductor current by 90°.

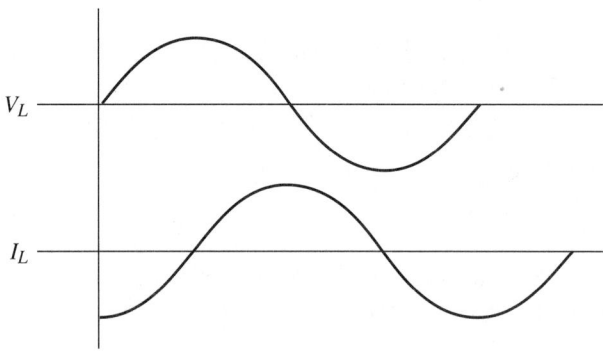

FIGURE 6.3 Inductor waveforms.

6.2 CAPACITORS

All components have some measure of capacitance, known as *stray capacitance.* Some capacitance will exist between the leads of components and the circuit ground; some capacitance will exist across an open circuit; some capacitance will exist between adjacent leads.

The effect that capacitors have on ac is called the *capacitive reactance*, X_C. The capacitive reactance can be found by the equation

$$X_C = \frac{1}{2(\pi)fC} \quad \pi = 3.14159... \tag{6.3}$$

Capacitor current leads capacitor voltage by 90°.

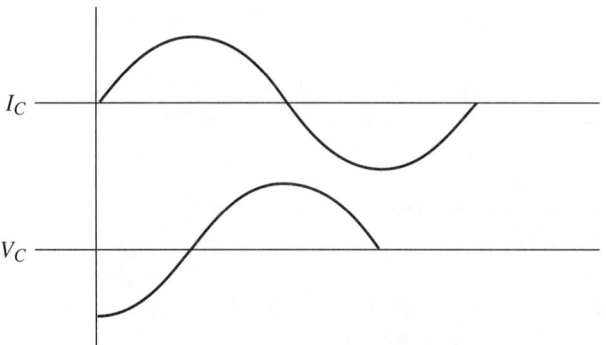

FIGURE 6.4 Waveform comparisons.

The capacitive reactance is a very high value when the circuit frequency is low. Therefore, the capacitor represents an open circuit to dc. Likewise, the reactance is very low when the frequencies are very high.

Due to this effect, the voltage across the capacitor has to build up over time, and the current leads the voltage by 90°. The waveform comparisons are shown in Figure 6.4.

6.3 *RC* TRANSIENT RESPONSE

When a square wave of voltage is applied to a resistor-capacitor network, the capacitor will charge toward the peak value of voltage. The actual waveform will be determined by the circuit *RC* time constant. The circuit will be an integrator (Figure 6.5) or a differentiator (Figure 6.6), depending on the placement of the capacitor in the circuit. The effect of the *RC* time constant is shown in Figure 6.7.

A special caution here: When using the oscilloscope, remember that the ground lead is connected to earth ground. As shown in Figure 6.8, you should reverse the component positions when taking measurements in order not to disturb the circuit.

CAUTION

Be careful using the oscilloscope.

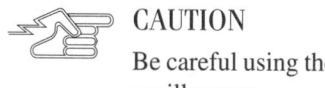

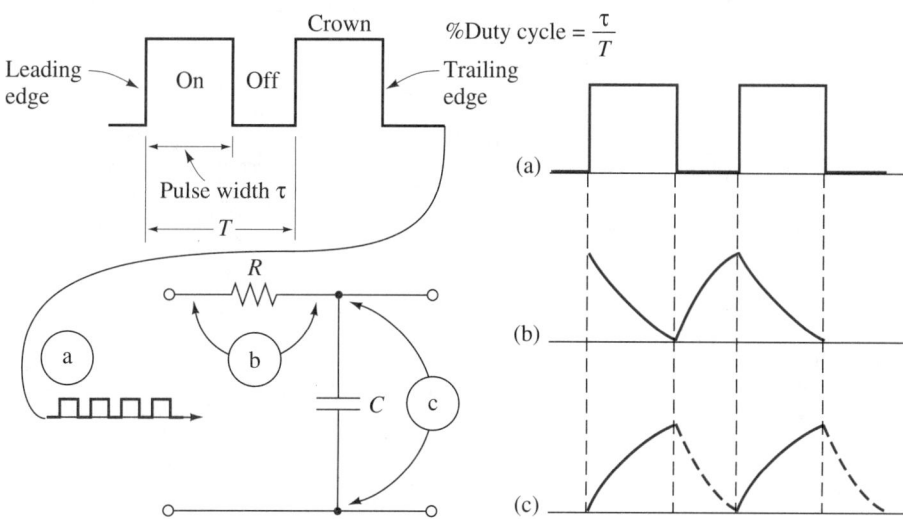

FIGURE 6.5 *RC* integrator.

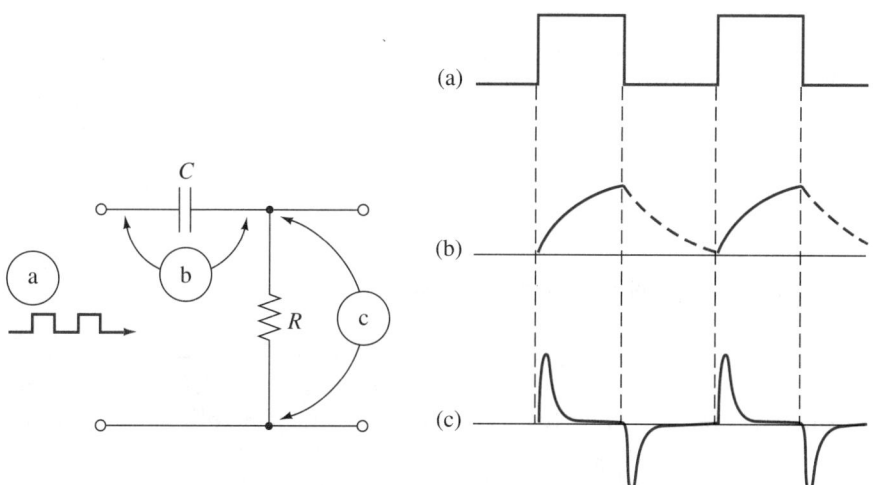

FIGURE 6.6 Differentiator.

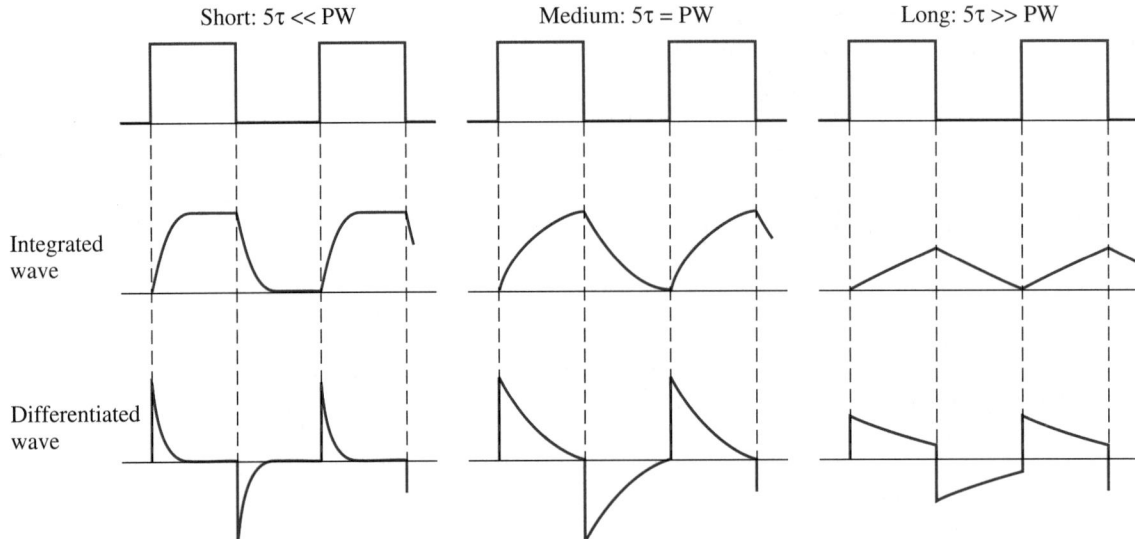

Short: 5τ << PW   Medium: 5τ = PW   Long: 5τ >> PW

Integrated
wave

Differentiated
wave

FIGURE 6.7 Time constants.

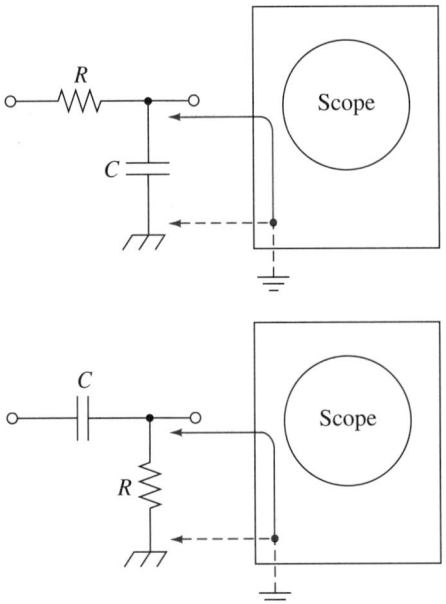

By changing the position of the components,
the ground is maintained.

FIGURE 6.8 Avoiding ground loops.

6.4 *RL* TRANSIENT RESPONSE

In an *RL* circuit like the one in Figure 6.9, the waveform across the resistor is an integration waveform because of the increasing and decreasing magnetic fields of the inductor. Due to the counterelectromotive force, CEMF, the inductor looks like an open circuit at the instant the voltage is applied. Over 5 *RL* time constants, the magnetic field builds up and current flows through the inductor and the resistor, producing the waveform of voltage shown.

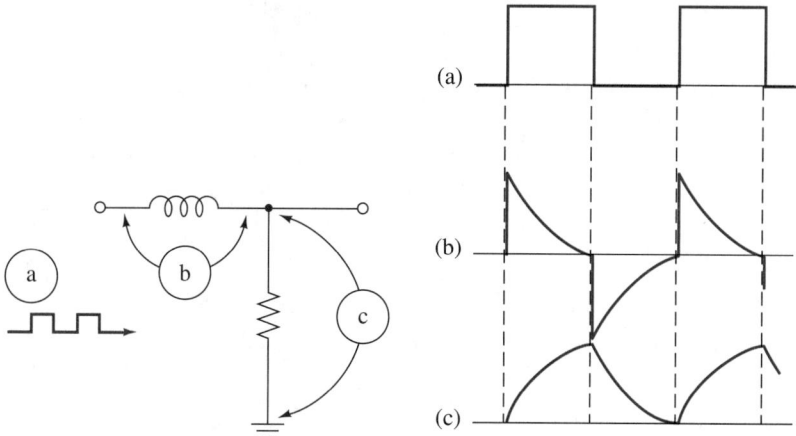

FIGURE 6.9 *RL* transient response.

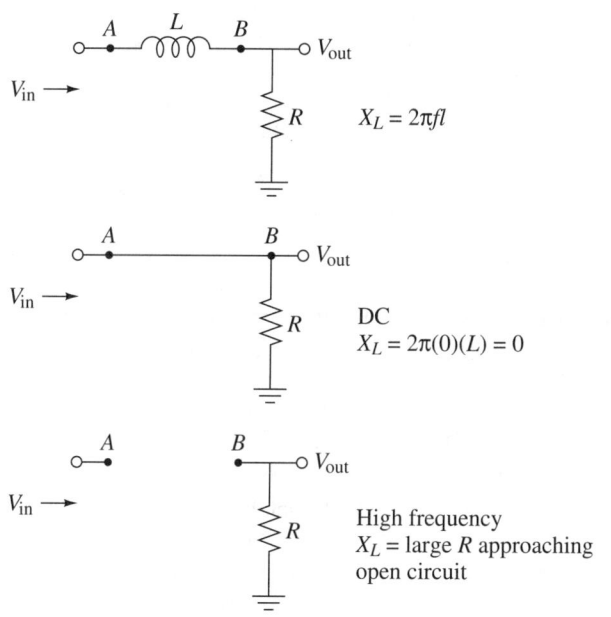

$X_L = 2\pi fl$

DC
$X_L = 2\pi(0)(L) = 0$

High frequency
$X_L =$ large R approaching
open circuit

FIGURE 6.10 Inductive coupling.

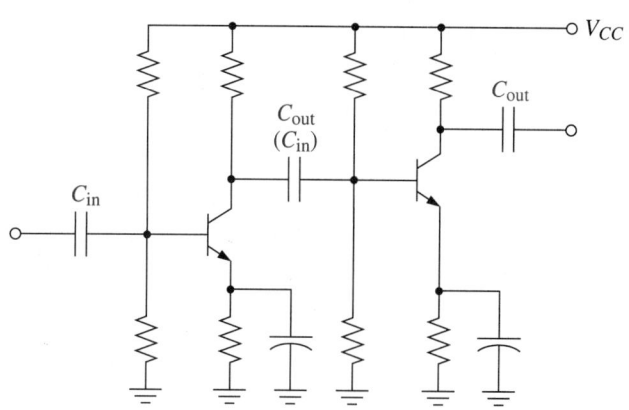

C_{out} of stage 1 is also C_{in} of stage 2.

FIGURE 6.11 Capacitive coupling.

6.5 CIRCUIT COUPLING

Inductive Coupling

Inductive coupling is used when it is desired to pass the low frequencies more easily than high frequencies. Remember, due to the inductive reactance, the inductor looks like a short to dc and an open to high ac. See Figure 6.10.

Capacitive Coupling

The capacitor blocks dc and passes ac. The transistor circuit in Figure 6.11 is using capacitive coupling to pass the ac signal and keep the dc biasing of each stage unaffected. The value of capacitance chosen for the job depends on the frequencies involved. Table 6.1 lists the most typical applications.

TABLE 6.1 Typical Applications

DC Blocking Capacitor Values	
Low frequency	1–10 μF electrolytic
Medium frequency	0.1–1 μF electrolytic
Low RF to 1 MHz	0.01–0.1 μF disk
High RF 1–100 MHz	0.001–0.01 μF ceramic or mylar

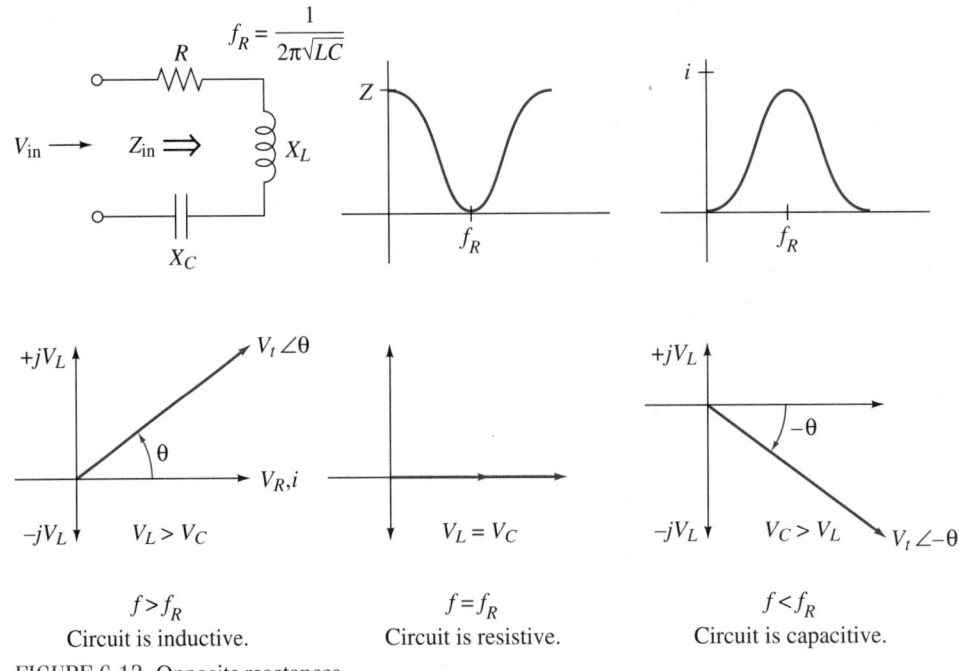

$$f_R = \frac{1}{2\pi\sqrt{LC}}$$

$f > f_R$	$f = f_R$	$f < f_R$
Circuit is inductive.	Circuit is resistive.	Circuit is capacitive.

FIGURE 6.12 Opposite reactances.

6.6 *RLC* CIRCUITS

When inductors and capacitors are combined in the circuit, the reactances have a cancellation effect. Figure 6.12 shows the vector waveform analysis of a typical circuit.

6.7 SERIES RESONANT CIRCUITS

When the frequency of the signal in Figure 6.13 is such that $X_L = X_C$, the circuit looks purely resistive and operates at maximum efficiency. The reactances cancel each other completely. The reactive energy is bouncing back and forth between the electric field of the capacitor and the magnetic field of the inductor. The voltages across the inductor and the capacitor are 180° out of phase. The circuit is said to be resonant. The characteristic curves are shown in Figure 6.14. This type of circuit is found in RF systems and would be tuned to a particular frequency.

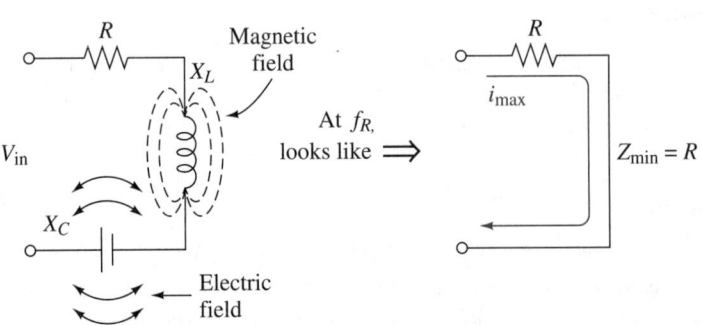

FIGURE 6.13 Series resonance.

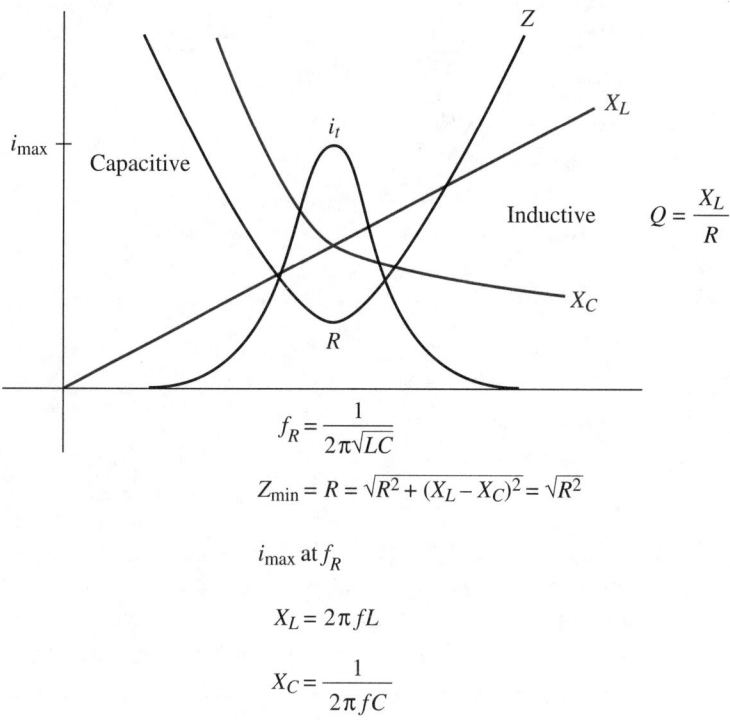

$$f_R = \frac{1}{2\pi\sqrt{LC}}$$

$$Z_{min} = R = \sqrt{R^2 + (X_L - X_C)^2} = \sqrt{R^2}$$

$$i_{max} \text{ at } f_R$$

$$X_L = 2\pi fL$$

$$X_C = \frac{1}{2\pi fC}$$

FIGURE 6.14 Series resonant circuit curves.

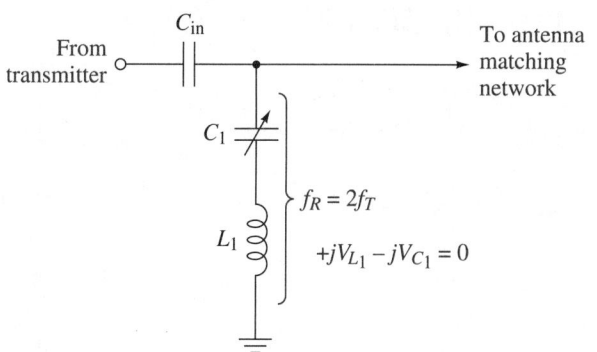

FIGURE 6.15 Frequency trap.

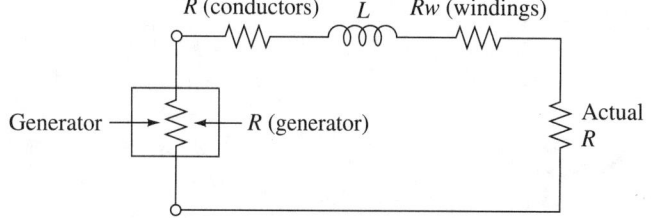

High-Q inductors have very small winding resistance.

$$Q_L = \frac{X_L}{Rw}$$

Measure the value of Rw with an ohmmeter.
Rw equals the resistance of the windings.

FIGURE 6.16 Simple RL circuit.

Figure 6.15 shows a frequency trap used in transmitters. This trap is tuned to the second harmonic of the transmitter frequency.

The overall quality of the circuit, the circuit Q, is determined by the total circuit resistance, which includes real resistors, the dc resistance of coil windings of the inductors, internal impedance of the generator, and resistance of the conductors. See Figure 6.16.

The Q of a coil is determined by the equation

$$Q = X_L/R \tag{6.4}$$

and high-Q inductors have very small coil resistances. The coil resistance can be measured with an ohmmeter and will vary with frequency. At radio frequencies and up, the

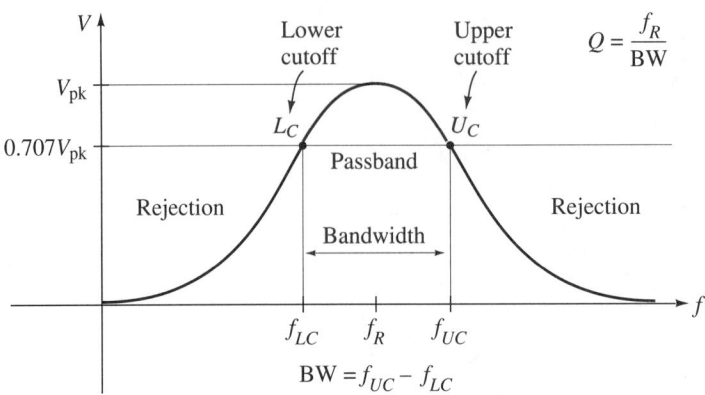

FIGURE 6.17   Relationship of bandwidth to circuit Q.

currents tend to travel along the surface of the wires (skin effect), and the effective resistance increases. Therefore, the coil Q will be lower at higher frequencies.

High-Q capacitors have highly conductive plates and very low leakage dielectrics. A high-Q series resistance circuit will have a high X_L at the resonant frequency, and both V_L and V_C are increased by the factor of Q. Thus,

$$V_C = EQ = V_L \qquad (6.5)$$

The circuit bandwidth in hertz is the band of frequencies (called the *midband*) where the output is greater than $0.707V_{max}$. As shown in Figure 6.17, the bandwidth is inversely related to the circuit Q. A high-Q circuit will have a narrower bandwidth.

6.8   TROUBLESHOOTING THE SERIES RESONANT CIRCUIT

Conditions can affect your measurements.

When testing resonant circuits you should remember that the following can upset your measurements:

1. Stray capacitance and inductance
2. Component tolerance
3. Temperature effects
4. Generator resistance
5. Component Q factors
6. Probe-loading effects
7. Capacitance of oscilloscope probes and cables
8. The input capacitance of your oscilloscope

Conditions 6, 7, and 8 are especially severe at frequencies above 1 MHz. These problems can detune a resonant circuit during measurements. You should use low-capacitance probes for your oscilloscope. The following steps should be used in the troubleshooting process.

Step 1.   Connect the generator as shown in Figure 6.18.
Step 2.   Connect scope across R.
Step 3.   Calculate the resonant frequency, $f_R = \dfrac{1}{2\pi\sqrt{LC}}$
Step 4.   Calculate the expected bandwidth, BW.
Step 5.   Sweep the generator output frequency. Locate the actual f_R.
Step 6.   Complete Table 6.2.

Table 6.2 lists the problems most likely to occur with the circuit in Figure 6.18.

TABLE 6.2 Likely Circuit Problems

Is There a Resonant Point?	f_R Value	BW	X_L	X_C	Condition
Yes	OK	OK	—	—	Normal
Yes	Low	—	X_L or X_C too large		Check L or C.
Yes	High	—	X_L or X_C too low		Check L or C.
Yes	OK	Too narrow	Q too high		Check R.
Yes	—	Too wide			
No		V_{out} drops off below f_R			Shorted L
No		V_{out} drops off above f_R			Shorted C

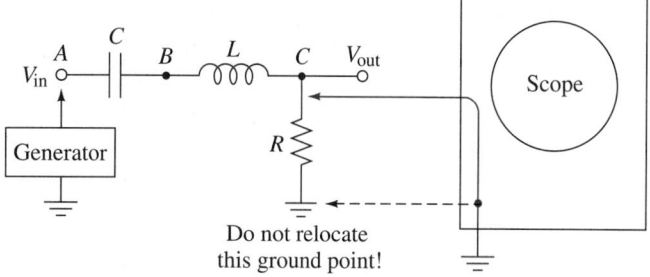

FIGURE 6.18 Troubleshooting series resonant circuits.

6.9 PARALLEL RESONANT CIRCUITS

When $X_C = X_L$, the circuit acts resistive with a power factor of 1. This circuit is used to select a particular frequency and reject others. It is extremely useful in communications circuits as oscillators, frequency multipliers, and tuned circuits. They are designed to resonate over a narrow band of frequencies.

When found in the class C circuit shown in Figure 6.19, the "tank" circuit rings at a given frequency determined by the component values. The characteristic curves are shown in Figure 6.20. Typical problems are identified in Table 6.3.

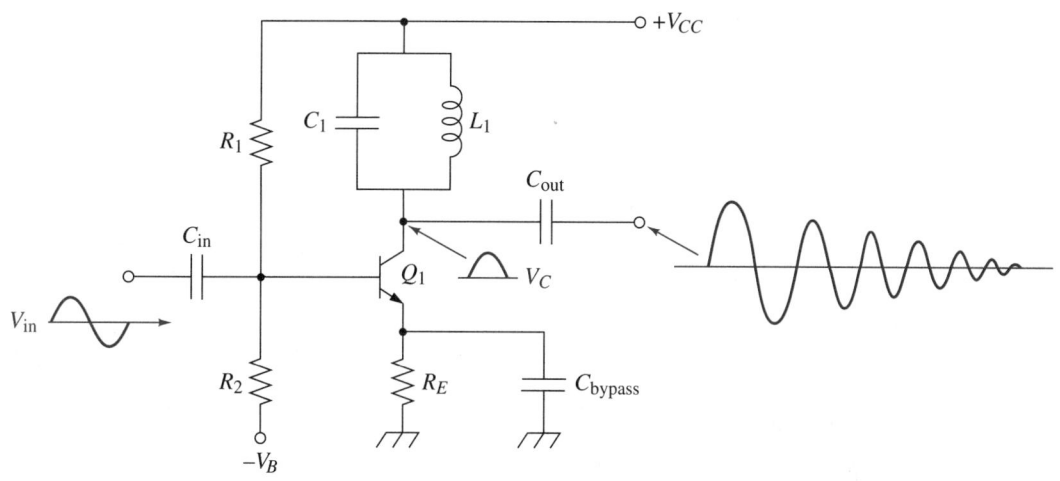

FIGURE 6.19 Class C circuit.

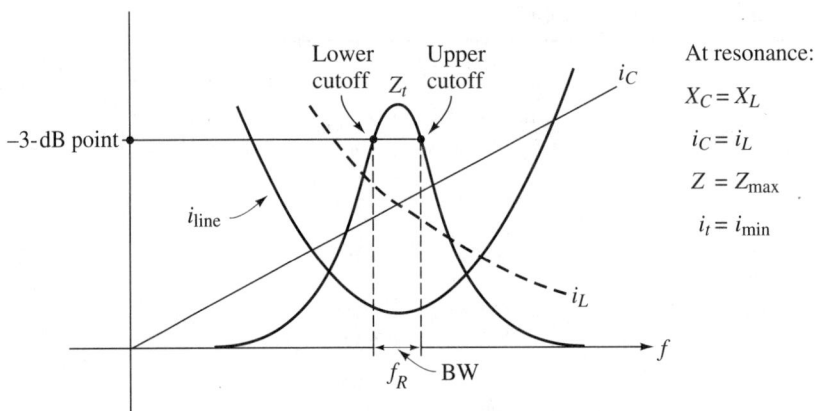

At resonance:

$$X_C = X_L$$

$$i_C = i_L$$

$$Z = Z_{max}$$

$$i_t = i_{min}$$

FIGURE 6.20 Parallel resonant circuit curves.

TABLE 6.3 Typical Problems

Is There a Resonant Point?	f_R	BW	Condition	
Yes	OK	OK	Normal	
Yes	Low	—	X_L too high X_C too low	Check L and C.
Yes	High	—	X_L too low X_C too high	Check L and C.
Yes	OK	Too narrow	Q too high	Add R_L.
Yes	OK	Too wide	Circuit Q too low— need higher Q coil.	
No	V_{out} decreases sharply at $f < f_R$			Open C
	V_{out} decreases sharply at $f > f_R$			Open L

6.10 PASSIVE FILTERS

The basic types of passive filters are

1. Low-pass
2. High-pass
3. Band-pass
4. Band-stop

Which one you use is determined by your application —whether you want to pass dc, block dc, or provide a capacitive load or an inductive load.

One very common application is to couple transistor stages together. The capacitor blocks dc and allows the ac signal to pass. A typical circuit is shown in Figure 6.21.

Several high-pass filters are given in Figures 6.22, 6.23, and 6.24.

The high-pass filter in Figure 6.25 is easy to troubleshoot; the results are shown in Table 6.4.

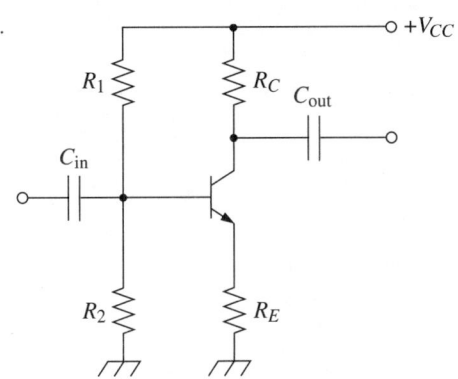

FIGURE 6.21 Capacitor coupling.

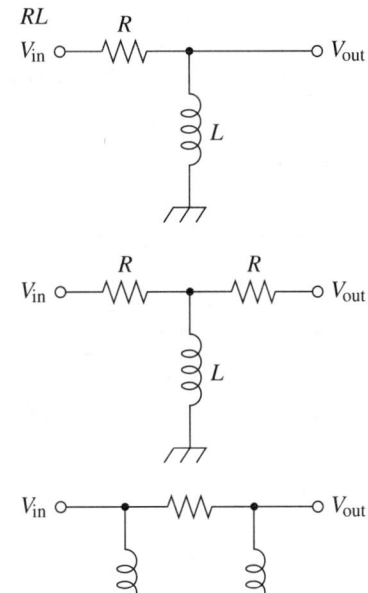

FIGURE 6.22 High-pass filters.

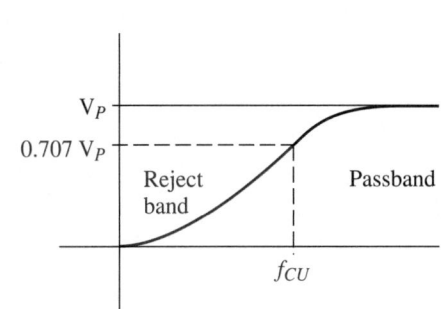

At the cutoff frequency (f_{CU}), $R = X_L$.

$$f_{CU} = \frac{R}{2\pi L}$$

$$L = \frac{R}{2\pi f_{CU}}$$

$$dB_{attenuation} = 20 \log\left(\frac{X_L}{\sqrt{R^2 + X^2_L}}\right)$$

At the cutoff frequency:

$$R = X_C$$

$$f_{CU} = \frac{1}{2\pi RC}$$

$$dB_{attenuation} = 20 \log\left(\frac{R}{\sqrt{R^2 + X^2_C}}\right)$$

FIGURE 6.23 High-pass filters.

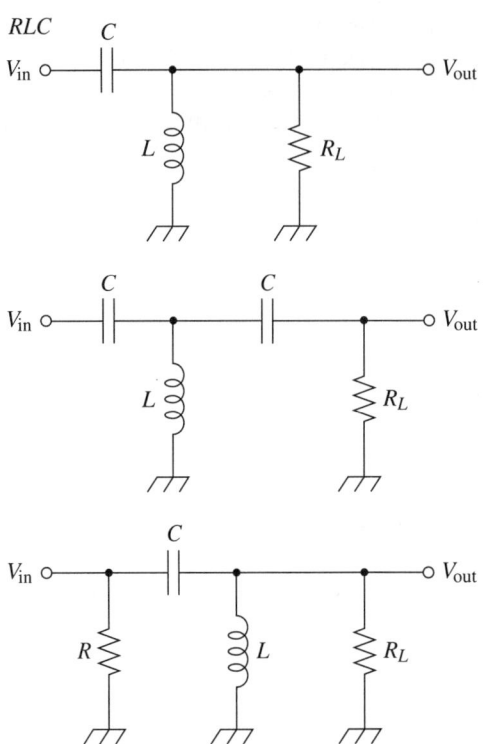

At the cutoff frequency:

$$f_{CU} = \frac{1}{2.8\pi\sqrt{LC}}$$

$$dB_{attenuation} = 20\log\left(\frac{jX_L \,||\, R_L}{jX_L \,||\, (R_L - jX_C)}\right)$$

FIGURE 6.24 High-pass filters.

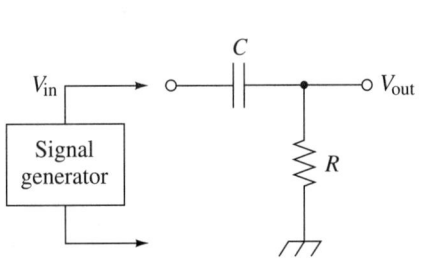

FIGURE 6.25 High-pass filters.

TABLE 6.4 Troubleshooting High-Pass Filters

Condition	Result
C open	$V_C = V_{in}$; $V_{out} = 0$
C short	$V_{out} = V_{in}$
R open	$V_{out} = V_{in}$
R short	$V_{out} = 0$
Vary the input frequency.	Determine $V_{out\,(max)}$.
Locate frequency where	$V_{out} = 0.707\,V_{out\,(max)}$
If C and R are good, but f_{CU} is not:	check the component value.

6.11 LOW-PASS FILTERS

How to troubleshoot the low-pass filter

The low-pass filter is used to pass frequencies from dc up to a set cutoff frequency determined by the component values. There are many applications for circuits like those in Figure 6.26.

1. Audio crossover networks for speakers
2. Passing dc
3. Blocking ac
4. Filtering in power supplies
5. Signal integration

Troubleshooting the low-pass filter (Figure 6.27) is similar to troubleshooting the high-pass filter; the results are shown in Table 6.5.

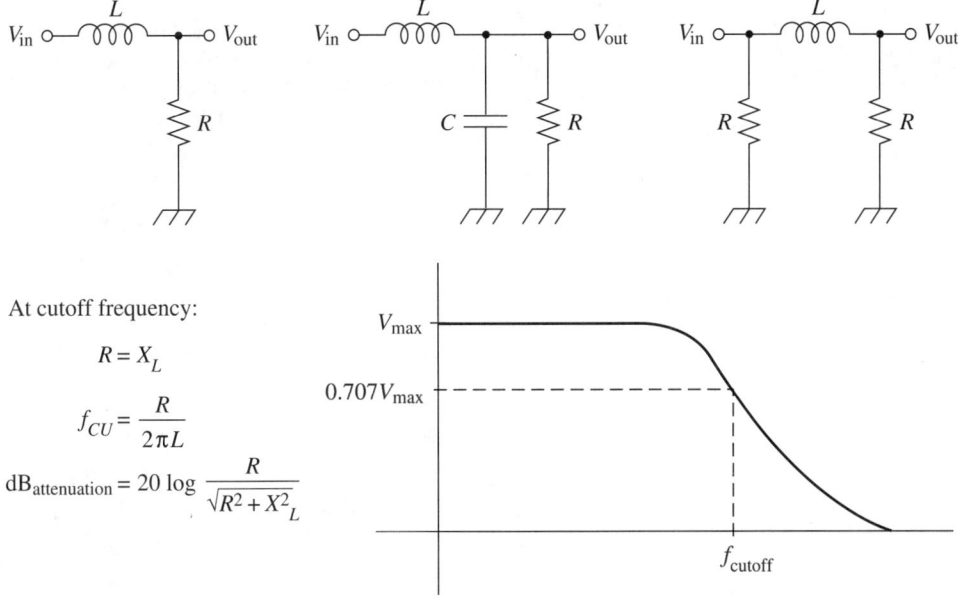

At cutoff frequency:

$$R = X_L$$

$$f_{CU} = \frac{R}{2\pi L}$$

$$dB_{attenuation} = 20 \log \frac{R}{\sqrt{R^2 + X_L^2}}$$

FIGURE 6.26 Low-pass filters.

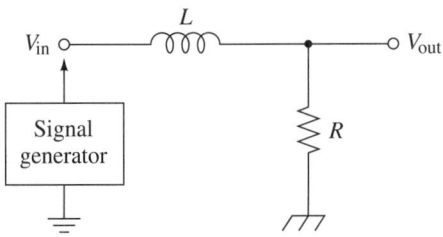

FIGURE 6.27 Low-pass filters.

TABLE 6.5 Troubleshooting Low-Pass Filters

Fault	Condition
Open L	$V_{out} = 0$
Shorted L	$V_{out} = V_{in}$
Open R	$V_{out} = V_{in}$
Shorted R	$V_{out} = 0$
Vary the frequency of V_{in} while monitoring the output. If the cutoff frequency is wrong:	check component value.

6.12   BAND-PASS AND BAND-STOP FILTERS

Band-pass and band-stop filters are combinations of high-pass and low-pass filters. Some typical circuits are given in Figures 6.28, 6.29, 6.30, 6.31, and 6.32.

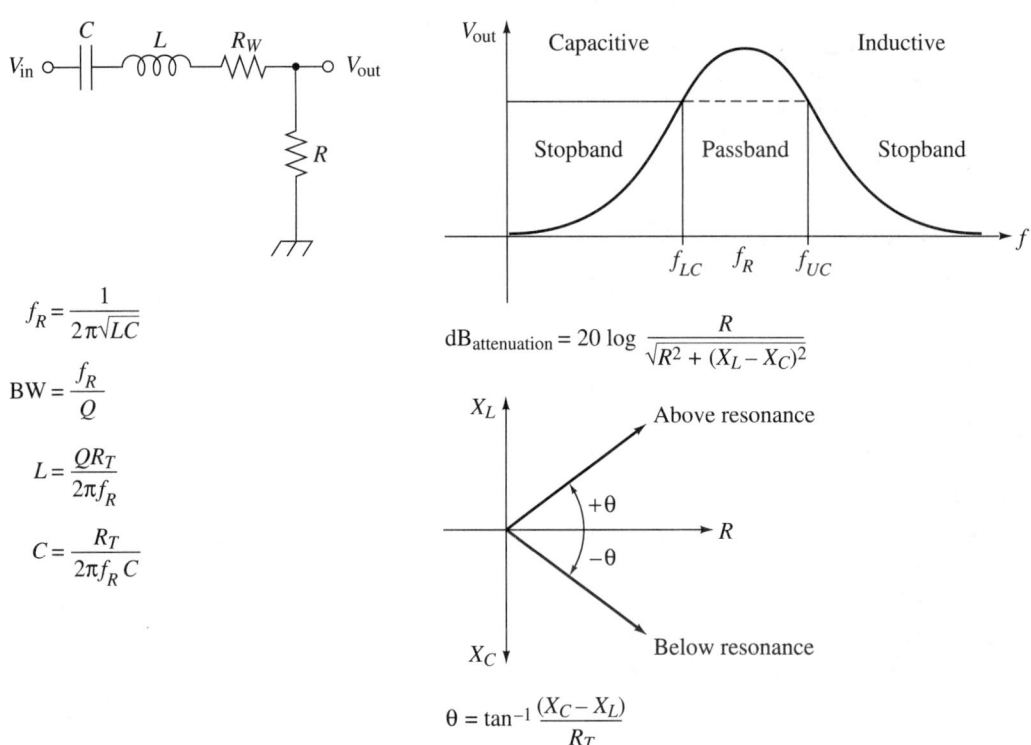

$$f_R = \frac{1}{2\pi\sqrt{LC}}$$

$$BW = \frac{f_R}{Q}$$

$$L = \frac{QR_T}{2\pi f_R}$$

$$C = \frac{R_T}{2\pi f_R C}$$

$$dB_{attenuation} = 20 \log \frac{R}{\sqrt{R^2 + (X_L - X_C)^2}}$$

$$\theta = \tan^{-1} \frac{(X_C - X_L)}{R_T}$$

FIGURE 6.28 Series resonant band-pass filter.

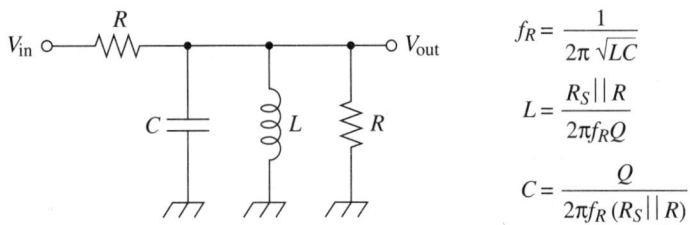

$$f_R = \frac{1}{2\pi\sqrt{LC}}$$

$$L = \frac{R_S || R}{2\pi f_R Q}$$

$$C = \frac{Q}{2\pi f_R (R_S || R)}$$

FIGURE 6.29 Parallel resonant band-pass filter.

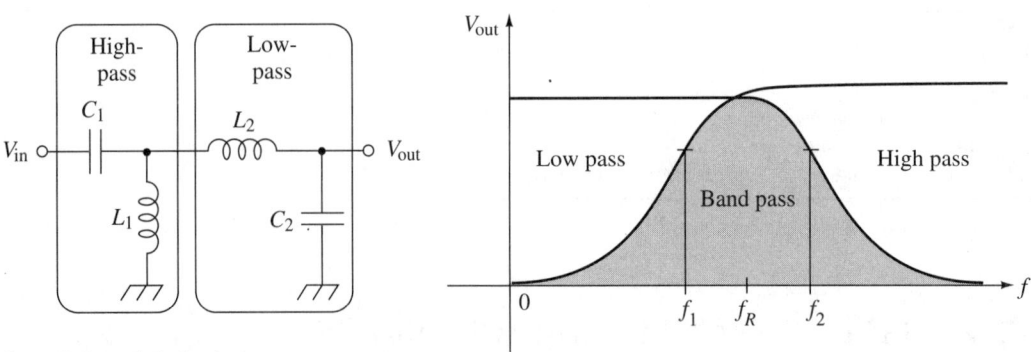

Low f: L_2 and C_2 dominate.

High f: L_1 and C_1 dominate.

FIGURE 6.30 High-pass low-pass combination.

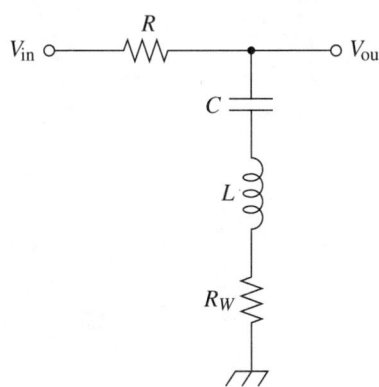

$R_t = R_S + R_W$

$$f_R = \frac{1}{2\pi\sqrt{LC}}$$

$$L = \frac{QR_t}{2\pi f_R}$$

$$C = \frac{1}{2\pi f_R QR_t}$$

$$dB_{attenuation} = 20 \log \frac{\sqrt{R^2_W + (X_L - X_C)^2}}{\sqrt{R^2_t + (X_L - X_C)^2}}$$

FIGURE 6.31 Band-stop filter.

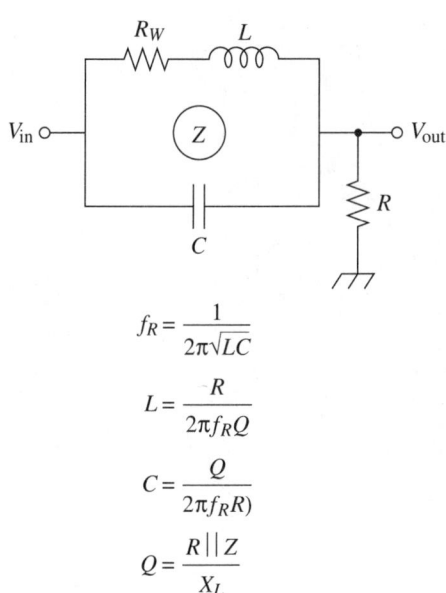

$$f_R = \frac{1}{2\pi\sqrt{LC}}$$

$$L = \frac{R}{2\pi f_R Q}$$

$$C = \frac{Q}{2\pi f_R R)}$$

$$Q = \frac{R \| Z}{X_L}$$

FIGURE 6.32 Parallel band-stop filter.

6.13 THE DIODE IN AC CIRCUITS

When a diode is used in ac circuits, you have to consider their ac resistance in addition to the dc bulk resistance of the *pn* material. The ac resistance is found by the following:

$$r'_{ac} = \frac{25 \text{ mV}}{I_{dc}} \qquad (6.6)$$

When analyzing a circuit, it often helps to use the second-approximation equivalent circuit shown in Figure 6.33.

The diode conducts only when forward-biased by the input signal. There are many applications, however; let's look at limiters, clampers, and voltage multipliers.

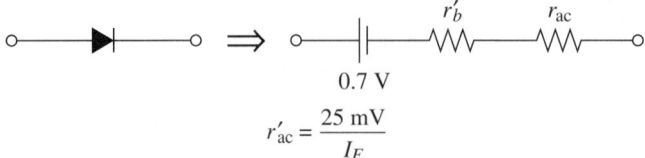

$$r'_{ac} = \frac{25\text{ mV}}{I_E}$$

FIGURE 6.33 Diode second-approximation circuit.

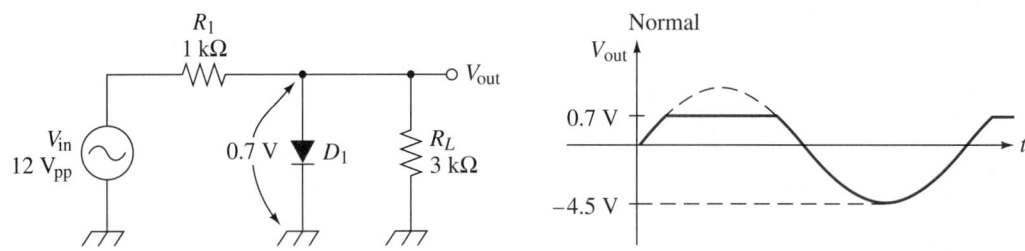

For positive V_{in}, D_1 is forward-biased and $V_0 = 0.7$ V.
For negative V_{in}, D_1 is reverse-biased and $V_0 = V_{in}$.

FIGURE 6.34 Positive limiter.

TABLE 6.6 Associated Problems

	V_{R_1}	V_{D_1}	V_{out}	Waveform
Normal	11.3 V$_{pp}$	0.7 V	0.7 V	See Figure 6.34.
D short	12 V$_{pp}$	0	0	0 ————
D open	3 V$_{pp}$	9 V$_{pp}$	9 V$_{pp}$	∿∿
R_L short	12 V$_{pp}$	0	0	0 ————
R_L open	1 V$_{pp}$	0.7 V	+0.7 V -6 V$_p$	+0.7 V -6 V$_p$
R_1 short	0	0.7 V	+0.7 V -12 V$_p$	+0.7 V -12 V$_p$
R_1 open	12 V$_{pp}$	0	0	0 ————

6.14 DIODE LIMITERS

Diode limiter circuits limit the amplitude of the load voltage. They can be either positive or negative. Figure 6.34 shows a typical positive limiter circuit, and Table 6.6 indicates its associated problems.

A negative limiter circuit and the waveforms for typical problems are illustrated in Figure 6.35.

6.15 DIODE CLAMPERS

Clampers are used to shift an ac waveform in the positive or negative direction. They do not change the peak-to-peak or rms values but rather the peak and the average values. Typical circuits are found in Figure 6.36, and various clamper faults for Figure 6.37 are summarized in Table 6.7.

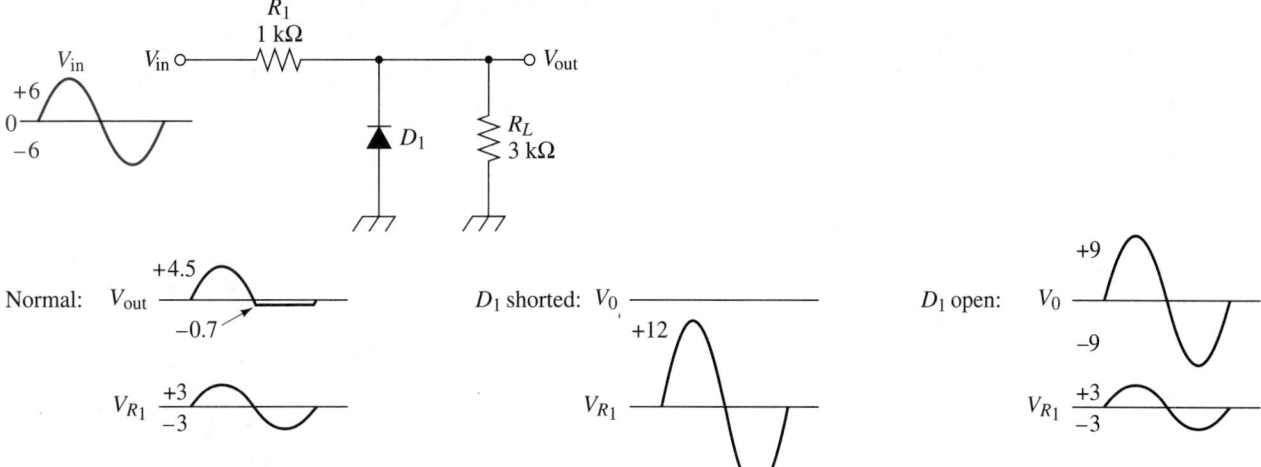

Normal: V_{out} — graph showing $+4.5$, -0.7

V_{R1} — graph showing $+3$, -3

D_1 shorted: V_0 — graph showing $+12$, -12

V_{R1}

D_1 open: V_0 — graph showing $+9$, -9

V_{R1} — graph showing $+3$, -3

FIGURE 6.35 Negative limiter.

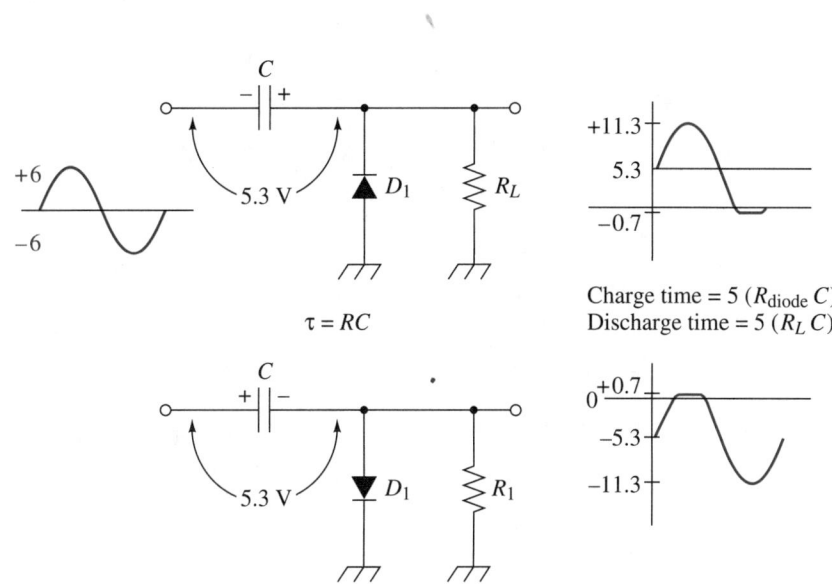

$\tau = RC$

Charge time $= 5\,(R_{diode}\,C)$
Discharge time $= 5\,(R_L\,C)$

FIGURE 6.36 Clampers.

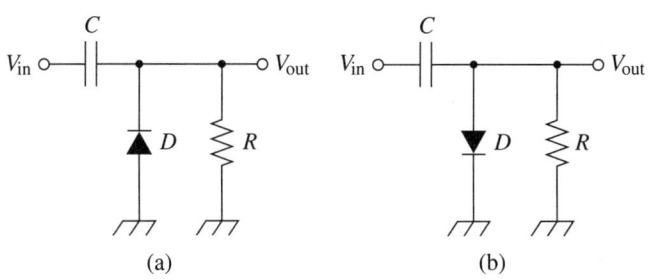

(a)   (b)

FIGURE 6.37 Clamper faults.

TABLE 6.7 Clamper Faults

C open		$V_{out} = 0$
C shorted	(a)	
	(b)	
C leaky		V_{out} distorted with changing dc reference.
D open		No clamping action
		V_{out} has $0\ V_{dc}$ reference
		$0\ V$
D shorted		Capacitor discharges rapidly through the diode.
		Possible damage to the source

6.16   VOLTAGE MULTIPLIERS

Voltage multipliers provide a dc output that is N times the peak value of the input. A voltage tripler is shown in Figure 6.38, and associated problems are identified in Table 6.8.

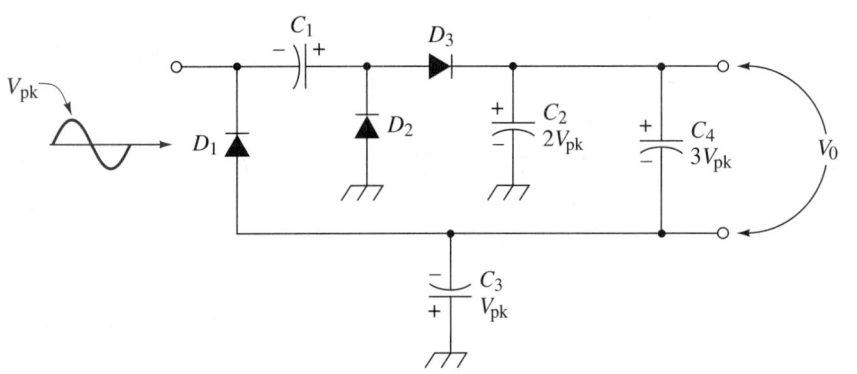

TABLE 6.8   Associated Problems

If $V_{out} \approx V_{pk}$:	check C_2.
If V_{C_2} is incorrect:	check V_{C_1}
$\quad V_{C_1}$ is good:	check D_3 and C_2.
$\quad V_{C_1}$ not good:	check D_2 and C_1.
If $V_{out} \approx 2\,V_{pk}$:	check D_1 and C_3.

V_0 should equal $3V_{pk} = (V_{C2}(2V_{pk}) + V_{C3}(V_{pk}))$

FIGURE 6.38   Voltage tripler.

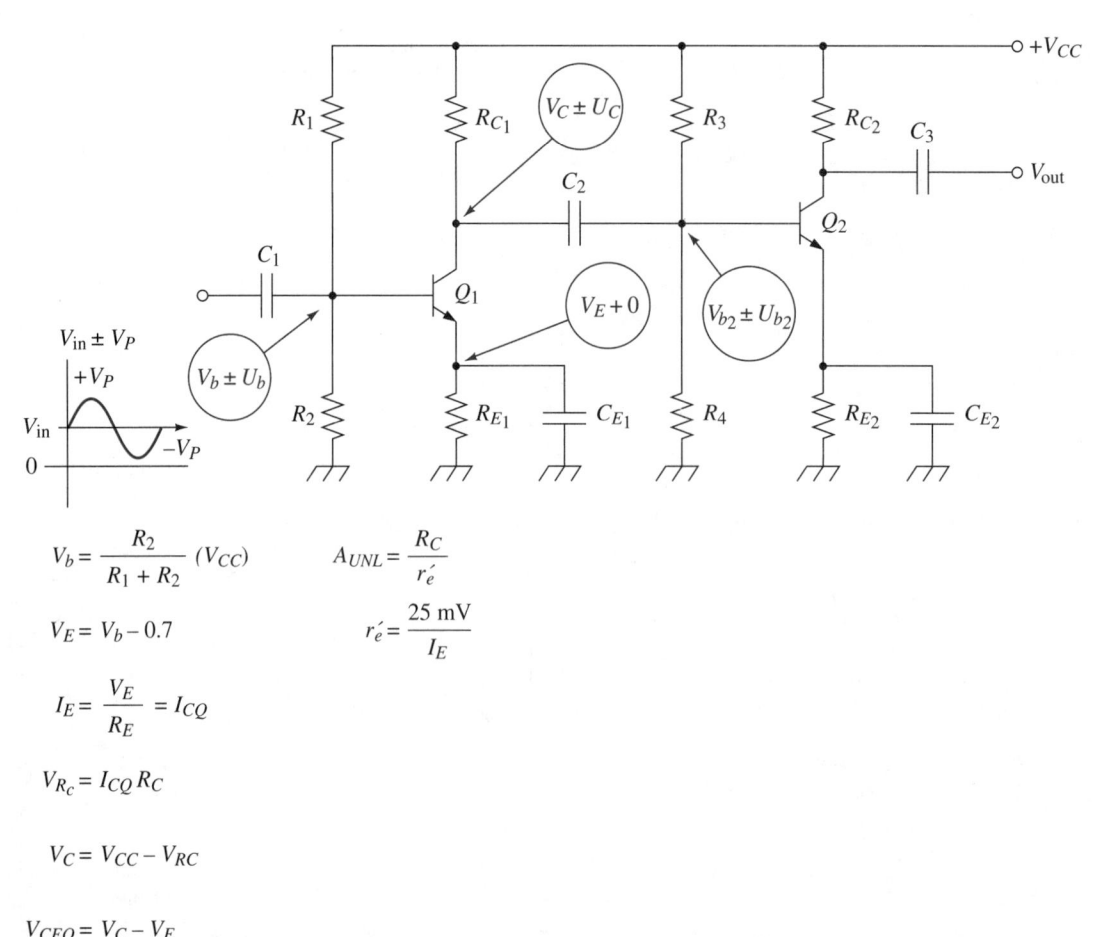

$$V_b = \frac{R_2}{R_1 + R_2}(V_{CC}) \qquad A_{UNL} = \frac{R_C}{r'_e}$$

$$V_E \approx V_b - 0.7 \qquad r'_e = \frac{25\ mV}{I_E}$$

$$I_E = \frac{V_E}{R_E} = I_{CQ}$$

$$V_{R_C} = I_{CQ}R_C$$

$$V_C = V_{CC} - V_{RC}$$

$$V_{CEQ} = V_C - V_E$$

FIGURE 6.39   Common-emitter amplifier.

6.17 CLASS A AUDIO AMPLIFIERS

The class A amplifier is one whose transistors conduct over a full 360° of the input cycle. They can be common-emitter, common-collector, or common-base circuits. Their characteristics are as follows:

1. Common emitter: voltage gain; 180° phase inversion; output taken off the collector
2. Common collector: large current gain; output taken off the emitter; collector tied to ground; used in impedance matching
3. Common base: voltage gain; no current gain; output taken off collector

Let's look at a two-stage common-emitter amplifier, as in Figure 6.39. In doing the dc analysis, note that C_1 and C_2 are coupling capacitors and act like open circuits. The bypass capacitor creates an ac ground at the emitter terminal and also is open to dc. The dc equivalent circuit is in Figure 6.40. The dc analysis allows us to determine the transistor's static operating point on the collector characteristic curves (Figure 6.41).

The ac input signal is processed as changes in the dc voltage levels, both positive and negative. The next step is to conduct an ac analysis and draw the ac equivalent circuit of Figure 6.42.

The second stage loads the first stage. Therefore, the ac resistance seen by the collector of Q_1 (r_C) is the parallel equivalent of R_C and Z_{in} of the second stage. Finally, the ac load line is superimposed on the collector curves and the ac analysis is finished.

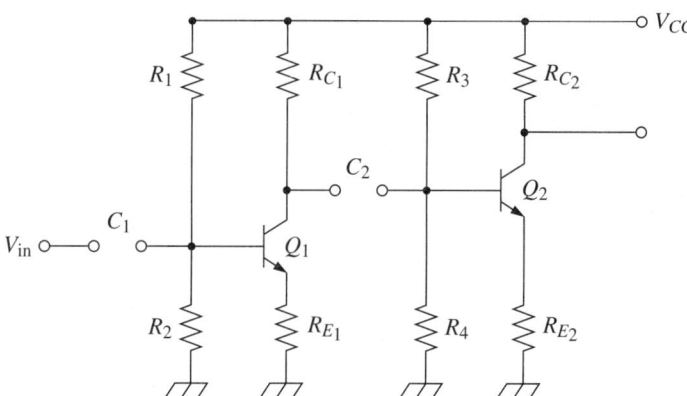

FIGURE 6.40 DC equivalent circuit.

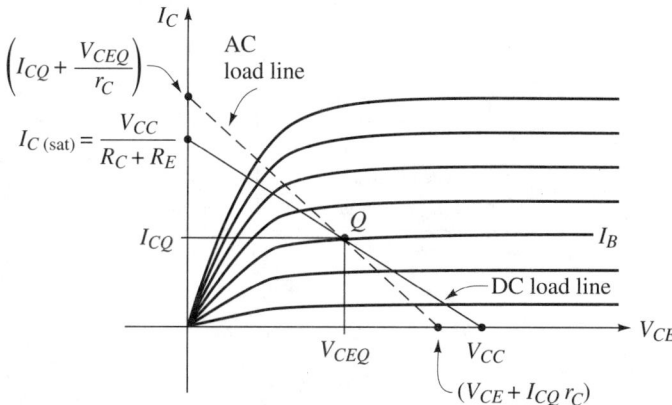

FIGURE 6.41 Static operating point (Q point).

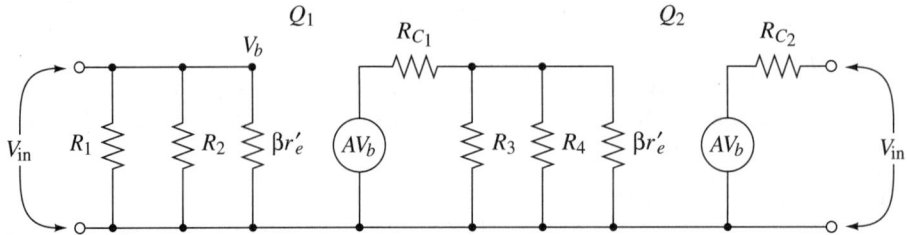

FIGURE 6.42   AC equivalent circuit.

6.18   THE SWAMPED EMITTER AMPLIFIER

Swamping is used for temperature stability by offsetting the temperature-sensitive resistance of the base-emitter diode. The presence of the large emitter resistor in Figure 6.43 causes an ac voltage drop. The resistance is also reflected back toward the base, showing up in the calculations for impedance. It also shows up in the voltage-gain equation and significantly reduces the gain.

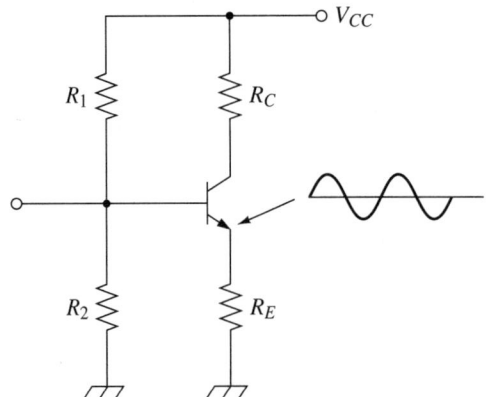

FIGURE 6.43   Swamped amplifier.

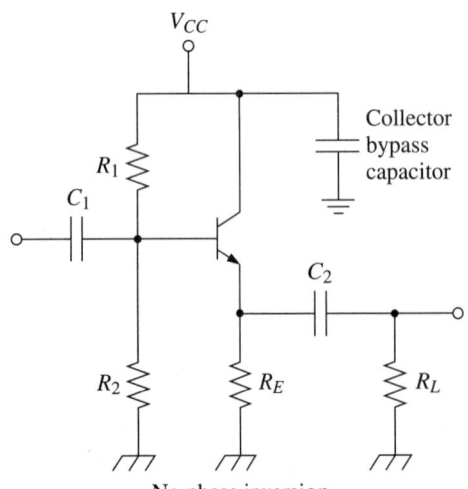

No phase inversion

FIGURE 6.44   Emitter follower.

TABLE 6.9   Common Problems

Condition	Problem
$V_{out} \approx V_{CC}$ or ground	Emitter-follower faulty
$V_E \approx V_{CC}$	R_2 open
$V_E \approx 0$	R_1 open
Emitter has no ac but has dc between 0 and V_{CC}. Signal present at base.	R_E faulty

6.19   THE EMITTER FOLLOWER

The common-collector circuit, better known as the emitter follower, is shown in Figure 6.44. The collector bypass capacitor may be required for decoupling purposes to prevent C_2 from charging or discharging through the power supply, thus causing changes in V_{CC} that result in a distorted output. Table 6.9 illustrates common problems with the emitter follower.

6.20   COMMON-BASE CIRCUIT

A typical common-base circuit is shown in Figure 6.45.

6.21   TROUBLESHOOTING A SINGLE STAGE

1. First you must recognize the type of stage:

   - CE—purpose is voltage gain with inversion
   - CC—output should match the input
   - CB—voltage gain with no inversion

It is important to recognize the type of stage.

2. This is a linear amplifier. The output should be a reproduction of the input. Any change is due to distortion. Nonlinear distortion is often due to the input voltage driving the base-emitter junction into its nonlinear region. See Figure 6.46.

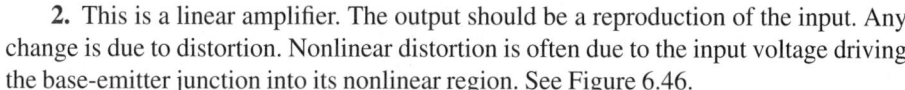

$$A_{UNL} = \frac{r_c}{r'_e} \approx 1$$

$$Z_{in} \approx r'_e \text{ (low)}$$
$$Z_{out} \approx R_C \text{ (high)}$$
No phase shift

FIGURE 6.45   Common-base circuit.

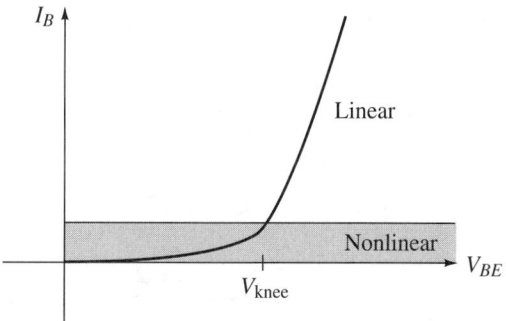

(a) Improper bias puts Q point near V_{knee}.

(b) V_{in} too large, driving V_{BE} to V_{knee}.

FIGURE 6.46   Base-emitter diode curve.

3. It is important to know exactly what you expect to see.

4. You can use a signal generator on the input to produce a known signal.

5. Then use your oscilloscope to analyze the stage.

6. Start at the output stage. If there is no output, check to see if the input to the stage is good. If the input is good, proceed with static testing of the components in the stage. If there is no input, go to the next stage and check its output. If you find a good signal there, the problem is in the coupling between the stages. Otherwise, you must continue looking for a good signal as you proceed back toward the generator.

6.22   THE CLASS B AMPLIFIER STAGE

The class B amplifier is widely used in audio and RF power circuits. It has a 50% duty cycle, and since each transistor conducts during only 180° of the input cycle, the power dissipation is very low. As shown in Figure 6.47, the operating point is biased near cut-off. Since the base-emitter diode is biased at almost 0.7 V, when the input signal exceeds 0.7 V, the transistor turns on and the output voltage goes from 12 V to 8.4 V.

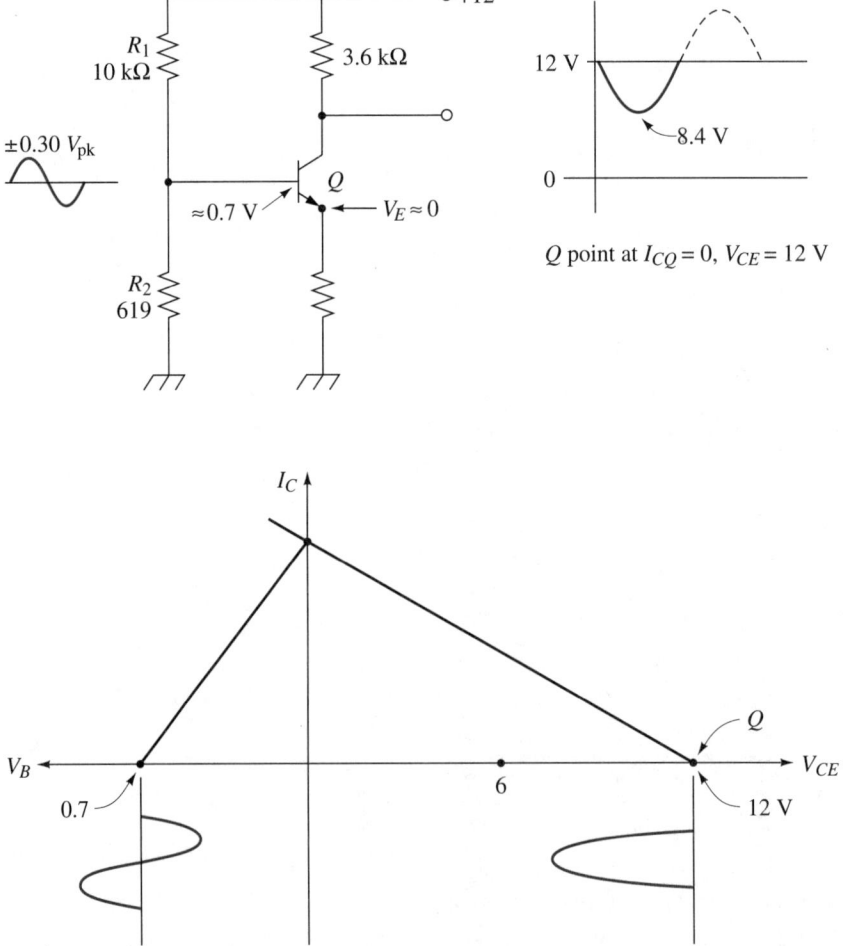

FIGURE 6.47 Class B amplifier stage.

6.23 COMPLEMENTARY PUSH-PULL STAGE

The push-pull circuit in Figure 6.48 incorporates two class B emitter-follower circuits, one using an *npn* and the other using a *pnp* transistor. Q_1 and Q_2 are complementary pairs such as the 2N3904 and 2N3906. The emitters are biased at $V_{CC}/2$. Diodes D_1 and D_2 hold the bases of Q_1 and Q_2 at the conduction threshold to prevent crossover distortion, as in Figure 6.49. The diodes and the base-emitter diodes must be perfectly matched. This improves the overall circuit stability.

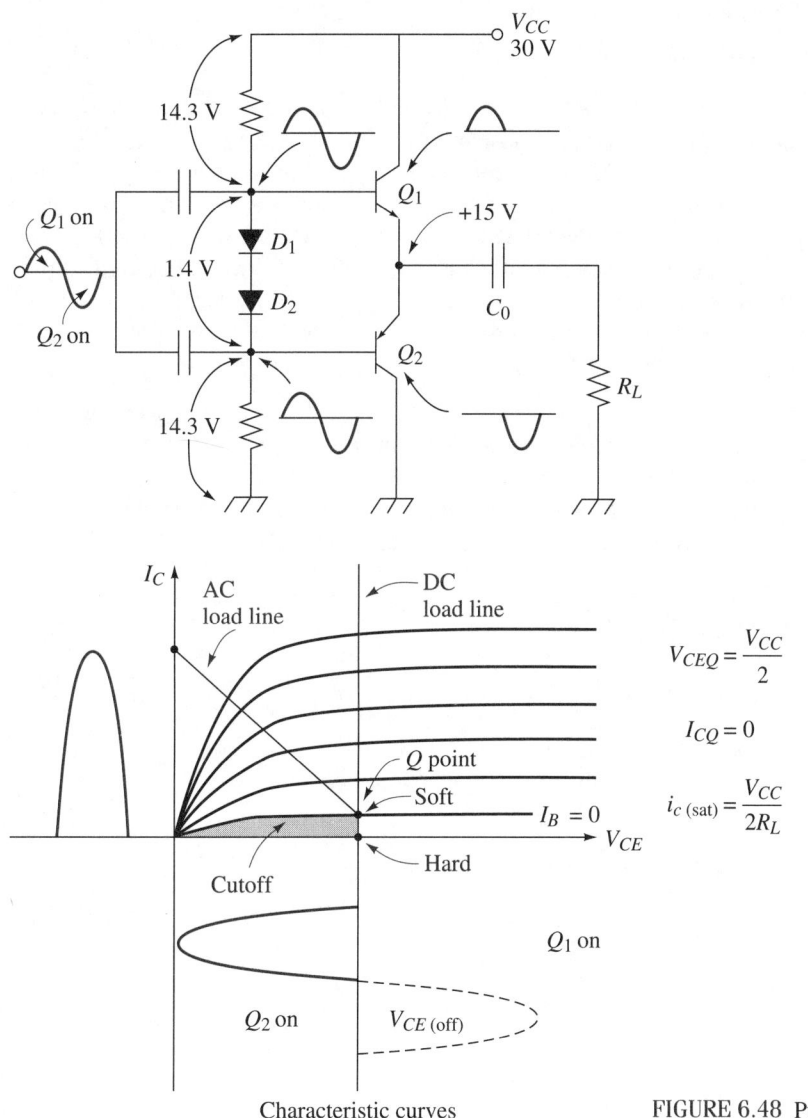

$$V_{CEQ} = \frac{V_{CC}}{2}$$

$$I_{CQ} = 0$$

$$i_{c\,(sat)} = \frac{V_{CC}}{2R_L}$$

Characteristic curves

FIGURE 6.48 Push-pull circuit.

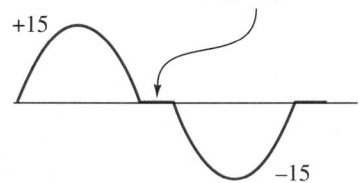

D_1 and D_2 hold Q_1 and Q_2 at conduction threshold to prevent crossover distortion.

FIGURE 6.49 Preventing crossover distortion.

6.24 TROUBLESHOOTING THE CLASS AB STAGE

Make sure that the problem is in the stage.

Of course, you know that any of the common transistor faults that we have already discussed can occur with either of the two transistors. When you are troubleshooting the class AB stage, you must first make certain that the problem is indeed in the stage. Here are some steps for you to follow:

1. Disconnect the load and check the output of the amplifier. If the output is normal, the load is faulty.

2. If the output is bad, make sure to check the input. If the input is normal, you will have to perform more tests on the amplifier.

3. Verify that the proper V_{CC} is present.

4. You can learn a lot about the problem by the appearance of the output waveform.
   a. Crossover distortion suggests a faulty bias diode.
   b. If the dc bias and operating point shift, the average output voltage will also shift. A shift down means that the lower transistor is conducting more than the upper one. Possibly, the driver transistor is conducting too much and increasing the bias on the lower transistor, resulting in a decreased bias of the upper transistor. This might be due to a change in resistor R_2. Other possible faults could be leaky C_1, thermal runaway, a short from the emitter of Q_2 to ground, or Q_1 shorted or even saturated.

After conducting these tests, if you are convinced that the problem is in the amplifier, you should disconnect the incoming signal. Once the input is separated from the amplifier stage, you are ready to perform static (dc) voltage tests, as we have already discussed.

When you have completed your static voltage tests and identified the most probable faulty component, the last step is to isolate the component and perform static tests on it.

6.25 JFET CIRCUITS

Common-Source Amplifier

All JFETs are voltage-controlled devices. The common-source amplifier is similar to the common-emitter amplifier in the following ways:

1. The output is inverted.
2. The voltage gain is greater than 1.

A typical common-source circuit and associated waveforms are shown in Figure 6.50.

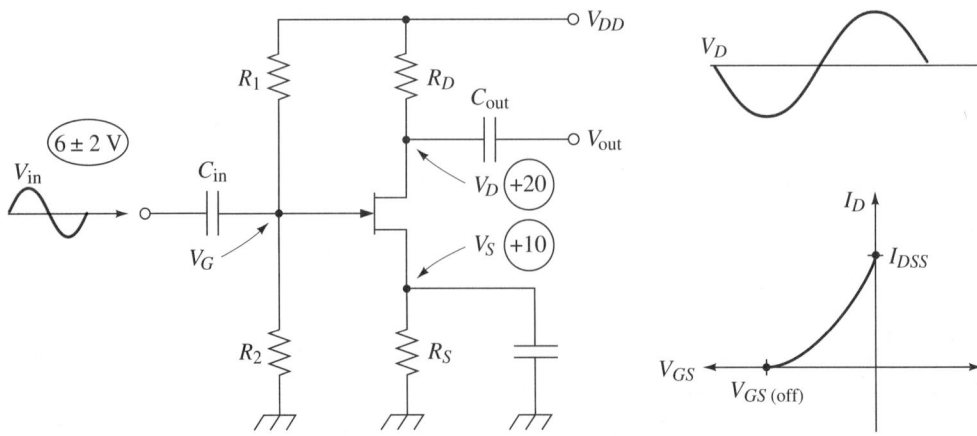

FIGURE 6.50 Common-source amplifier and waveforms.   Conditions: $V_{GS} = -4$ V

Any increase in the gate voltage will cause a decrease in V_{GS}, thereby increasing the drain current. When I_D increases, the drain voltage, V_D, decreases. It follows, then, that any decrease to the gate voltage will cause the drain voltage to increase.

Common Drain

The common-drain circuit (also known as a source follower) is similar to the emitter-follower amplifier. See Figure 6.51. It has the following characteristics:

1. No phase shift
2. Very high input impedance
3. Very low output impedance
4. Voltage gain equal to 1

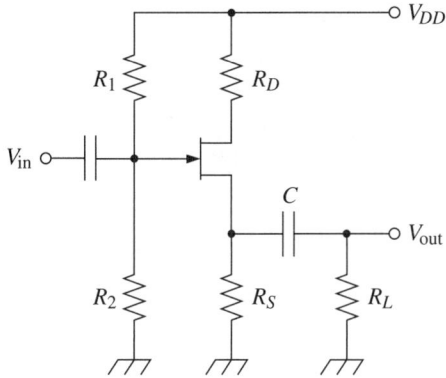

FIGURE 6.51 Common-drain circuit.

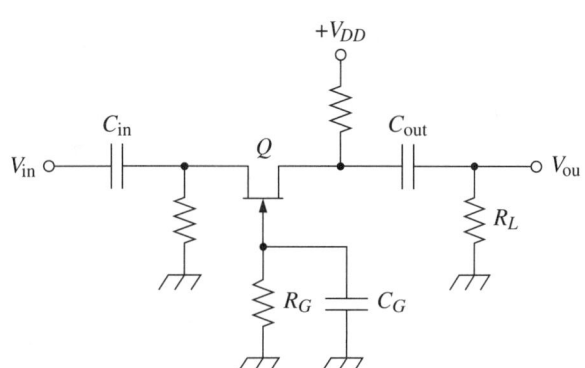

FIGURE 6.52 Common-gate circuit.

Common Gate

The common-gate circuit (Figure 6.52) is similar to the common-base transistor circuit. Its characteristics are

1. Low input impedance
2. High output impedance
3. Voltage gain greater than 1

6.26 TROUBLESHOOTING JFET CIRCUITS

What to check if the amplifier has no output voltage

If the amplifier has no output voltage,

1. Check the input voltage
2. Check the supply voltages
3. Check all ground connections

If all these are normal, you can assume that the amplifier is faulty. Study Figure 6.53. You can see that if R_2 opens, the gate voltage will rise and forward-bias the junction. This may open the junction. With the junction open, the depletion region will disappear and the drain current will increase to the level of

$$I_D = \frac{V_{DD}}{R_D + R_S + r_D}$$

The increasing drain current will cause the voltage across R_S to increase and V_{GS} to decrease.

A positive V_{GS} indicates an open junction.

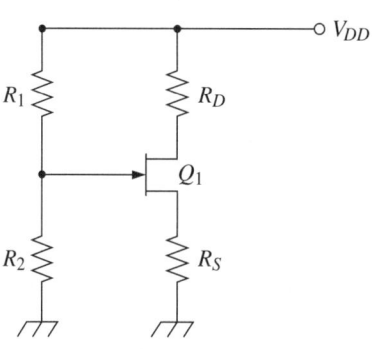

FIGURE 6.53

6.27 THE TRANSFORMER-COUPLED CLASS A STAGE

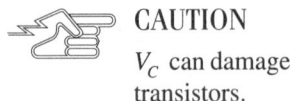
You will find circuits like the one in Figure 6.54 used extensively in RF communications circuits as a tuned amplifier with a specific power gain over a given bandwidth. When the input voltage goes negative, the collector current decreases and the magnetic field associated with the transformer primary coil collapses. The collapsing field generates a CEMF and V_C exceeds V_{CC}.

Be careful—if R_L is disconnected, the reflected impedance (Z_1) will be very high. V_C can become large enough to damage the transistors.

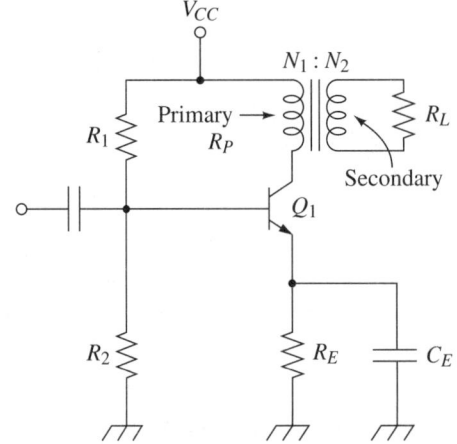

$$\frac{N_1}{N_2} = \frac{V_1}{V_2} = \frac{I_2}{I_1} = \sqrt{\frac{Z_1}{Z_2}}$$

R_P = primary resistance—very low

$$V_{CEQ} = V_{CC} - I_{CQ}(R_P + R_E)$$

$$Z_1 = \left(\frac{N_1}{N_2}\right)^2 Z_2 = \left(\frac{N_1}{N_2}\right)^2 R_L$$

$$Z_1 = \frac{\Delta V_{CE}}{\Delta I_C}$$

$$I_{C\,(max)} = I_{CQ} + \Delta I_C$$

$$V_{CE\,(max)} \approx 2V_{CEQ} \approx 2V_{CE}$$

max n = 50%

$$P_D = V_{CC}\,I_{CQ}$$

$$PP = V_{CE\,(max)} - V_{CE\,(min)} = \text{primary } V$$

$$V_{PP} = \frac{N_2}{N_1}\,PP = \text{secondary voltage}$$

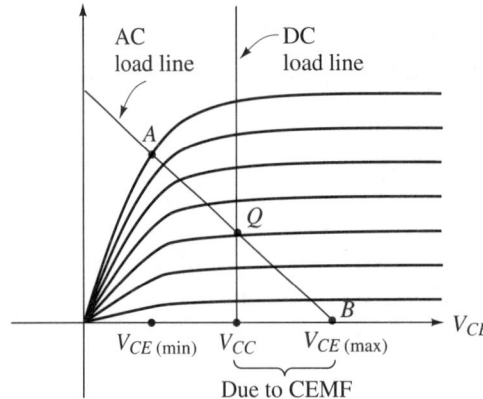

FIGURE 6.54 Transformer-coupled RF amplifier.

6.28 TROUBLESHOOTING THE TRANSFORMER-COUPLED CLASS A STAGE

How to troubleshoot the transformer-coupled Class A stage

The transformer can develop open turns and shorted turns. If the primary opens, the collector voltage will be zero. An open secondary will result in the voltage across the load being equal to zero, but the collector voltage may be normal.

If the primary or secondary becomes shorted, the reflected impedance will be zero, and the collector voltage will also be zero.

6.29 OPERATIONAL AMPLIFIERS

The operational amplifier (op amp) is a high-gain dc amplifier with very high input impedance and low output impedance. The symbol for the 741 op amp is given in Figure 6.55. The heart of the op amp is the differential amplifier, shown in Figure 6.56.

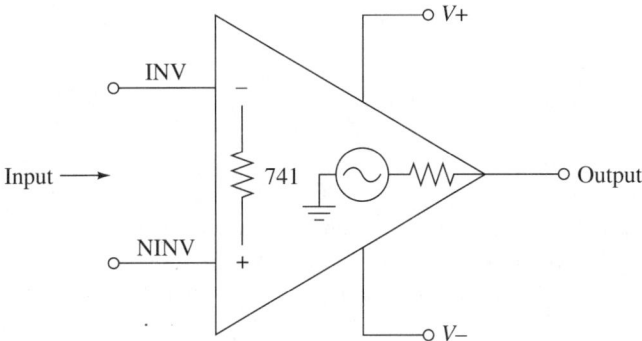

FIGURE 6.55 741 op amp.

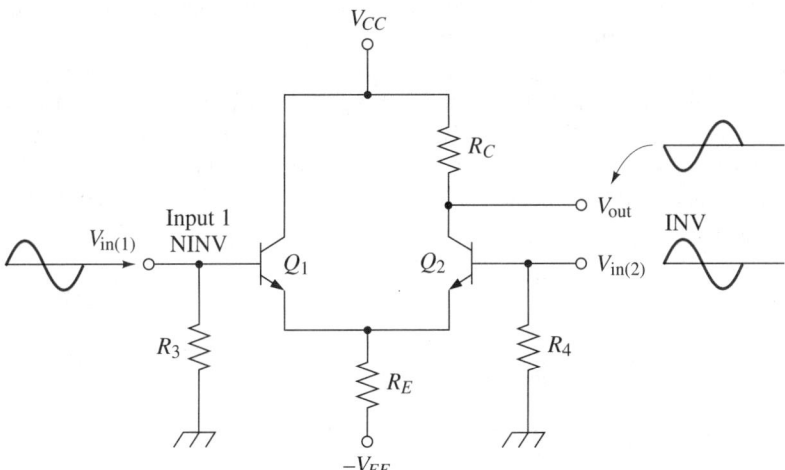

FIGURE 6.56 Differential amplifier.

The differential amplifier circuit accepts two input signals and produces an output signal that is proportional to the difference between the two. The maximum gain of the circuit is the open-loop gain, A_{OL}, which is, typically, 100,000 for the 741. The differential input is called the *error voltage*. As you can see from the figure, one input is designated as the *inverting input* and the other is called the *noninverting input*. A signal on the inverting input results in an output signal that is 180° out of phase with the input. One of the most important characteristics of the circuit is that the amplifier tends to reject any signals that appear on both input terminals simultaneously. This is also known as the common-mode rejection ratio, or CMRR, of the amplifier. It is the CMRR of the op amp that makes it so valuable in the design of instrumentation amplifiers. The basic differential amplifier is added with additional circuitry, including a push-pull amplifier, and is made available in an eight-pin IC.

There are many applications that use an op amp in the open-loop mode. Due to the extremely high gain, the op amp immediately saturates when the error voltage is different from zero. Thus the output will be a square wave, as shown in Figure 6.57.

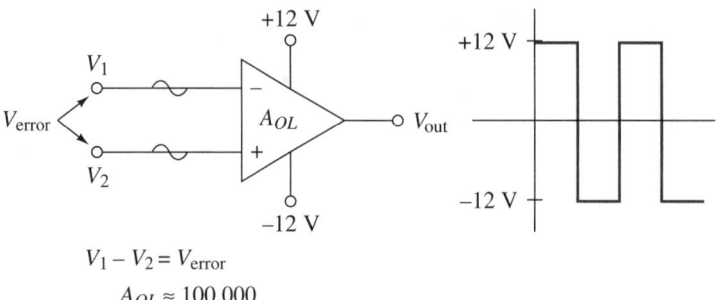

$$V_1 - V_2 = V_{error}$$
$$A_{OL} \approx 100{,}000$$

FIGURE 6.57 Open-loop operation.

If you are interested in amplifying the input signal and retaining its shape, the gain of 100,000 is of no use to you. So, you have to do something to control the gain. This requires a negative feedback path from the output to the input. Let's look at the most common circuits.

6.30   INVERTING AMPLIFIERS

If you sample the output voltage and feed this sample back to the input through a feedback resistor, you will have the circuit shown in Figure 6.58. This is called an *inverting amplifier* and is the op-amp equivalent of the common-emitter and common-source amplifiers. Since the input signal is fed into the inverting amplifier, the output will be inverted. If this circuit is operating properly, the two input terminals will be within microvolts of each other. This is important when troubleshooting a linear circuit.

The negative feedback will reduce the gain to a usable value. In fact, the gain is found by the equation

$$A_V = R_F/R_1 \tag{6.7}$$

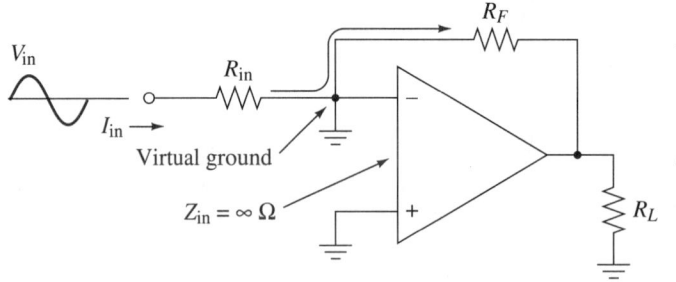

$$V_{in} = I_{in} R_{in} \qquad Z_{in\,(stage)} = R_{in}$$
$$V_{out} = I_{in} R_F \qquad Z_{out} = R_F \,||\, r_{out}$$
$$A_V = -\frac{R_F}{R_i} \qquad CMRR = \frac{A_{CL}}{A_{CM}}$$

Note: CMRR = Common-mode rejection ratio

FIGURE 6.58 Inverting amplifier.

6.31   NONINVERTING AMPLIFIERS

If the input signal is fed into the noninverting terminal, the result is the noninverting amplifier of Figure 6.59. The closed-loop gain is found by

$$A_V = \frac{R_F}{R_1 + 1} \tag{6.8}$$

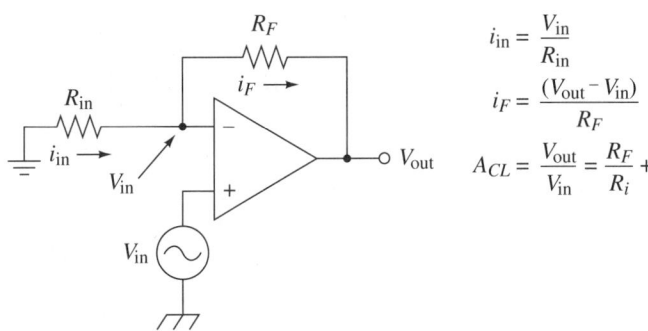

$$i_{in} = \frac{V_{in}}{R_{in}}$$

$$i_F = \frac{(V_{out} - V_{in})}{R_F}$$

$$A_{CL} = \frac{V_{out}}{V_{in}} = \frac{R_F}{R_i} + 1$$

FIGURE 6.59 Noninverting amplifier.

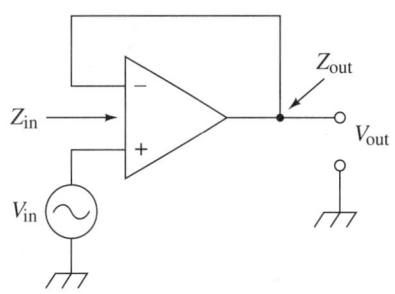

High Z_{in}
Low Z_{out}
$A_V \approx 1$
No phase shift

FIGURE 6.60 Voltage follower.

6.32 VOLTAGE FOLLOWERS

The op-amp equivalent of the emitter follower and the source follower is the voltage follower shown in Figure 6.60.

6.33 TROUBLESHOOTING BASIC LINEAR OP-AMP CIRCUITS

It is easy to troubleshoot the op-amp circuit because there are only a few components to worry about. Each component develops definite symptoms when faulty.

Open Feedback Resistor The feedback loop is lost and the gain of the stage goes up to the open-loop value. The output will saturate to $\pm V_{CC}$, as shown in Figure 6.61.

Checking the components of an op-amp circuit.

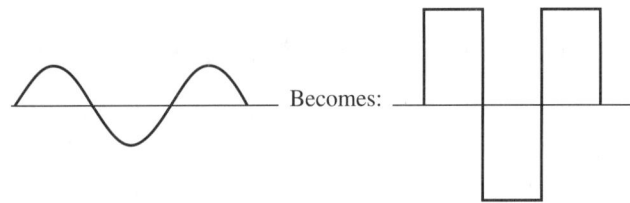

FIGURE 6.61 Saturated output.

Open Series Resistor If the inverting terminal is positive when R_i opens, the output will become positive. The positive voltage is fed back to the inverting terminal, and the output becomes negative. The result is that the circuit oscillates.

If R_2 in Figure 6.62 opens, the circuit acts like a voltage follower with a gain of 1, and the output will look like the input.

Summary

There are three basic problems: no output, low output, or distorted output.

No Output • Check V_{CC}.
• Check V_{in}.
• If the output is saturated, the input resistor may be shorted, the feedback resistor may be open, or there may be an open ground.
• The op amp may be bad.

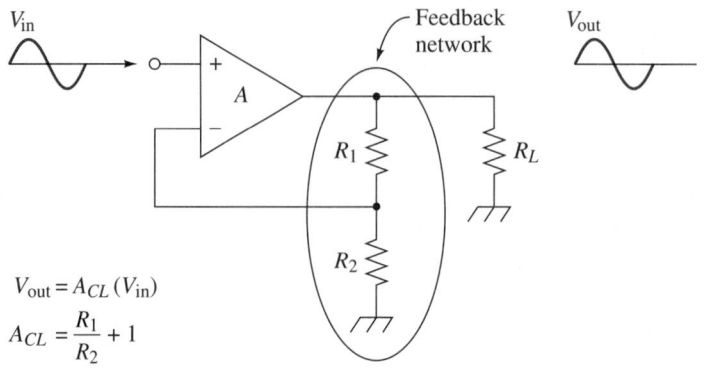

$V_{out} = A_{CL} (V_{in})$

$A_{CL} = \dfrac{R_1}{R_2} + 1$

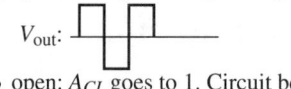

R_1 open: Gain goes to open-loop value.

V_{out}:

R_2 open: A_{CL} goes to 1. Circuit becomes a voltage follower.

FIGURE 6.62 Troubleshooting the feedback network.

Low Output
- V_{CC} may be low.
- V_{in} may be low.
- Check the resistor values.
- The op amp may be bad.

Distorted Output
- Check V_{CC}.
- Check V_{in}.
- Bad op amp.

6.34   TROUBLESHOOTING COMPARATOR CIRCUITS

Some of the most common problems with the comparator circuit

The comparator is an op-amp circuit using open-loop gain to compare the two input voltage levels. This type of circuit is very common in digital systems. The circuit in Figure 6.63 is often referred to as a level detector, and the polarity of its dc output indicates which of the two inputs is the larger. The bypass capacitor is used to short to ground any changes in the input voltage that might upset the reference level voltage. Table 6.10 lists some of the most common problems with the comparator circuit.

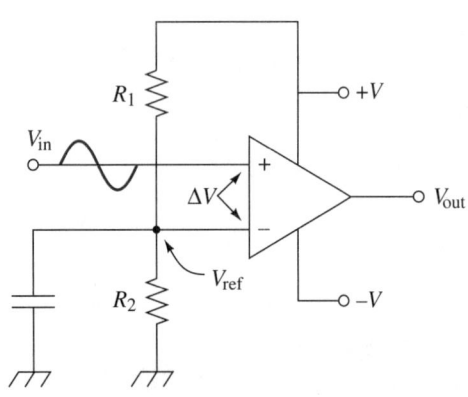

FIGURE 6.63 Comparator.

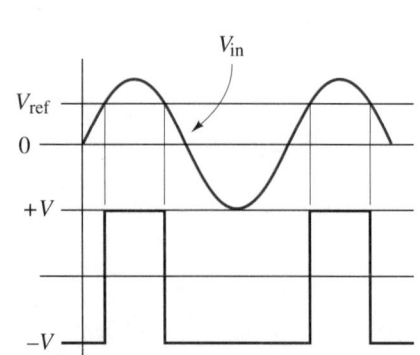

For $V_{in} > V_{ref}$, $V_{out} = +V$.

For $V_{in} < V_{ref}$, $V_{out} = -V$.

TABLE 6.10 Common Problems

Most Common Problems	Causes
No output	No input
	$+V$ or $-V$ faulty
	Op amp faulty
V_{out} changes levels at the wrong reference point. V_{ref} is too high.	R_1 shorted
	R_2 has wrong value
V_{out} changes levels at wrong reference point. V_{ref} is too low.	R_1 open
	R_2 wrong value
	Capacitor shorted

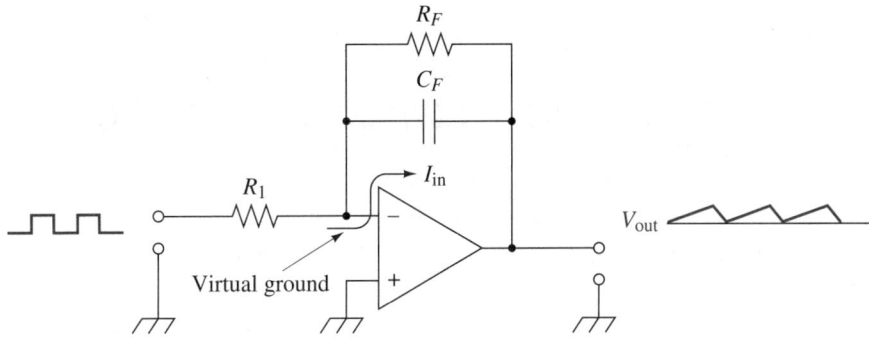

FIGURE 6.64 Integrator.

6.35 TROUBLESHOOTING INTEGRATOR CIRCUITS

The mathematical concept of integration is one of dividing a given signal waveform into many small segments and then adding the areas under each segment. Therefore, the output of the integrator in Figure 6.64 is proportional to the area under the input curve. The capacitor charges at a constant rate and, since the input is applied to the inverting input, the output is 180° out of phase. Thus the output will be a negative-going ramp voltage when the input is a positive pulse. The feedback resistor helps to eliminate any offset voltage so that the output is centered around 0 V. We know that the RC network will have a critical frequency given by

$$f_C = \frac{1}{6.28 R_F C_F} \qquad (6.9)$$

As X_C approaches the value of R_F, the integration action stops. Thus the capacitive reactance should be kept less than $0.1R_F$.

As the frequency of operation increases, the capacitive reactance decreases until the feedback resistor is shorted out. Therefore, the maximum operating frequency is 10 times f_C.

Table 6.11 summarizes the most common trouble spots with the op-amp integrator circuit.

Some of the most common problems with the op-amp integrator circuit

TABLE 6.11 Common Problems

Integrator Troubles	Results
R_F open	V_{out} offset around a level other than 0 V.
	Low cutoff frequency decreases.
R_F shorted	Loss of integration.
	Circuit gain lost.
	Output drops.
C_F open	Circuit acts like an inverting amplifier.
	$A_{CL} = R_F / R_L$
	Output is a square wave.
C_F shorted	Same as for a shorted R_F.
	Test C_F and R_F individually.

6.36 TROUBLESHOOTING DIFFERENTIATOR CIRCUITS

The most common problems in differentiator circuits

By changing the position of the capacitor, the circuit in Figure 6.65 becomes a differentiator. The output voltage is proportional to the rate of change of the input voltage.

Like the integrator, the differentiator will have a maximum frequency, which is given by

$$f_{max} = 0.1 f_C \qquad (6.10)$$

where

$$f_C = \frac{1}{6.28 RC} \qquad (6.11)$$

If the signal frequency exceeds this limit, the output waveform will be distorted.
See Table 6.12 for a discussion of the most common faults in this circuit.

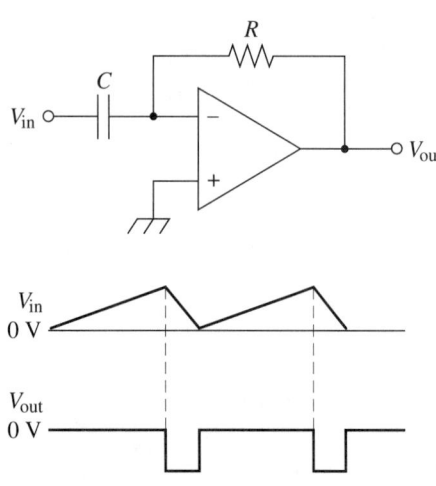

FIGURE 6.65 Differentiator.

TABLE 6.12 Common Faults

Differentiator Troubles	Results
R open	V_{out} will be zero.
R shorted	Increased gain and there will be loss of differentiating action.
C open	Gain increases to open-loop value.
C shorted	Circuit gain drops to zero. V_{out} drops to zero.

6.37 PROBLEMS

1. What is the resonant frequency of the circuit in Figure 6.66?

2. If Figure 6.67 is the response curve for the circuit in Figure 6.66, what is the most likely problem?

3. What happens to the gain of the common-emitter stage in Figure 6.68 if the by-pass capacitor opens?

4. Given the voltage waveforms in Figure 6.69, what are the possible problems?

5. What is meant by crossover distortion?

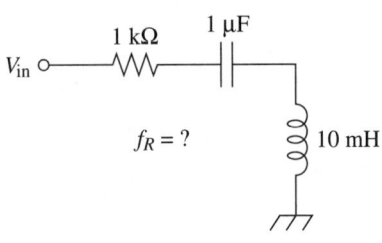

FIGURE 6.66 Problem 1.

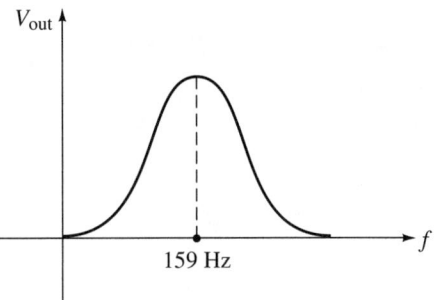

FIGURE 6.67 Problem 2.

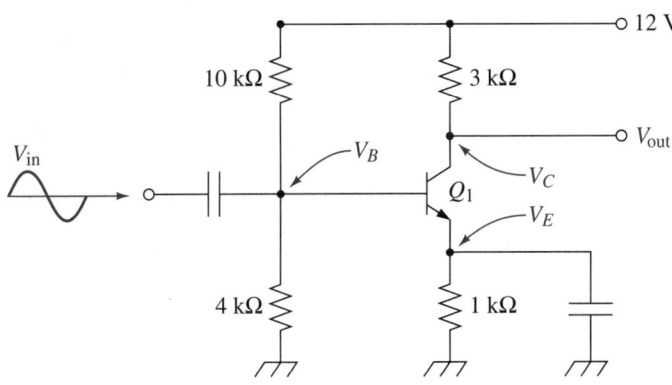

FIGURE 6.68 Problem 3.

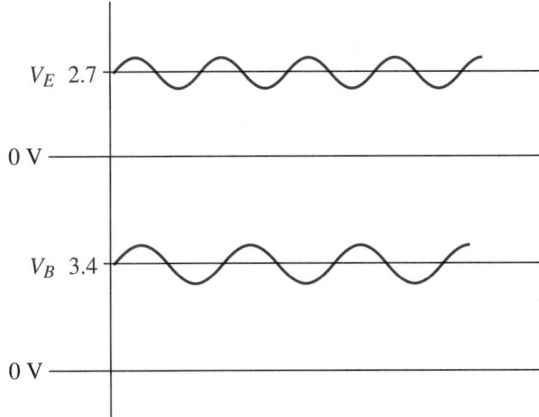

FIGURE 6.69 Problem 4.

6. Identify the troubles in the class AB circuit of Figure 6.70.

7. What type of JFET circuit is Figure 6.71?

8. List the most common transformer faults.

9. Calculate the closed-loop gains of the op-amp circuits in Figure 6.72.

10. Identify the possible trouble for the circuit in Figure 6.73.

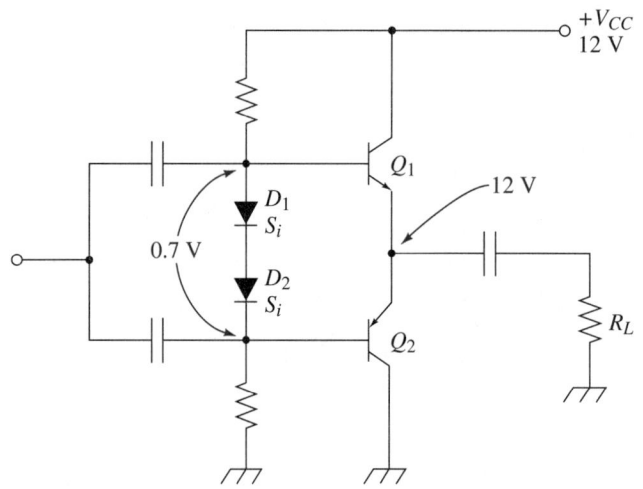

FIGURE 6.70 Problem 6.

FIGURE 6.71 Problem 7.

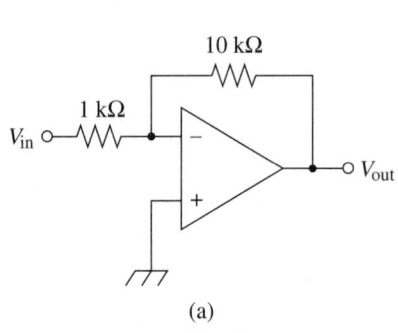

(a)

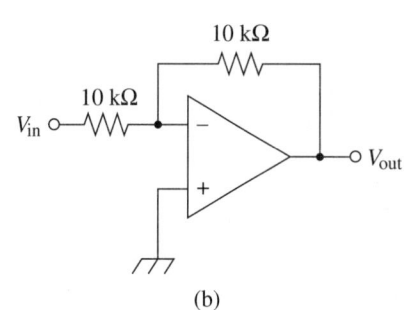

(b)

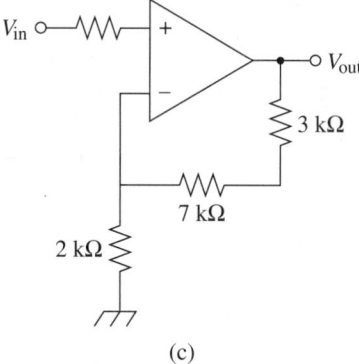

(c)

FIGURE 6.72 Problem 9.

FIGURE 6.73 Problem 10.

11. What should the output of the comparator in Figure 6.74 look like?

12. What characteristic of the op amp is valuable in the measurement of very small signals?

13. Determine the value of the reference voltage for the comparator in Figure 6.75.

14. Determine the frequency at which the circuit in Figure 6.76 will start to lose its linear output characteristics.

15. The circuit in Figure 6.77 has the readings indicated. Determine the possible cause(s) of the problem.

16. The circuit in Figure 6.78 has the readings indicated. Determine the cause of the problem.

17. When troubleshooting an integrator circuit, what symptoms would indicate an open feedback resistor?

18. The output of your inverting amplifier suddenly becomes a square wave with a level equal to the saturation voltage of the circuit. What do you think the problem is?

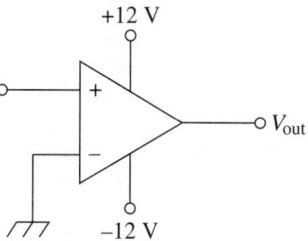

FIGURE 6.74 Problem 11.

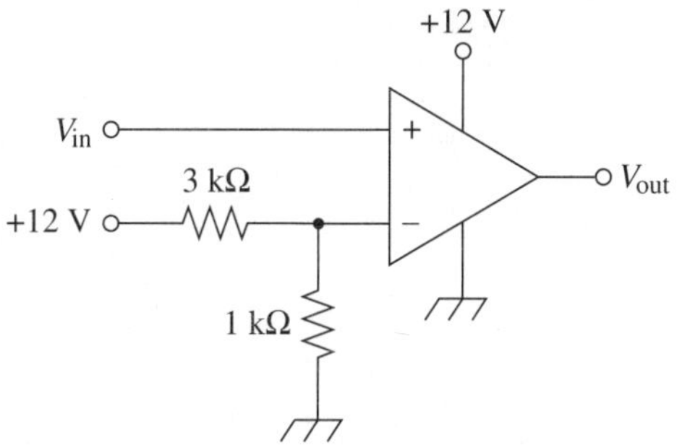

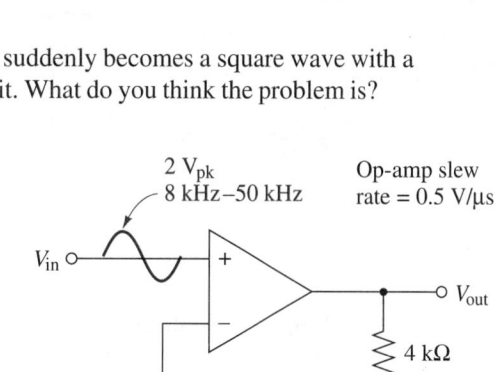

FIGURE 6.75 Problem 13.

FIGURE 6.76 Problem 14.

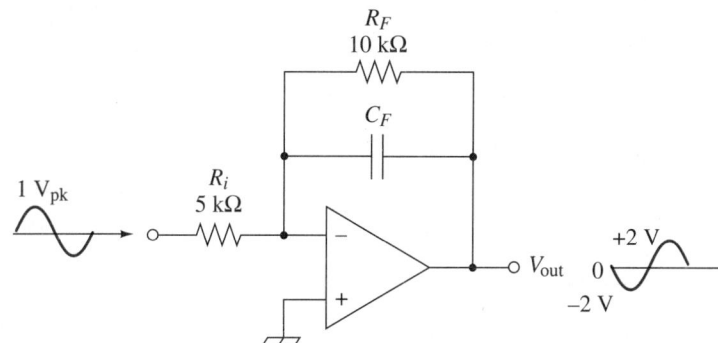

FIGURE 6.77 Problem 15.

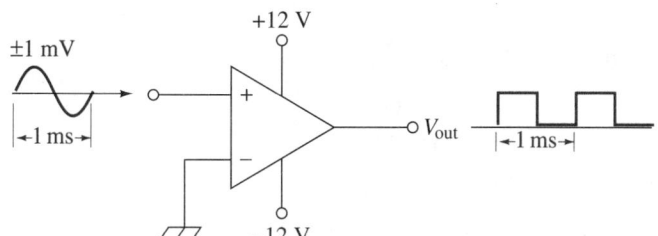

FIGURE 6.78 Problem 16.

19. The output of your nonlinear amplifier is too low. You measured the voltage on the input terminals and found the difference to be 800 mV. What happened?

20. What is meant by virtual ground?

6.38 TROUBLESHOOTING EXAMPLE 1

You have the circuit in Figure 6.79 working and have made the measurements listed in Table 6.13. Your calculations provide the normal values listed in the table.

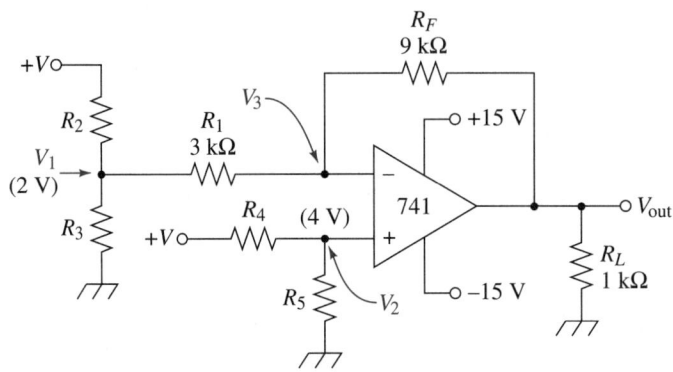

FIGURE 6.79 Troubleshooting Example 1.

TABLE 6.13 Troubleshooting Example 1

	V_1	V_2	V_{out}	V_3
	Calculated Values			
	2.00 V	4.00 V	10.00 V	4.00 V
	Measured Values			
a.	2.02	3.00	9.98	3.00
b.	2.02	3.38	7.50 no load	3.38
c.	2.02	3.38	4.83 with load	3.38
d.	2.02	3.99	7.50 no load 4.83 with load	3.37
e.	2.02	3.99	9.98	3.99

By superposition, the contribution by the inverting terminal should be

$$(+2)\left(\frac{-9\,k\Omega}{3\,k\Omega}\right) = -6\,V$$

The contribution by the noninverting terminal should be

$$(+4)\left(\frac{-9\,k\Omega}{3\,k\Omega + 1}\right) = 16\,V$$

The total V_{out} should be

$$+16 + (-6) = +10\,V$$

After the circuit was constructed, the measurement values listed in part d of Table 6.13 were obtained. Something went wrong.

The problem was identified as

1. V_{error} too large; in fact, V_{error} should always be nearly 0 V for linear operation.
2. V_{out} was too low.

3. V_{out} varies with load.

Follow the troubleshooting steps given here and make sure you understand the reasons for the conclusion.

Step 1. Set $V_1 = 2.02$ V and $V_2 = 3.00$ V, and check to see that $V_3 = V_2$ or $V_{error} = 0$. For linear operation, V_{error} will always be close to 0 V.
Step 2. Measure V_{out}. The circuit worked fine. See part a of Table 6.13.
Step 3. Set level of V_2 to be 3.38 V. At this point the error voltage $(V_3 - V_2)$ was not 0 V. This indicated a problem.
Step 4. V_{out} measured 7.5 unloaded and dropped when loaded. See parts b and c of Table 6.13.
Step 5. The op amp functions fine until $V_2 = 3.38$ V. At that point the op-amp output remains at 7.50 V unloaded.
Step 6. The op amp was replaced.
Step 7. The new op amp provided the voltage measurements in part e of Table 6.13.

Follow the steps and make your conclusion.

Troubleshooting Problem 1

Study the comparator circuit in Figure 6.80 and the associated measurements listed in Table 6.14. Calculate the normal values and enter them in the table. List the troubleshooting steps that you would take to find the circuit fault. Can you identify the problem with the circuit?

Once you feel that you have the answer, breadboard the circuit and inject the fault you listed. How do your measurements agree with those in Table 6.14? Then, conduct your troubleshooting following the steps you indicated before. Did you arrive at the same conclusions?

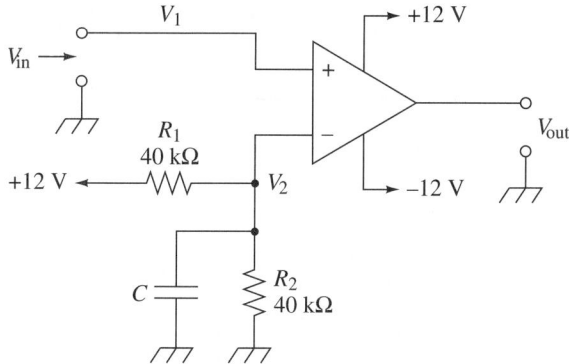

FIGURE 6.80 Troubleshooting Problem 1.

TABLE 6.14 Troubleshooting Problem 1

Calculated Values			Normal Waveforms
V_1	V_2	V_{out}	

Troubleshooting Problem 2

The limiter circuit in Figure 6.81 is faulty. The waveforms are shown and voltage measurements listed in Table 6.15. What do you think the problem is? Calculate the normal voltage levels and then breadboard the circuit. Inject the fault and take the measurements indicated. Troubleshoot the circuit, recording each step. What are your conclusions?

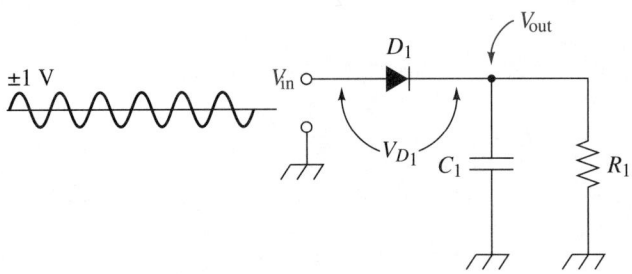

This is the familiar limiter (clipper) circuit found in AM radios as the detector stage.

FIGURE 6.81 Troubleshooting Problem 2.

TABLE 6.15 Troubleshooting Problem 2

	V_{in}	V_{D_1}	V_{out}
Possible Fault			
Measured		0.7 V	−0.3 V
Normal Calculated			
Your Measurements			

TROUBLESHOOTING UNTUNED AMPLIFIER CIRCUITS

7.1 INTRODUCTION

Most electronic circuits contain untuned amplifiers in one form or another. The audio amplifier is a very common example. When troubleshooting an untuned amplifier, it is customary to first inject a known signal into the front stage of the amplifier and then trace this signal through the system with an oscilloscope until a point is reached where the signal either becomes distorted or lost. By tracing the signal with your oscilloscope, you can see what is happening to the signal and quickly identify any undesirable conditions, such as the following:

How to troubleshoot an untuned amplifier

1. Hum on the waveform
2. Distortion, such as clipping
3. Unwanted noise spikes
4. Excess ac ripple voltage
5. Ringing effects

Once the defective stage has been identified, the next logical step is to isolate the stage and perform the static tests that we have already discussed. Static testing should help you to isolate the component or components that are faulty.

The final step is to perform individual component testing on the suspected faulty item to verify your conclusions. We can summarize these steps as follows:

1. Identify the problem.
2. Inject a good signal.
3. Trace the signal stage by stage.
4. Isolate the faulty stage.
5. Perform static voltage measurements on the faulty stage.
6. Isolate the faulty component.
7. Perform component tests to verify condition.
8. Identify the component.
9. Replace the component.

Inherent in the previous discussion is the assumption that you know what the normal (good) signal should look like. It is easy to troubleshoot when the signal is missing. If there is a signal present, the big question is this: Is it the right one?

Therefore, let's discuss some of the most important amplifier parameters, how they affect the output, and, also, how to measure them.

7.2 FREQUENCY RESPONSE

The frequency response of an amplifier tells you how well the amplifier performs over a given range of frequencies. Most amplifiers should have a fairly constant gain over the range of frequencies; this gain is referred to as the amplifier's *bandwidth*.

Generally, the frequency response curve is a graph of output voltage versus frequency (for a given level of input voltage) plotted on semilog paper. The bandwidth is determined by the circuit components. A typical frequency response curve is shown in Figure 7.1.

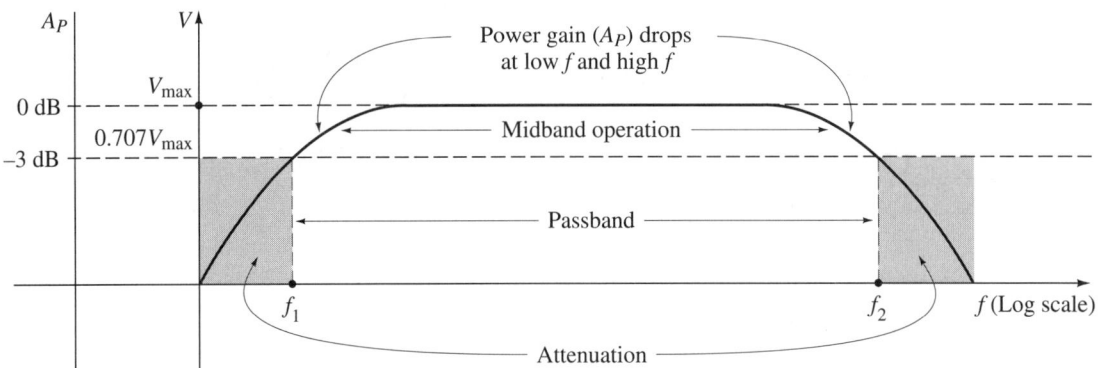

FIGURE 7.1 Frequency response.

The bandwidth is defined as the range of frequencies between the lower critical frequency and the upper critical frequency. The critical frequency is defined as the frequency at which the output of the amplifier is equal to 0.707 of the midband value. This is also known as the half-power (or −3-dB) point.

It is possible to measure these −3-dB points. You should take the following steps to establish the critical frequencies of the circuit in Figure 7.2.

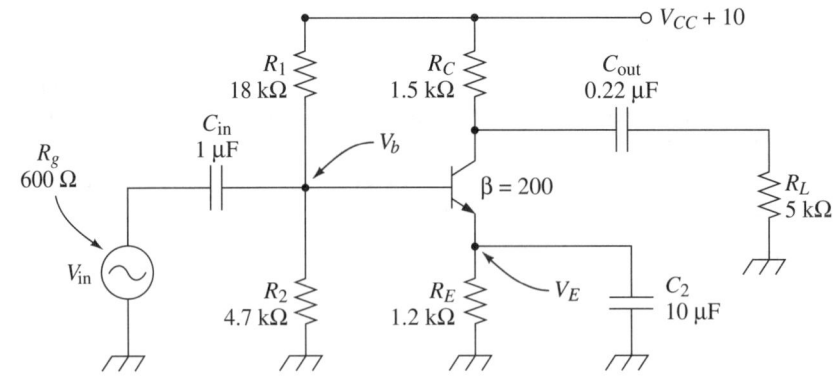

AC equivalent circuit

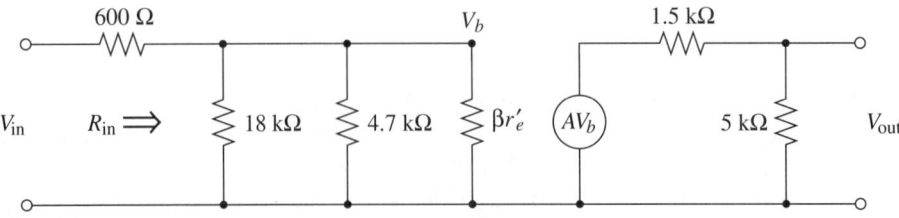

FIGURE 7.2 AC equivalent circuit.

1. Set the amplifier to the maximum undistorted output at approximately the mid-frequency of operation.

2. Hold the input-voltage level constant. To find the lower critical frequency, decrease the signal frequency until the output drops to 0.707 times the midband value.

3. Hold the input-voltage level constant. To find the upper critical frequency, increase the signal frequency until the output drops to 0.707 times the midband value.

The low-frequency response of the amplifier in Figure 7.2 is determined by the input and output coupling capacitors and by the emitter-bypass capacitor.

Example 1

Calculate the lower critical frequency for the circuit in Figure 7.2.

Step 1. Calculate the critical frequency of the input coupling network.

$$R_{in} = R_G + (R_1 \| R_2 \| \beta r_e^1)$$
$$= 600 + (3727 \| 4386)$$
$$= 600 + 2015$$
$$= 2617 \, \Omega$$

$$C_{in} = 1 \, \mu F$$

$$f_C = \frac{1}{2\pi RC} = \frac{1}{2\pi(2617)(1)(10^{-6})} = \frac{1}{0.016435} = 60.85 \, Hz$$

The dc analysis is as follows:

$$V_b = \frac{4.7 \, k\Omega}{22.7 \, k\Omega} (10) = 2.07 \, V$$

$$V_E = 1.37 \, V$$

$$I_E = \frac{1.37}{1.2 \, k\Omega} = 1.14 \, mA$$

$$r_e^1 = \frac{25 \, mV}{1.14 \, mA} = 21.93 \, \Omega$$

$$\beta r_e^1 = (200)(21.93) = 4386 \, \Omega$$

Step 2. Calculate the critical frequency of the output coupling network.

$$R = R_C + R_L = 6.5 \, k\Omega$$

$$C = C_{out} = 0.22 \, \mu F$$

$$f_C = \frac{1}{2\pi RC} = \frac{1}{2\pi 6500(0.22)(10^{-6})} = 111.35 \, Hz$$

Step 3. Calculate the critical frequency of the emitter-bypass circuit.

$$R = \frac{R_G \| R_1 \| R_2}{\beta} + r_e^1 = 517 + 22 = 539 \, \Omega$$

$$C = 10 \, \mu F$$

$$f_C = \frac{1}{2\pi 539(10)(10^{-6})} = 29.54 \, Hz$$

Step 4. The lower critical frequency is the largest. So, $f_C = 111.35 \, Hz$.

The high frequency response is determined by factors that are most difficult to measure. For example,

1. A transistor has interelectrode capacitance that is not important at the lower frequencies.
2. Stray capacitance throughout the circuit is important.
3. Stray lead inductance also can be very important.

It is possible to determine the frequency response using a signal generator and your oscilloscope.

7.3   AMPLIFIER TESTING: SINE-WAVE METHOD

Our objective is to determine the lower and upper critical frequencies of the circuit in Figure 7.3. The signal generator's output impedance will have to be matched as closely as possible to the input of the amplifier. It is important to use a load resistor whose value is as close to the normal value as possible. Be sure to use a low-capacitance probe with the oscilloscope.

Preset the frequency of the generator to the approximate midband of the amplifier. Now, vary the generator frequency positive and negative until the output drops to 0.707 times the midband value. Read these two frequencies as the lower and upper critical values. See Figure 7.4.

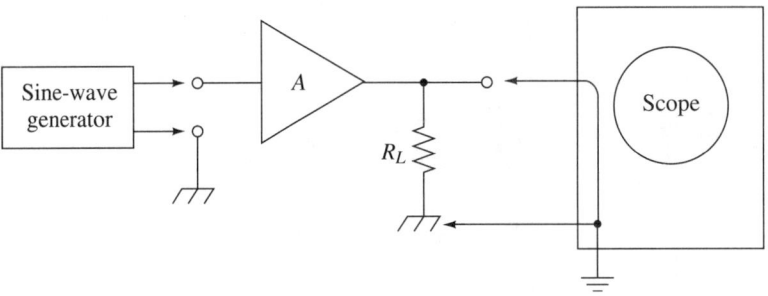

FIGURE 7.3   Sine-wave testing of an amplifier.

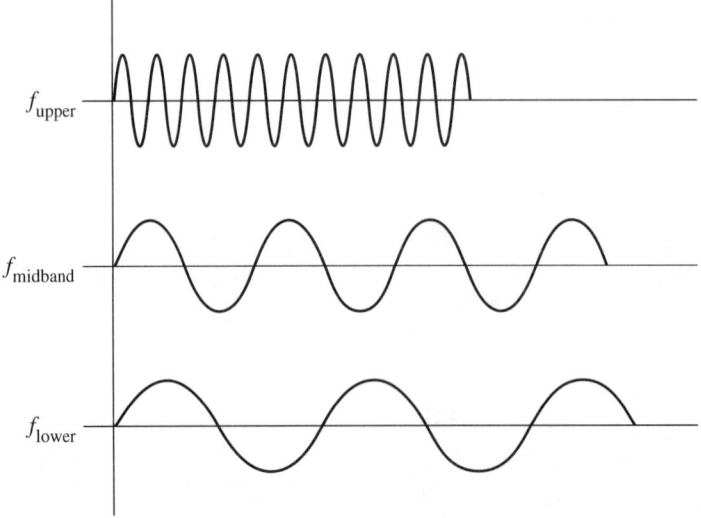

FIGURE 7.4   Critical frequency values.

7.4 AMPLIFIER TESTING: SQUARE-WAVE METHOD

The amplifier frequency response can be determined by observing its reaction to a (frequency-rich) square-wave input signal. See Figure 7.5. By measuring the rise time (t_1), the time between 10% and 90% of the output, the cutoff frequency can be calculated as follows:

$$f_{UC} = 0.35/t_1 \qquad (7.1)$$

This frequency, known as the *upper cutoff (UC) frequency,* is the frequency at which the amplifier output drops to 70.7% of its level at the midband operation frequency.

By measuring the fall time of the output, the *lower cutoff (LC) frequency* can be determined as follows:

$$f_{LC} = 0.35/t_1 \qquad (7.2)$$

The lower cutoff frequency is that frequency where the output again drops to 70.7% of its midband level.

The rise time and sharpness of the trailing edge are determined by the high-frequency response, whereas the fall time and sharpness of the leading edge are determined by the low-frequency response.

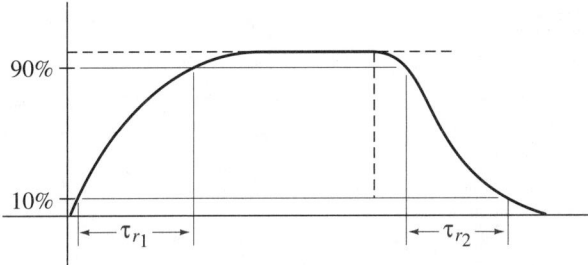

Lower cutoff frequency $= \dfrac{0.35}{\tau_{r2}}$

Upper cutoff frequency $= \dfrac{0.35}{\tau_{r1}}$

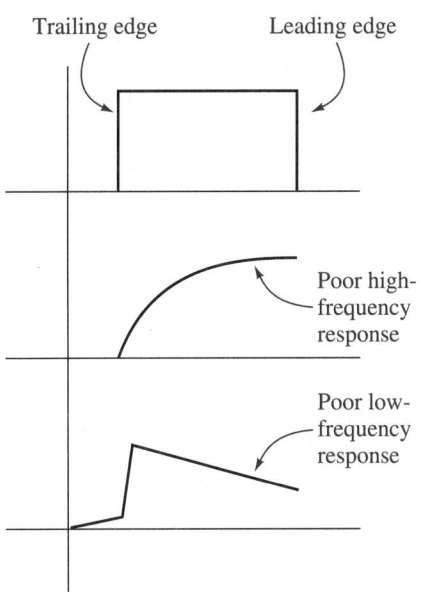

FIGURE 7.5 Determining amplifier frequency response.

7.5 OUTPUT POWER VERSUS LOAD

It is easy to see from Figure 7.6 that the power across the load resistor is maximum when the amplifier output impedance is matched to the value of the load resistor. At this particular frequency, the amplifier output impedance is known as the *dynamic output impedance*.

The power output is found by calculating the load power across the entire midband frequency range using a test circuit, as in Figure 7.6. In some cases a minimum power output is necessary for a given value of input voltage (input sensitivity requirement). The range of frequencies within which a given power level must be maintained is called the *power bandwidth* and is given by:

$$P_L = \frac{V_{out}^2}{R_L} \tag{7.3}$$

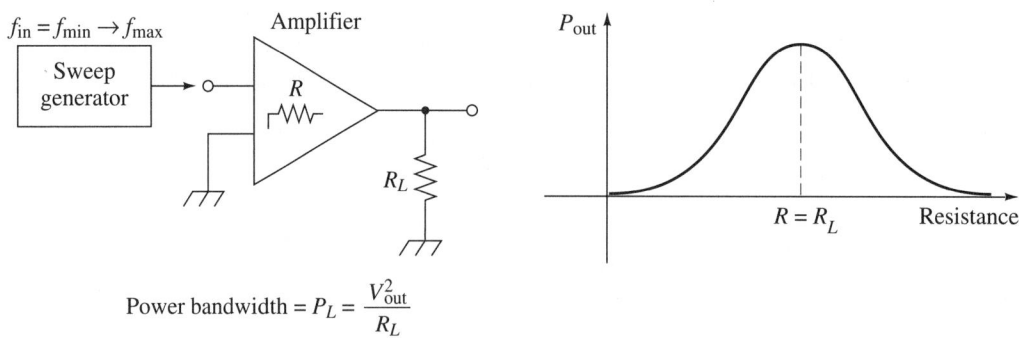

Power bandwidth = $P_L = \dfrac{V_{out}^2}{R_L}$

FIGURE 7.6 Output power versus load.

7.6 AMPLIFIER DISTORTION

The output of a linear amplifier is supposed to be an amplified inverted, or noninverted, reproduction of the input signal. When the output does not look like the input, the signal has distortion.

Distortion in an amplifier is due to the creation of harmonics of the input signal. Regardless of the amplifier circuit type, odd and even harmonics can exist. In single-ended amplifiers the harmonics are predominantly even, whereas push-pull types generate odd harmonics.

Let's assume that our amplifier has a pure 400-Hz sine-wave input signal. The output would be as in Figure 7.7. Any generated harmonics are effectively added to the 400-Hz signal, giving the outputs shown in Figure 7.8.

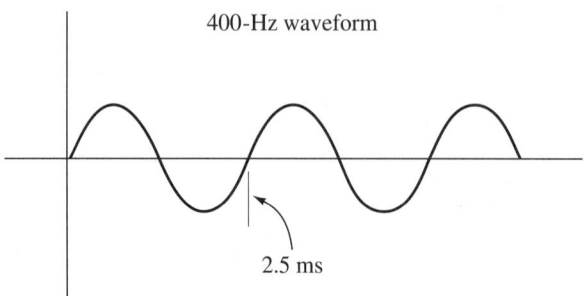

FIGURE 7.7 Output for 400-Hz waveform.

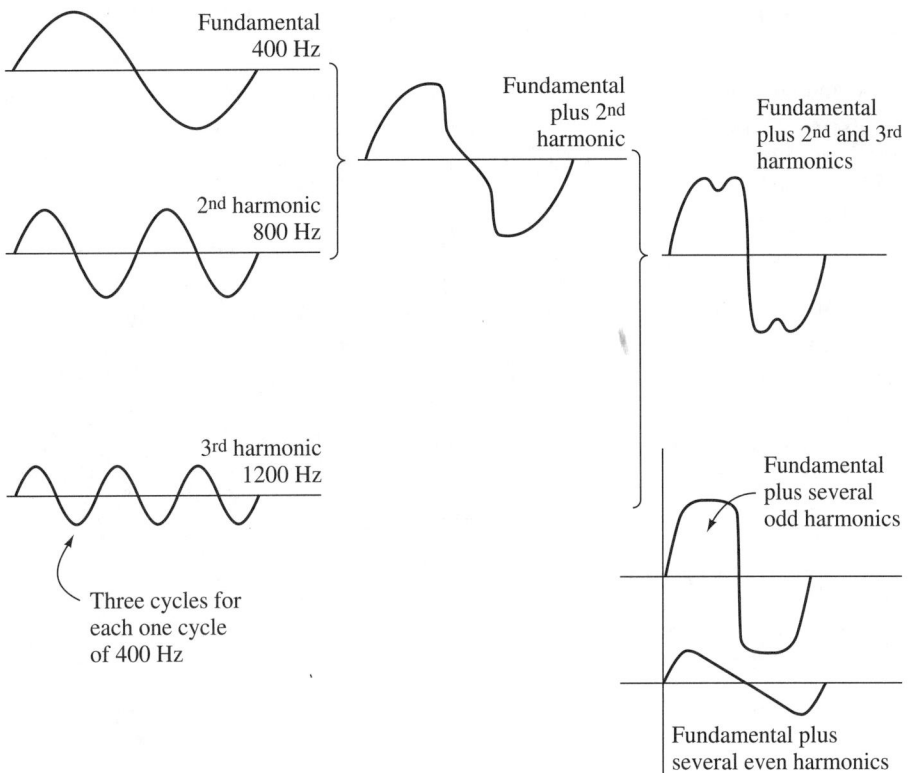

FIGURE 7.8 Creation of a complex waveform.

The ratio of total RMS values of all harmonics to that of the fundamental frequency is known as the *total harmonic distortion factor* (THD). THD decreases as the power output level is reduced.

Refer to Figure 7.9. The first step is to measure or calculate the RMS voltage (total amplifier output) at point A.

The second step is to measure or calculate the RMS voltage of the harmonics without the fundamental at point B.

Calculate the %THD as follows:

$$\%\text{THD} = \left(\frac{B}{A}\right)100\% \tag{7.4}$$

By using the oscilloscope, it is possible to determine more closely which frequencies give the most distortion.

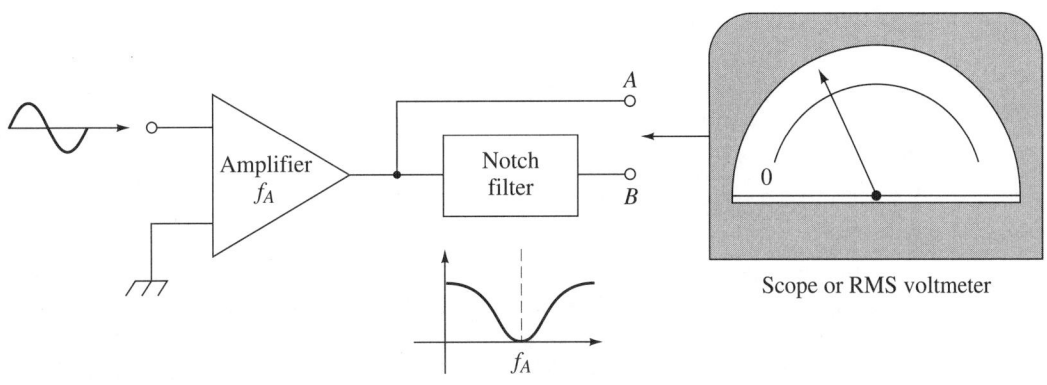

FIGURE 7.9

7.7 DISTORTION ANALYSIS BY SQUARE WAVES

It's common practice to analyze amplifier distortion by observing its response to a square-wave input signal, as shown in Figure 7.10. Square waves have a very high odd harmonic content. Fourier analysis tells us that a square wave can be expressed as the sum of n-odd harmonics as follows:

$$V = \frac{4\,V}{\pi}\left(\sin \omega t + \frac{1}{3}\,\sin 3\,\omega t + \frac{1}{5}\,\sin 5\,\omega t + \cdots\right) \tag{7.5}$$

If the amplifier is provided a square-wave input of frequency f_X and the output is a clean reproduction, then the frequency response is good to about $10f_X$. By monitoring the input on one channel and the output on the second channel, any distortion will be obvious.

Some of the most common waveforms are shown in Figure 7.11.

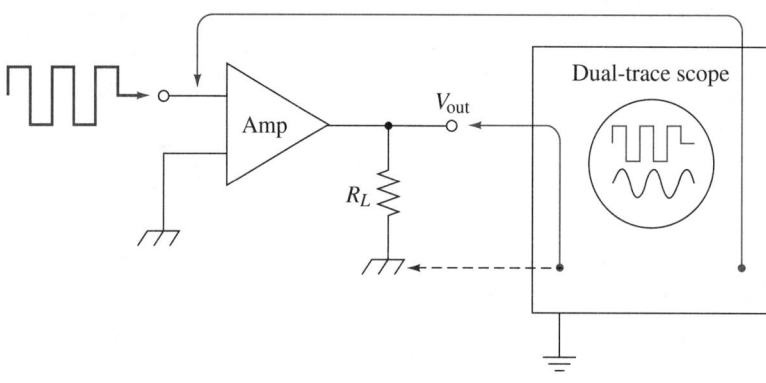

FIGURE 7.10 Response to square-wave input signal.

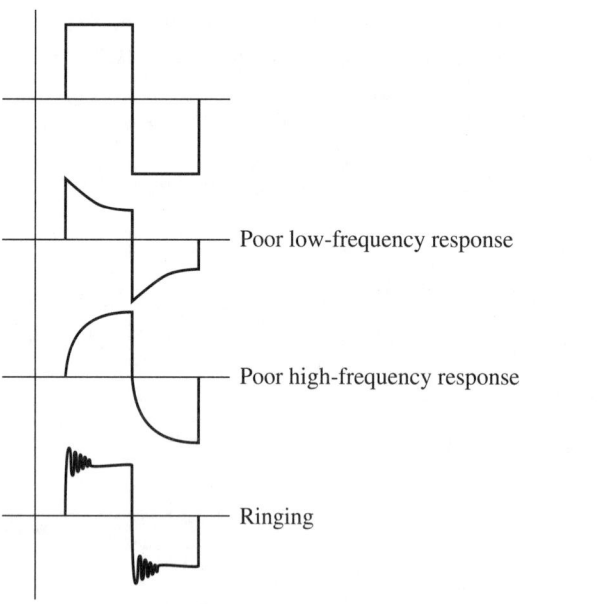

FIGURE 7.11 Waveform distortion.

7.8 MEASURING NOISE IN THE AMPLIFIER

The oscilloscope is the best tool to use here when looking for noise, hum, oscillations, etc. By monitoring the amplifier output with no input signal applied, the trace on the scope will represent the noise signal. See Figure 7.12. The noise may be inductively coupled in the leads of the amplifier. If you are looking for the presence of 60-Hz line hum, set the scope syncronization control to line and look for a stationary sine-wave pattern. If

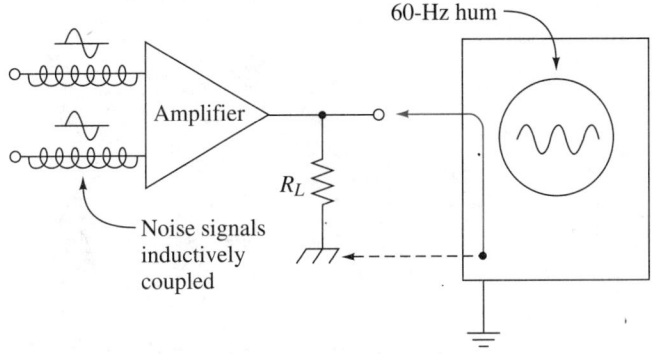

FIGURE 7.12 Observing an amplifier's noise level.

the trace is not stationary, then the hum is due to oscillation frequencies other than the line frequency of 60 Hz. If the trace disappears when the amplifier's input terminals are shorted, the problem is due to stray pickup. If not, the problem is in the amplifier circuits.

7.9 TROUBLESHOOTING FEEDBACK AMPLIFIERS

Troubleshooting feedback amplifiers is quite tricky. You cannot open the loop because the circuit gain will revert to the open-loop level and the op amp will saturate. Figure 7.13 is a typical feedback stage. If Q_1 is distorting the signal, the feedback signal to the emitter of Q_1 will also be distorted, but with a phase shift. The end result will be distortion within the loop but a mostly undistorted output for Q_2. Therefore, the negative feedback loop not only minimizes distortion but also helps stabilize the overall circuit gain.

Problems in troubleshooting feedback amplifiers

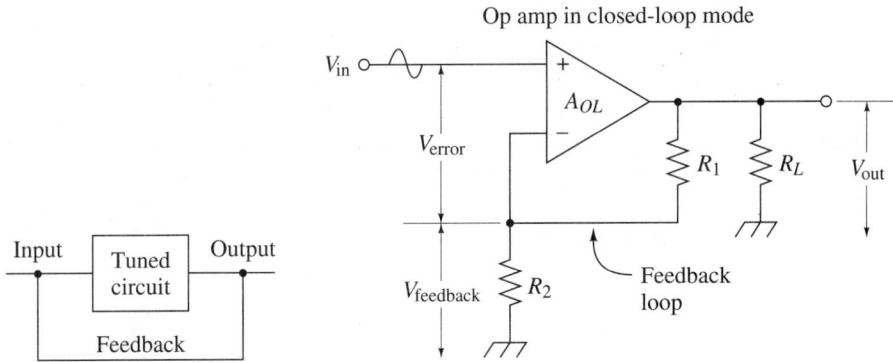

FIGURE 7.13 Feedback circuit.

7.10 CAUSES OF DISTORTION

1. One major cause of distortion is the stage being overdriven by too large an input signal.

2. Distortion will occur if the Q point is too far toward the saturation point on the collector curves. See Figure 7.14.

3. Distortion will occur if the Q point is too far toward the cutoff point on the collector curves.

4. The inherent nonlinearity of the amplifier can also cause distortion.

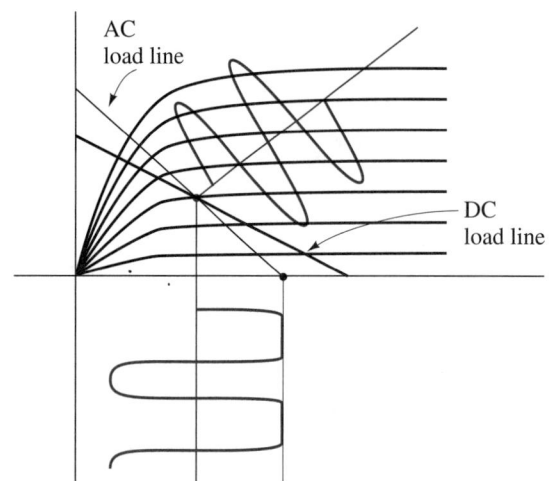

FIGURE 7.14 Distortion due to cutoff point.

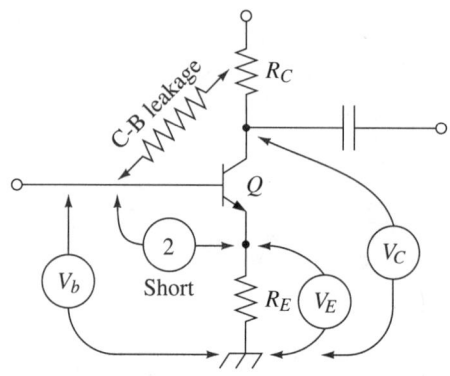

FIGURE 7.15 Transistor leakage paths.

5. Transistor leakage from the collector to base may also cause distortion by reducing the gain of the stage due to increasing the forward-bias voltage on the base.

Refer to Figure 7.15. To check for collector-base leakage:

1. Check the value of V_C.
2. Short the base to emitter to turn off the transistor.
3. Recheck the value of V_C.
4. V_C should equal V_{CC}. If there is a collector-to-base leakage path, the value of V_C will be less than V_{CC}.

Important Points to Remember

- Don't open the loop; the circuit will lose its closed-loop gain.
- In Figure 7.16, V_1 and V_2 are not the same. V_2 will include the effects of the feedback voltage and be more useful in computing circuit voltage gain.
- Make sure that you do not connect the ground lead of the oscilloscope to the base or emitter terminal. This will set up ground loops and large circulating currents, which could destroy the circuit Q.
- Be careful of any confusion due to voltmeter loading.

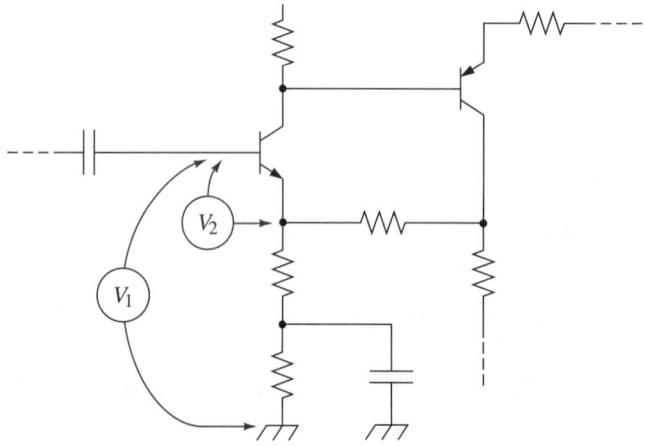

FIGURE 7.16

7.11   TROUBLESHOOTING EXAMPLE 1

Figure 7.17 a, b, and c is a complete two-stage amplifier.

1. Perform complete dc and ac analyses of the circuit.
2. Given the voltage waveforms, troubleshoot the circuit and identify the problem.

A troubleshooting problem

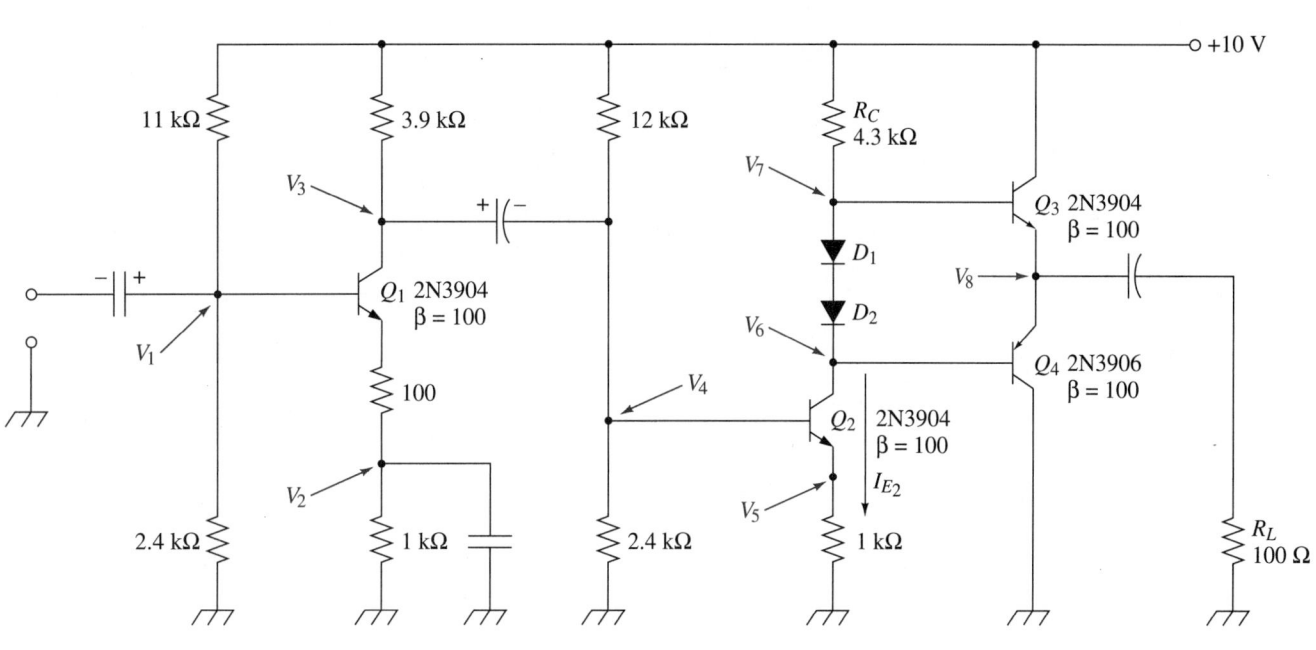

DC equivalent circuit

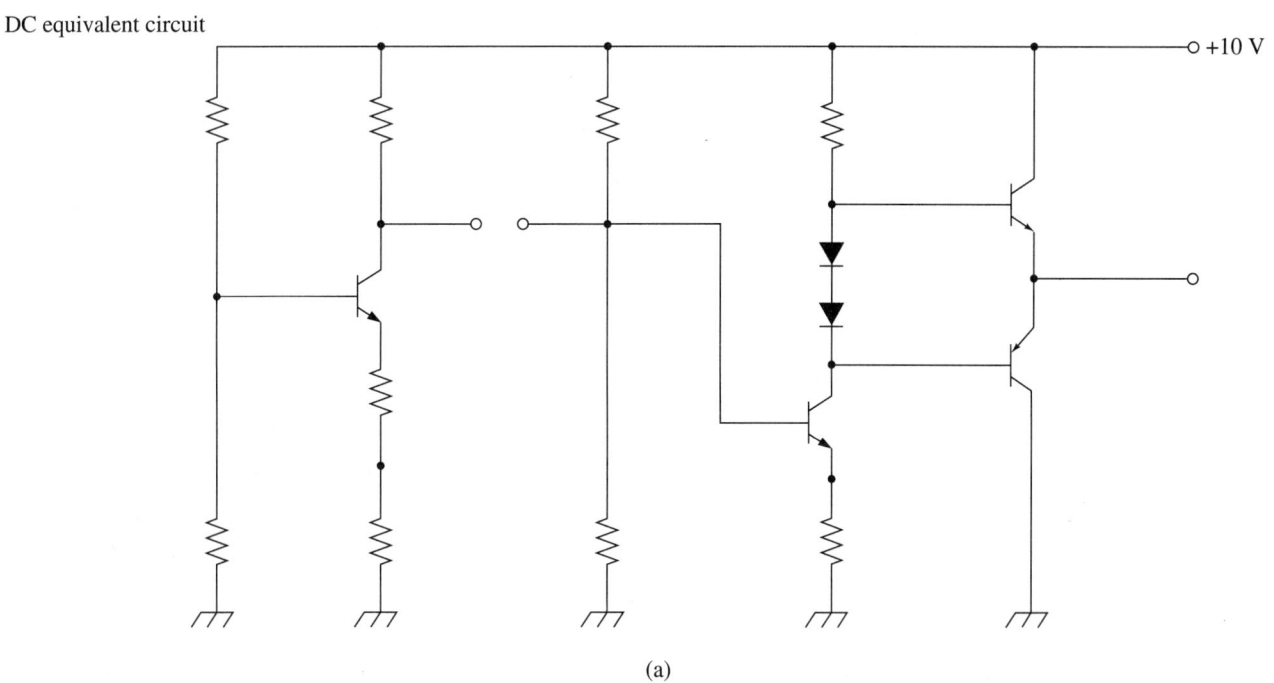

(a)

FIGURE 7.17 Two-stage amplifier analysis.

DC analysis

$$V_1 = \frac{2.4 \text{ k}\Omega}{13.4 \text{ k}\Omega}(10) = 1.79 \text{ V}$$

$$V_2 = 1.79 \text{ V} - 0.7 \text{ V} = 1.09 \text{ V} \approx 1.1 \text{ V}$$

$$I_{E1} = I_{CQ1} = \frac{1.1 \text{ V}}{1.1 \text{ k}\Omega} = 1.0 \text{ mA}$$

$$V_{C1} = V_3 = 10 - (1 \times 10^{-3})(3.9 \times 10^3) = 10 - 3.9 = 6.1 \text{ V}$$

$$V_4 = \frac{2.4 \text{ k}\Omega}{14.4}(10) = 1.666 \text{ V}$$

$$V_5 = 1.666 - 0.7 = 0.966 \text{ V} \approx 1.0 \text{ V}$$

$$I_{E2} = \frac{1 \text{ V}}{1 \text{ k}\Omega} = 1.0 \text{ mA}$$

$$V_{RC} = (I_{E2})(R_C) = (1 \times 10^{-3})(4.3 \times 10^3) = 4.30 \text{ V}$$

$$V_6 = V_{CC} - V_{RC} - V_{D1} - V_{D2} - V_5 =$$

$$V_6 = 10 - 4.3 - 0.7 - 0.7 - 1.0 = 4.3 \text{ V}$$

$$V_7 = 4.3 + 1.4 = 5.7 \text{ V}$$

$$V_8 = 5 \text{ V}$$

$$V_{E3/4} = 5 \text{ V}$$

$$I_{E3/4} = \frac{5 \text{ V}}{100 \text{ }\Omega} = 50 \text{ mA}$$

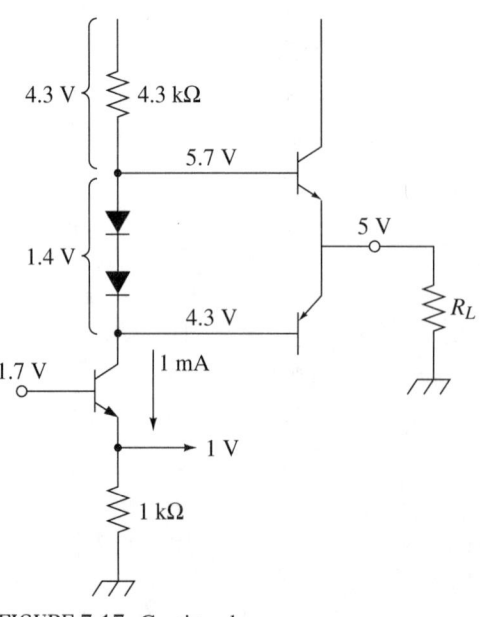

FIGURE 7.17 Continued.

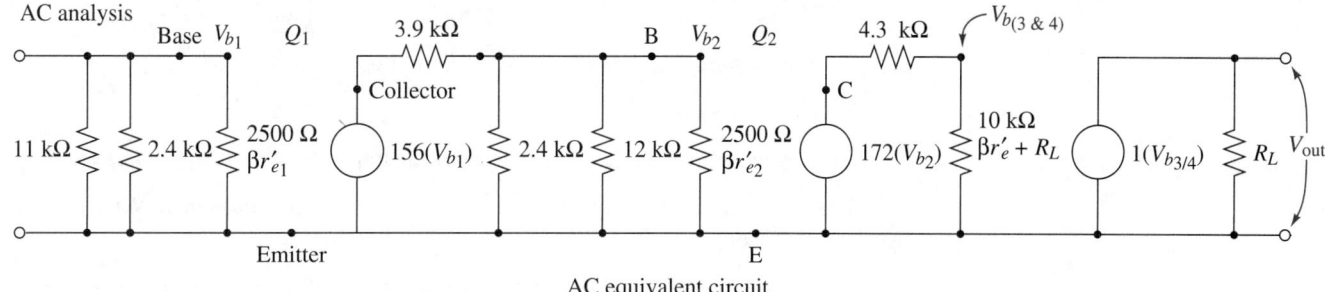

AC equivalent circuit

$$\text{For } Q_1: r'_{e_1} = \frac{25 \text{ mV}}{I_{E_1}} = \frac{25 \text{ mV}}{1 \text{ mA}} = 25 \, \Omega \qquad\qquad \text{For } Q_2: r'_{e_2} = 25 \, \Omega$$

$$\beta r'_{e_1} = (100)(25) = 2500 \, \Omega \qquad\qquad\qquad \beta r'_{e_2} = 2500 \, \Omega$$

$$A_{V(\text{unl})} = \frac{R_C}{r'_{e_1}} = \frac{3900}{25} = 156 \qquad\qquad\qquad A_{V(\text{unl})} = \frac{4300}{25} = 172$$

$$\text{For } Q_3 \text{ or } Q_4: r'_e = \frac{25 \text{ mV}}{50 \text{ mA}} = 0.5 \, \Omega$$

$$\beta(r'_e + R_L) = 10 \text{ k}\Omega$$

$$V_{b(3\ \&\ 4)} = (0.7)(172)(.22)(156) \, V_{b1}$$

$$= (4132.13) \, V_{b1}$$

$$V_{b2} = (0.22)(156) \, V_{b1}$$

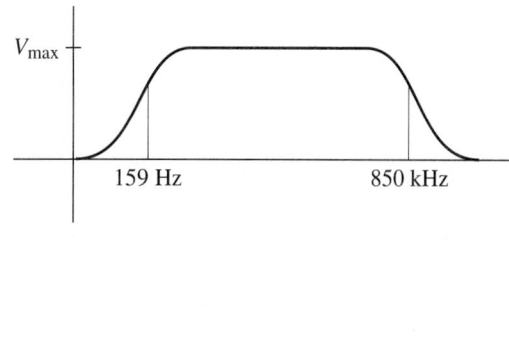

Total output: $V_{\text{out}} = 5 \pm (4132.13)V_{b1}$ V

(c)

FIGURE 7.17 Continued.

7.12   PROBLEMS

1. What is meant by the bandwidth of an amplifier?

2. What is the bandwidth of the amplifier in Figure 7.18?

3. Discuss the procedure that you would use to determine the critical lower and upper frequencies of the circuit in Figure 7.18.

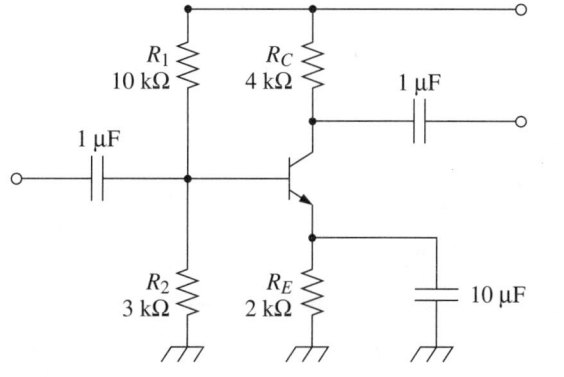

FIGURE 7.18 Problem 2.

4. What determines the low-cutoff frequency of an amplifier?

5. What is the second harmonic of a 3-kHz signal?

6. What is distortion?

7. Why are square waves used to analyze an amplifier's frequency response?

8. Figure 7.19 is the output of an amplifier with a square input signal. What type of distortion is present?

9. Why is the oscilloscope so important when troubleshooting noise in an amplifier?

10. Looking at the collector curves in Figure 7.20, would you expect the output to be distorted?

11. What is meant by the −3-dB points?

12. You are troubleshooting an amplifier using the square-wave method. The output is shown in Figure 7.21. Calculate the critical frequency.

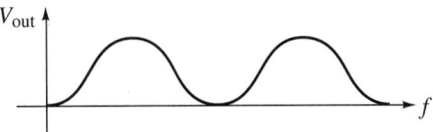

FIGURE 7.19 Problem 8.

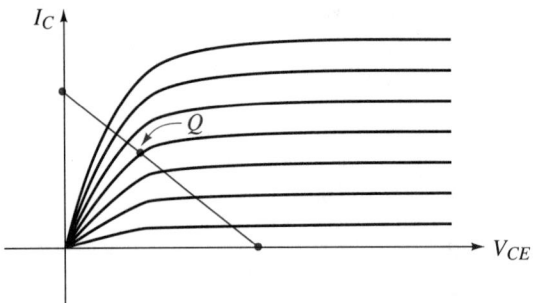

FIGURE 7.20 Problem 10.

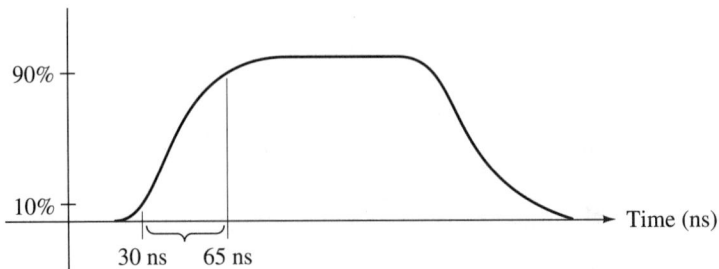

FIGURE 7.21 Problem 12.

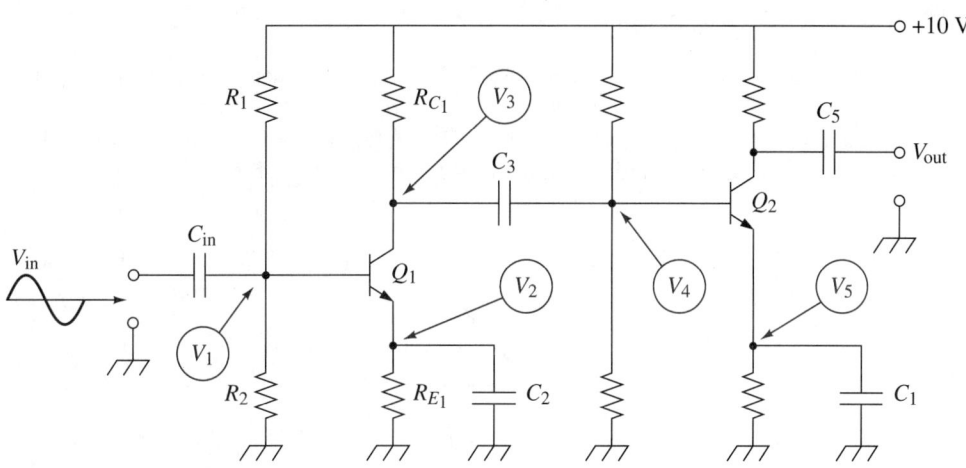

FIGURE 7.22 Problems 14–20.

TABLE 7.1 Problems 14–20

Condition	V_1	V_2	V_3	V_4	V_5	V_{out}
Normal	1.7 V	1 V	5.6 V	1.8 V	1.1 V	6 V
A	1.7 V	1 V	10 V	1.8 V	1 V	0
B	1.7 V	1 V	5.6 V	1.8 V	1.1 V	0
C	1.7 V	1.7 V	5.6 V	1.8 V	1.1 V	0

13. What type of distortion does Figure 7.21 represent? Use Figure 7.22 and Table 7.1 for Problems 14–20.

14. Using the waveforms for condition A, identify the fault.

15. You made measurements and they are listed in condition B. Identify the problems.

16. The circuit has no output. Circuit conditions are shown in C. What is wrong?

17. If the input level is too high, sketch the output.

18. Capacitor C_1 is shorted. What will the output look like?

19. Capacitor C_1 is open. What will the output look like?

20. Transistor Q_2 is shorted. What will the output look like?

Troubleshooting Problem 1

Breadboard the circuit shown in Figure 7.23. Make your measurements at the points listed in Table 7.2 and compare them with the normal waveforms shown. Once you are satisfied that your circuit is operational, inject the faults listed, one by one, while measuring the voltage levels and sketching the waveforms in the table.

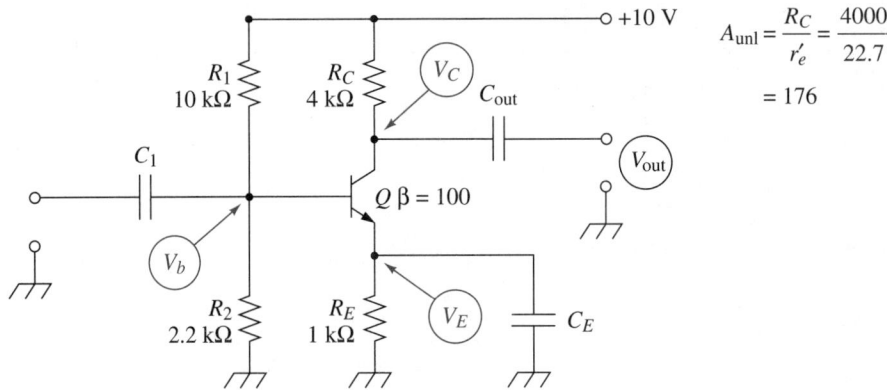

$$A_{unl} = \frac{R_C}{r'_e} = \frac{4000}{22.7}$$

$$= 176$$

FIGURE 7.23 Troubleshooting Problem 1.

TABLE 7.2 Troubleshooting Problem 1

	Normal			
	V_b	V_E	V_C	V_{out}
Faults	1.9 V 1.7 V 0 〰	1.1 V 0 ⎯	4.5 V 0 ⊓⊔	⊔⊓
Open CE				
Shorted R_C				
Open R_1				

8

TROUBLESHOOTING TUNED AMPLIFIERS

8.1 INTRODUCTION

The tuned amplifier is a major block in any communications system. A tuned amplifier is designed to provide a specific output over a limited bandwidth. See Figure 8.1. A signal whose frequency lies between the lower cutoff and upper cutoff points will be amplified, whereas one with frequency outside the band will be attenuated. Unlike other amplifiers, the tuned amplifier is designed to have a narrow (high-Q) frequency bandwidth. Normally, the more narrow bandwidth is the desirable one. The roll-off rate becomes very important in critical applications. Therefore, the Q of the circuit (and, as a result, the roll-off rate) determine the steepness of the "skirts" of the response curve, as shown in Figure 8.2.

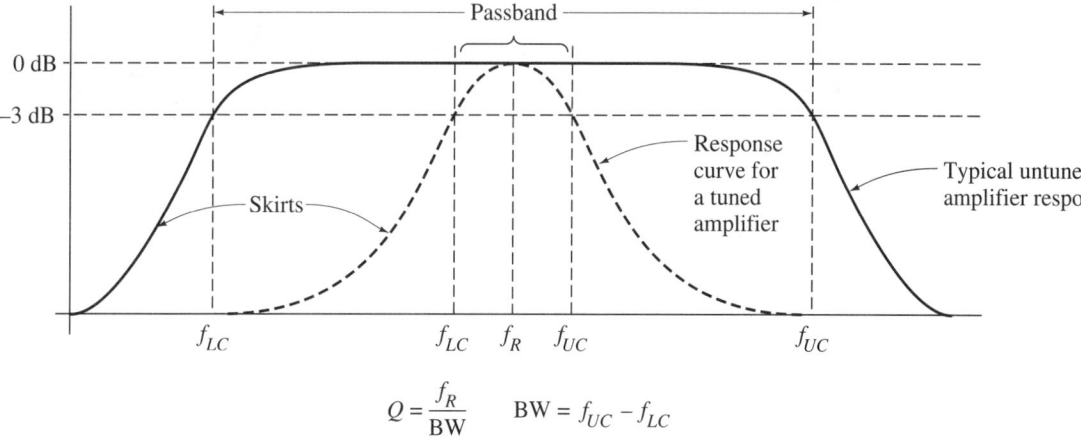

$$Q = \frac{f_R}{\text{BW}} \qquad \text{BW} = f_{UC} - f_{LC}$$

FIGURE 8.1 Amplifier response curves.

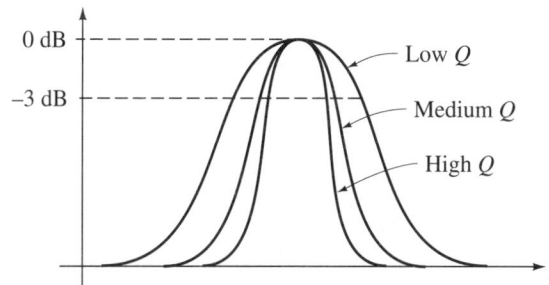

FIGURE 8.2 Tuned circuit response curves.

8.2 DISCRETE TUNED AMPLIFIERS — THE CLASS C CIRCUIT

The class C amplifier circuit (Figure 8.3) is very common in RF circuits. The transistor is biased well in the cutoff region. The value of V_{BB} is set so that the transistor turns on only at the positive peaks of the input voltage. Since the transistor turns on with each positive peak, the collector current, I_C, will also pulse. Each pulse sustains the flywheel oscillations of the tank circuit at the resonant frequency. The tank circuit is tuned to the fre-

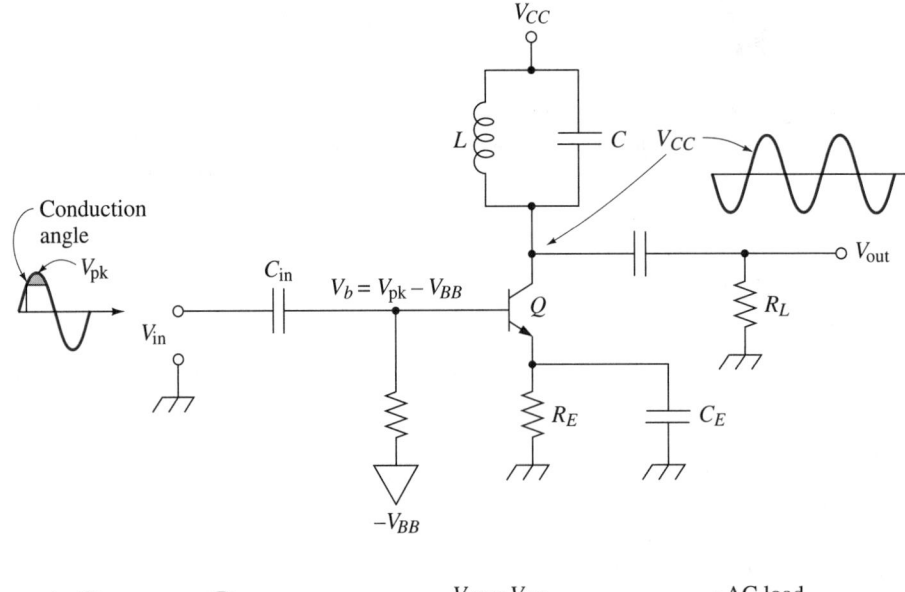

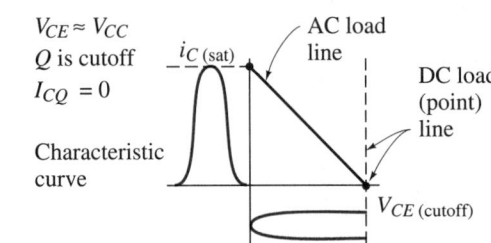

FIGURE 8.3 Class C amplifier circuit.

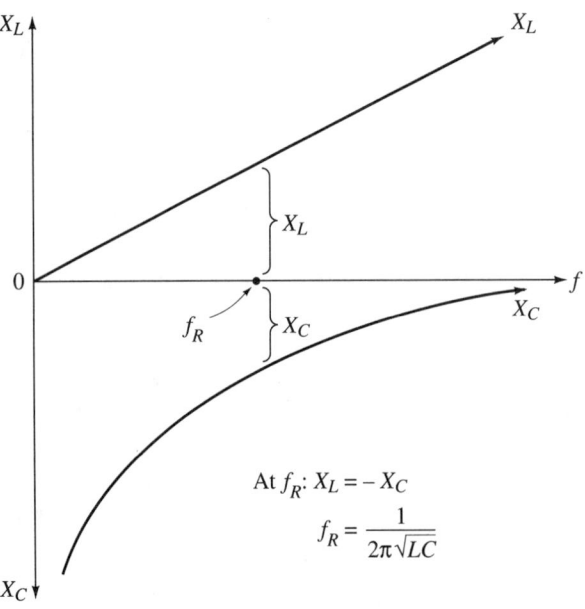

FIGURE 8.4 Reactance curves.

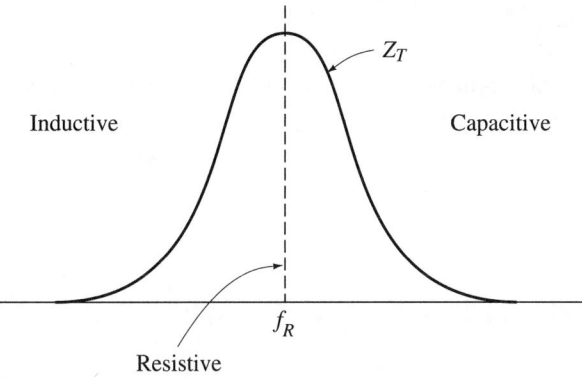

FIGURE 8.5 Tank circuit response curve.

quency of the input voltage or a harmonic of this frequency. When the output frequency is equal to a harmonic, the circuit is a frequency multiplier.

From Figures 8.4 and 8.5, you can see the following:

1. When the input frequency is less than the resonant frequency of the tank circuit, the capacitive reactance exceeds the inductive reactance and most of the tank circuit current passes through the inductance. The circuit looks inductive.

2. When the input frequency equals the resonant frequency, the reactances cancel each other out, and the circuit looks like R_L only—purely resistive.

3. When the input frequency is greater than the resonant frequency, the inductive reactance exceeds the capacitive reactance, and the circuit looks capacitive.

8.3 TUNING THE CIRCUIT

When tuning the circuit, place your ammeter as shown in Figure 8.6 to measure the collector current with no applied input signal. This is the value of I_{CQ}.

Next, apply the input signal and adjust the variable inductance until you observe a dip in the collector current. At this point, the tank circuit is tuned to the input frequency.

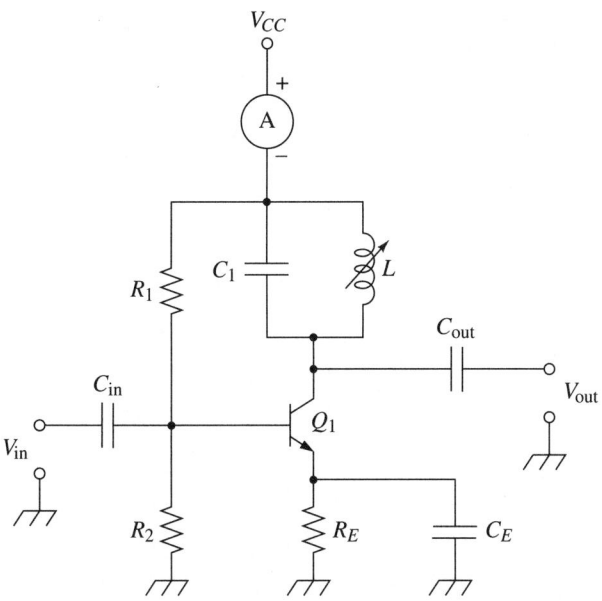

FIGURE 8.6 Tuning the circuit.

8.4 CLASS C STAGE COMMON FAILURE

The most common problem with this circuit is a drift in the resonant frequency. This drift is usually due to component aging and can best be solved by retuning the amplifier.

8.5 TROUBLESHOOTING THE CLASS C AMPLIFIER

Some common faults in the class C amplifier

As a reminder: If the resonant frequency is not as expected, this might be due to the tolerances of the inductors and capacitors. It might also be due to stray capacitance and inductance. Junction capacitance can also disturb the circuit.
    Here are some common faults:

1. Open inductor. The capacitor blocks dc; therefore, $V_C = 0$.
2. Shorted inductor. $V_C = V_{CC}$.
3. Open capacitor. The circuit is inductive and will pass high frequency.
4. Shorted capacitor. $V_C = V_{CC}$.

These should be followed up with static tests of the inductor and the capacitor.

8.6 IC CIRCUITS—OP-AMP ACTIVE FILTERS

Figure 8.7 shows an op-amp circuit that is the equivalent of the discrete tuned amplifier. These are known as active filters and are designed to act as shown in Figure 8.8, including

1. Low-pass filters
2. High-pass filters
3. Band-pass filters
4. Notch filters

Examples of these circuits are shown in Figures 8.9, 8.10, 8.11, and 8.12.

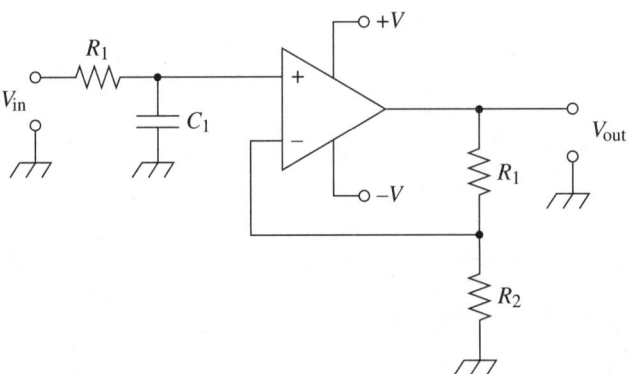

FIGURE 8.7 Op-amp circuit.

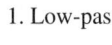

1. Low-pass

2. High-pass

3. Band-pass

4. Band-stop
   (notch)

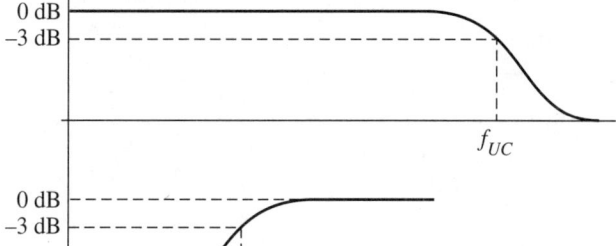

FIGURE 8.8 Active filters.

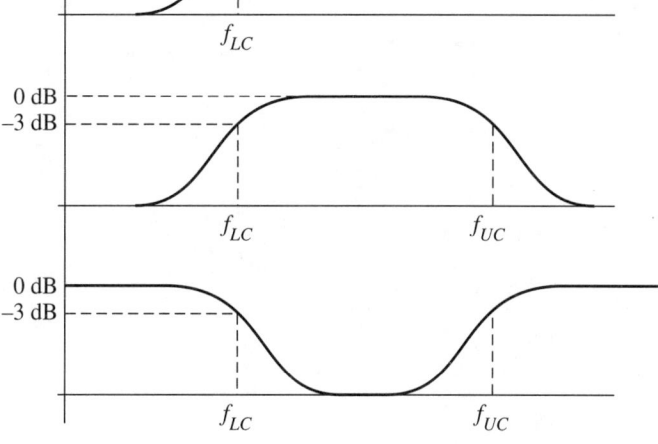

The number of RC circuits (poles) determine the roll-off rate and steepness of the "skirts."

$$f_{UC} = \frac{1}{2\pi RC}$$

$$A_{CL} = \frac{R_1}{R_2} + 1$$

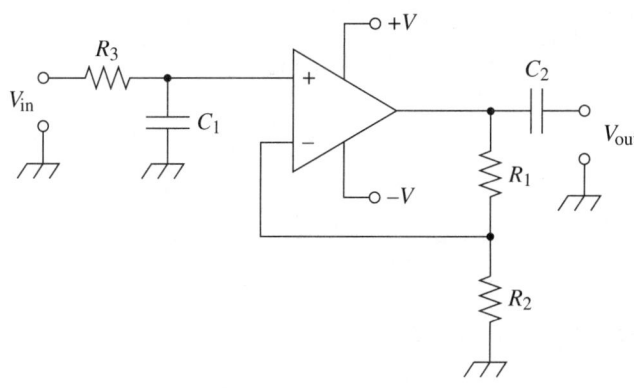

FIGURE 8.9 Low-pass filter.

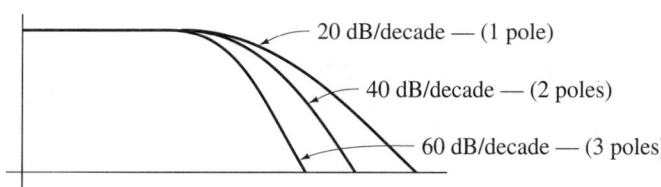

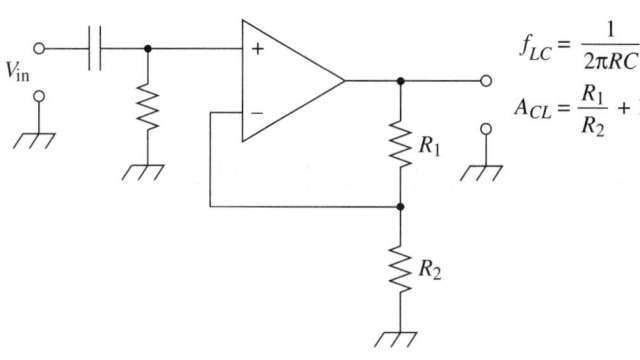

$$f_{LC} = \frac{1}{2\pi RC}$$

$$A_{CL} = \frac{R_1}{R_2} + 1$$

FIGURE 8.10 High-pass filter.

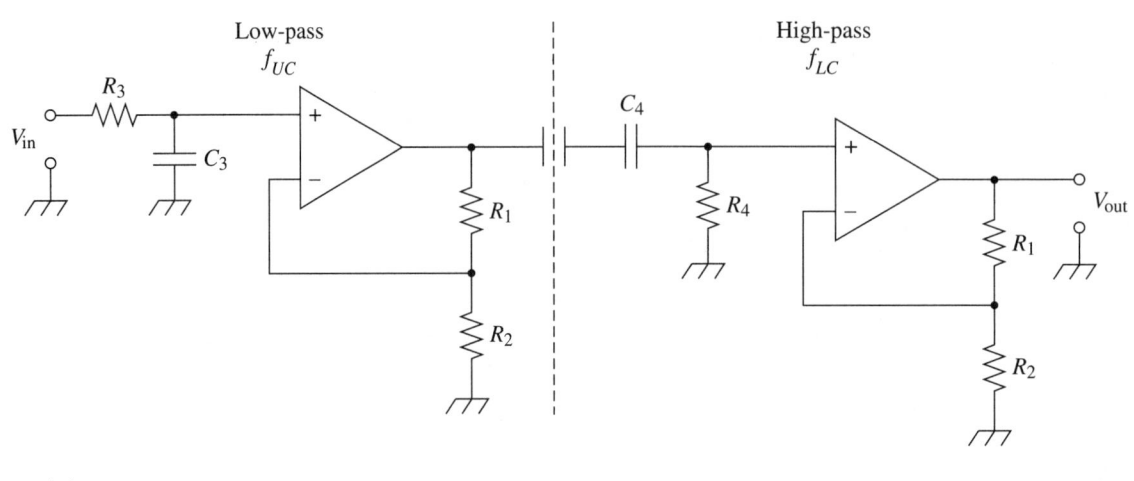

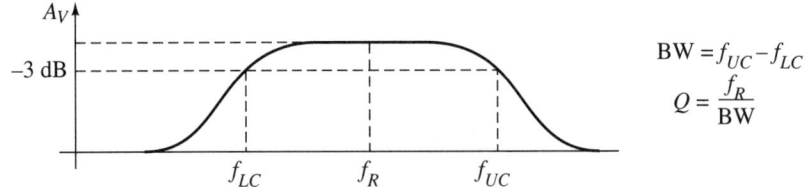

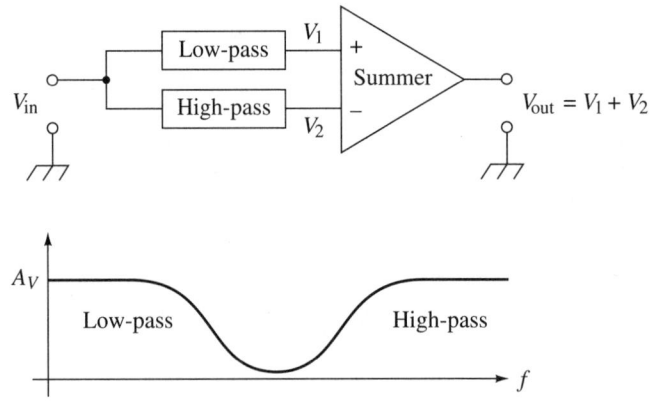

FIGURE 8.11 Band-pass filter incorporates a low-pass and a high-pass filter.

$$\text{BW} = f_{UC} - f_{LC}$$
$$Q = \frac{f_R}{\text{BW}}$$

FIGURE 8.12 Notch filter.

8.7 COMMON FAULTS OF LOW-PASS FILTERS

1. R_1 open: The closed-loop gain equals 1.
2. R_2 shorted: The closed-loop gain equals 1.
3. R_2 open: The closed loop gain is lost and the circuit goes to open loop. The output is clipped.
4. R_3 open: The output voltage equals 0.
5. C_1 shorted: The output voltage equals 0.
6. C_1 open: The upper cutoff frequency is bad and the roll-off rate is disturbed.

8.8 COMMON FAULTS OF HIGH-PASS FILTERS

1. R_1 open: The closed-loop gain equals 1.
2. R_2 shorted: The closed-loop gain equals 1.
3. R_2 open: The circuit goes into open-loop gain.
4. R_3 shorted: The output voltage equals 0.

5. C_3 open: The output voltage equals 0.
6. R_1 shorted: The output voltage equals 0.
7. C_3 shorted: The lower cutoff frequency is bad and the roll-off rate is bad.

8.9 COMMON FAULTS OF BAND-PASS FILTERS

1. R_1 open: The circuit goes to open-loop gain.
2. R_3 shorted: The circuit goes to open-loop gain.
3. R_2 open: The closed-loop gain doubles.
4. R_3 open: The output voltage equals 0.
5. C_3 open: The output voltage equals 0.
6. C_4 shorted: The circuit becomes a low-pass filter.
7. C_3 shorted: The circuit becomes a high-pass filter.

8.10 GENERAL TROUBLESHOOTING

1. Determine the filter type.
2. Check the input voltage.
3. Disconnect the load and check the output voltage.
4. Check the supply voltages.
5. Perform a static test of the components.
6. Replace the op amp.

Some general troubleshooting tips

8.11 OSCILLATORS

An oscillator circuit is one that produces an output signal without any apparent input signal. The circuit takes power from the dc supply and converts it to an ac output.

As you can see from Figure 8.13, the key to the circuit's operation is the positive feedback loop. The feedback voltage is applied to the front of the amplifier, amplified, and returned to the front, in phase, to be amplified even more.

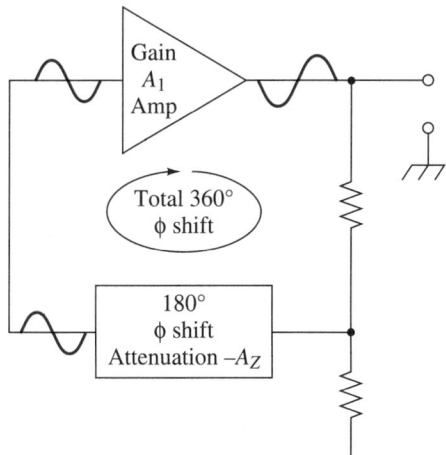

Thermally generated signal is fed back to the front, amplified, fed back, and amplified, over and over. The signal continues to build in value.

FIGURE 8.13 Oscillator circuit.

In the oscillator circuit, thermally generated signals are fed back to the amplifier input to be amplified again. If the closed-loop gain is greater than 1, the signal will continue to increase in value. At some value of the output, if the loop gain can be made equal to 1, the oscillations will be sustained. This is called the *Barkhausen criterion*. In terms of gain

$$A_1 A_2 = 1 \qquad (8.1)$$

If the total loop gain is less than 1, the oscillations will fade. If the total loop gain is greater than 1, the oscillator will become saturated.

The stability of the oscillator is its ability to maintain a constant-output amplitude and frequency.

8.12 THE WIEN-BRIDGE OSCILLATOR

The Wien-bridge oscillator (Figure 8.14) is extensively used for frequencies up to 1 MHz. It has very low distortion and uses an RC feedback network. At the resonant frequency of the RC network, the feedback network has an attenuation factor of $B = \frac{1}{3}$. Also at the resonant frequency, the op amp has a gain of 3. Thus the total gain, AB, is equal to the product of the two values, or 1, and the net phase shift is 0.

In the circuit, A is the positive feedback path. R and C constitute a low-pass filter, whereas R_1 and C_1 form a high-pass filter. Both have the same critical frequency. Therefore, the Wien bridge is a lead-lag network (Figure 8.15). At low frequencies, C_1 looks like an open circuit and the output voltage is 0. At high frequencies, C_2 looks like a short circuit and the output voltage is 0. The resonant frequency is

$$f_R = \frac{1}{6.28RC} \qquad (8.2)$$

The closed loop gain of the amplifier circuit is

$$A_{CL} = \frac{2R'}{R'} + 1 = 3 \qquad (8.3)$$

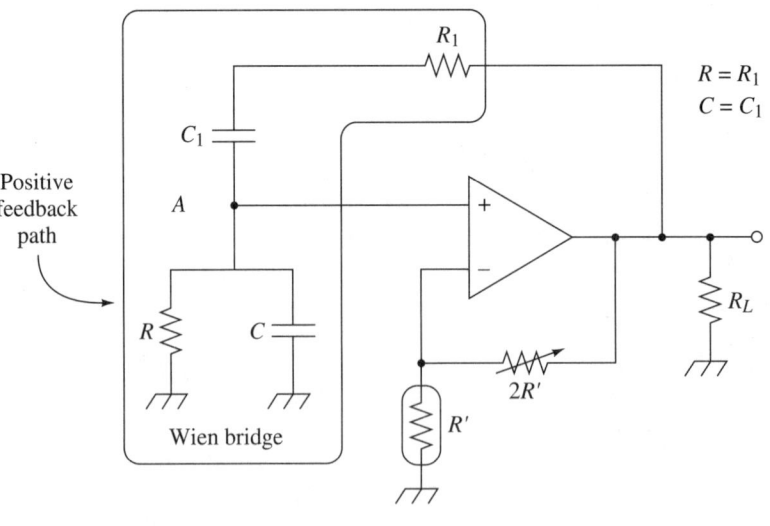

R and C are low-pass filters.
R_1 and C_1 are high-pass filters.
Both have the same f_C.

FIGURE 8.14 Wien-bridge oscillator.

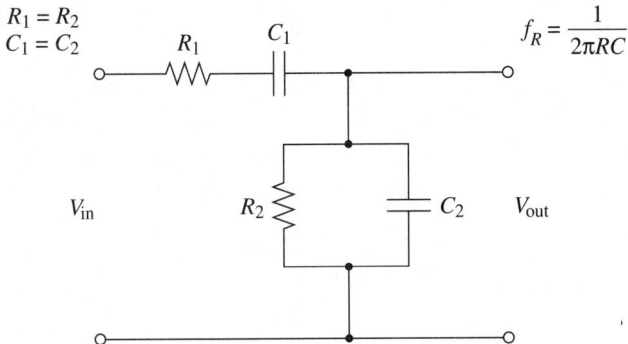

$R_1 = R_2$
$C_1 = C_2$

$f_R = \dfrac{1}{2\pi RC}$

At low frequencies, C_1 looks like an open circuit and $V_{out} = 0$.

At high frequencies, C_2 looks like a short circuit and $V_{out} = 0$.

FIGURE 8.15 Wien bridge.

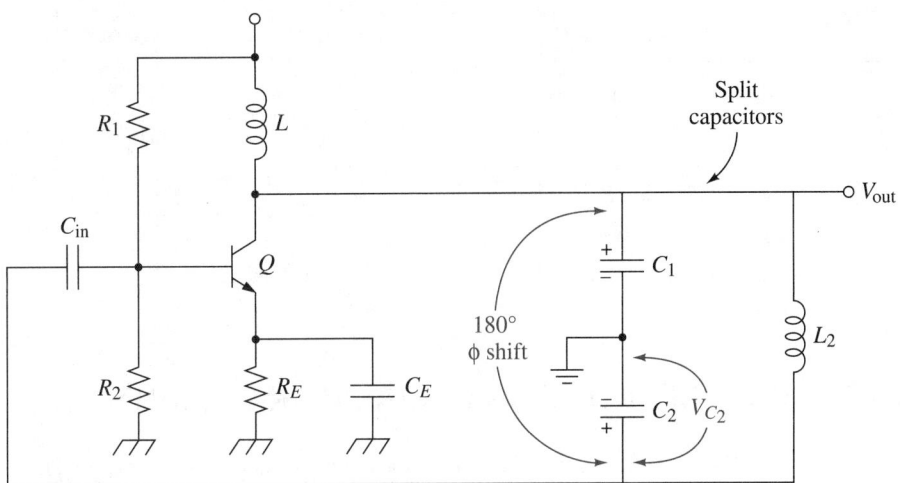

$$f_R = \dfrac{1}{2\pi\sqrt{LC_T}}, \qquad C_T = \dfrac{C_1 C_2}{C_1 + C_2}$$

V_{C_2} is the feedback voltage.

V_{C_2} and V_{C_1} are 180° out of phase.

V_{in} equals V_{C_2}.

FIGURE 8.16 Colpitts oscillator.

8.13   THE COLPITTS OSCILLATOR

The Colpitts oscillator (Figure 8.16) is recognized by its split capacitors. The voltage across C_2 is the feedback voltage. The voltage across C_2 is 180° out of phase with the voltage across C_1. The voltage across C_2 acts as the input voltage to the amplifier.

8.14   CRYSTAL-CONTROLLED OSCILLATORS

Figure 8.17 is the symbol and equivalent circuit for a quartz crystal. The crystal is made of silicon dioxide (SiO_2) and vibrates naturally at a constant rate when subjected to an electric field. The frequency of vibrations is determined by the physical dimensions of the crystal. As you can see, the quartz acts like a series and parallel resonant circuit. These crystals are incorporated in oscillator circuits to obtain extremely stable output signals. See Figure 8.18.

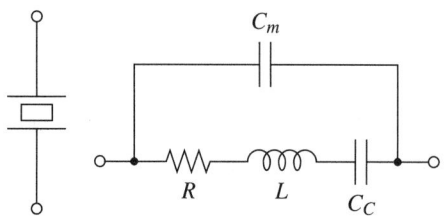

FIGURE 8.17 Quartz crystal.

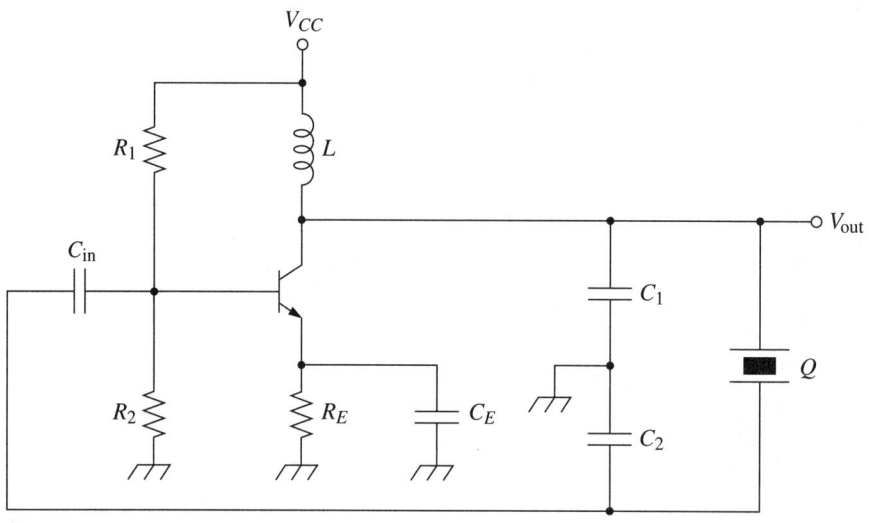

FIGURE 8.18 Crystal oscillator.

8.15 TROUBLESHOOTING OSCILLATOR CIRCUITS

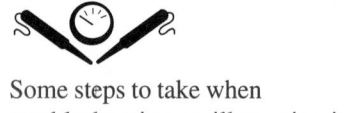

Some steps to take when troubleshooting oscillator circuits

1. In an oscillator circuit, every component (except the resistors used for dc biasing) is involved in the formation of the output signal.

Prior to troubleshooting, it is important to make certain that the problem is with the oscillator and not the surrounding stages.

2. Check the power supply to verify its operation.

3. It is difficult to troubleshoot oscillators due to the many schemes available. In some audio oscillator bias arrangements, the transistor is operating in class A mode, but RF oscillators are usually operated in class C mode. The circuits are likely to be emitter followers, source followers, or voltage followers and can be static-tested the same as any follower circuit.

4. First you should check the bias conditions to determine the class of operation.

  a. If the circuit is class A, the base-emitter junction must be forward-biased.
  b. When the oscillator is class C, any forward bias on the base-emitter junction indicates a problem.

5. Measure the amplitude and frequency of the output signal. Use your oscilloscope with an RF probe to observe and measure the output. Verify the frequency with your frequency counter.

6. If the frequency is acceptable, but the amplitude is low:

  a. Retune the circuit and check the output again.
  b. Check for excessive leakage of the transistor.
  c. Verify that the IC chips are normal.
  d. Make sure that the dc bias is correct so that the circuit gain will be proper.

e. Perform the dc voltage measurements and compare them with normal values or calculated values.

f. Perform resistance tests on the capacitors and inductors.

7. If the frequency is only slightly low or high, it is necessary to retune the circuit.

8. If the frequency is very low or very high, you should suspect the frequency-controlling components.

Here is a quick test of your oscillator circuit: Place a small capacitor from base to ground or from the collector to ground and measure the collector voltage and the emitter voltage. The capacitor should stop circuit oscillations, and you should see a change in the collector and emitter voltages.

A quick test

8.16 SOLID-STATE SWITCHING CIRCUITS

Solid-state circuits are designed to operate in two states. They are either on or off. The states are also referred to as

1. High or low
2. True or false
3. Yes or no
4. Logic 1 or logic 0

Solid-state switching circuits are the fundamental building blocks of computer systems.

8.17 THE TRANSISTOR SWITCH

The most basic switch is the transistor switch in Figure 8.19. When the input voltage goes high, the transistor saturates and the output goes to $V_{CE(sat)}$. When the input voltage goes low, the transistor cuts off and the output goes to V_{CC}. Therefore, the Q point has two locations on the dc load line in Figure 8.19(b): the transistor operator as a switch, either shorted (on) or open (off). There are some requirements:

1. The input low must be low enough to turn the transistor off.
2. The input high must be positive enough to saturate the transistor.
3. The output voltage high is normally equal to V_{CC}.
4. The base resistor is such that

$$I_B \gg \frac{I_{C(sat)}}{B_{dc}} \tag{8.4}$$

and

$$I_B = \frac{V - V_{BE}}{R_B} \tag{8.5}$$

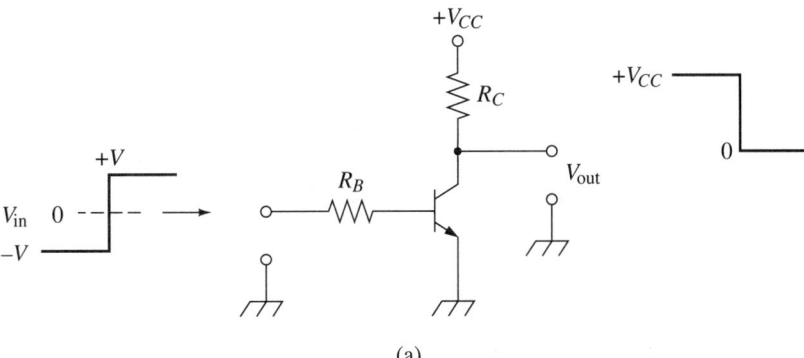

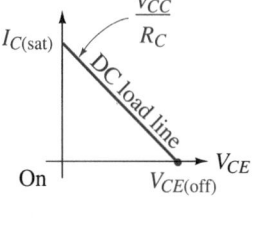

(a)

(b)

FIGURE 8.19 Transistor switch.

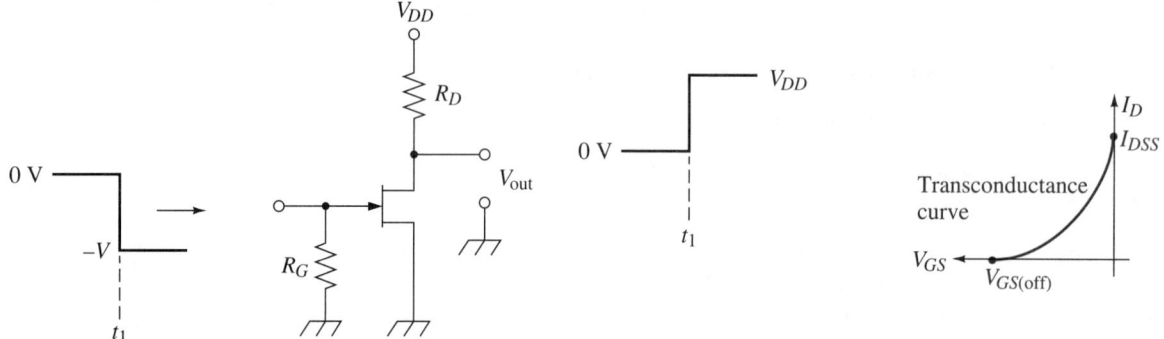

FIGURE 8.20 JFET switch.

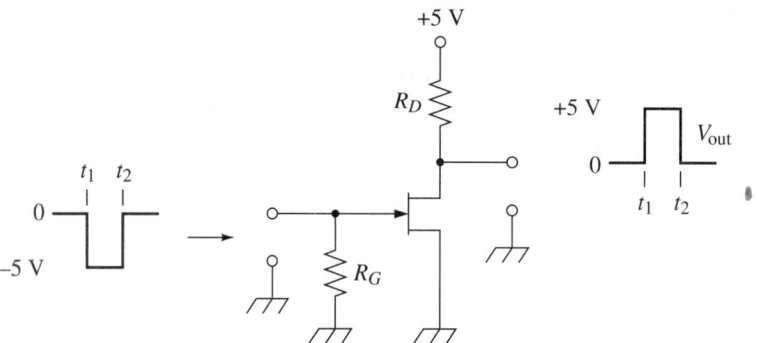

FIGURE 8.21 Using a standard
TTL 5 V logic level.

8.18   THE JFET SWITCH

Figure 8.20 shows the basic JFET switch. Remember that the gate of the JFET must not
go positive. Thus the input voltage is either 0 V or $-V_{GS(off)}$. Again, the transistor is either
saturated or cut off and the output is either approximately 0 V (V_{DS}) or V_{DD}.

   JFETs can switch very fast and can be controlled by standard logic voltage levels, as
in Figure 8.21.

8.19   THE MOSFET SWITCH

The MOSFET in Figure 8.22 is an excellent switch. MOSFETs are used extensively in
analog switching circuits. They can be controlled by the application of either positive or
negative gate voltages. The output will be $(V_{DD} - I_{DSS}R_D)$ when the input voltage is 0 V.
When the input goes to $+V$, the transistor saturates and the output is equal to $V_{DS(on)}$ (almost 0 V).

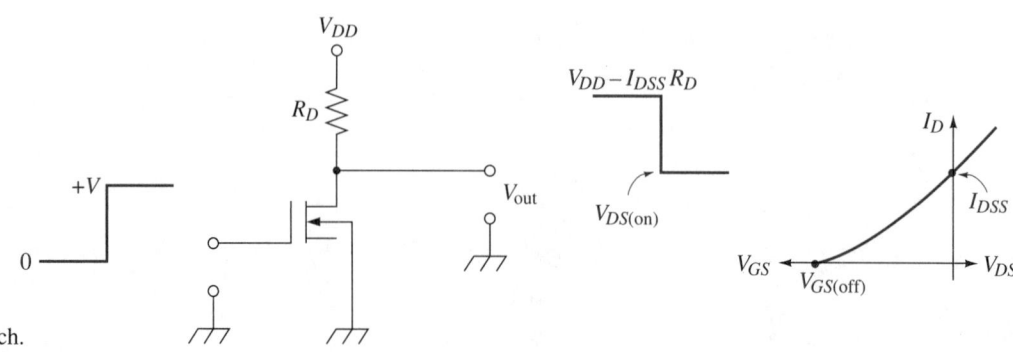

FIGURE 8.22 MOSFET switch.

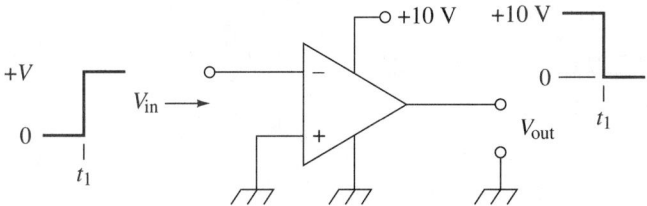

FIGURE 8.23 Op-amp switch.

8.20   THE OP-AMP SWITCH

As you can see from Figure 8.23, the op-amp switch is really a comparator with a 0-V reference. Since this in an inverter circuit, the output will be $-V_{sat}$ when the input voltage is $+V$. Therefore, the output switches between $+V_{sat}$ and $-V_{sat}$.

8.21   THE OUTPUT PULSE

The characteristics of a square-wave pulse are shown in Figure 8.24. The pulse width and space width are measured at the half-power (-3-dB) points. Ideal square waves have a 50% duty cycle. The duty cycle is critical when dealing with ICs. The propagation delay between input square-wave transition and the output transition is due to

1. The time required for the transistor to change states (known as the *delay time*)
2. The *rise time,* which is the time to go from cutoff to saturation
3. The *storage time,* which is the time required to come out of saturation
4. The *fall time,* which is the time required to go from saturation into cutoff
5. The *slew rate* of an op amp

The upper cutoff frequency of the device can be found from

$$f_2 = 0.35/t \qquad (8.6)$$

To pass a square wave with no distortion, the upper cutoff frequency of the circuit must be about 10 times the frequency of the square waves.

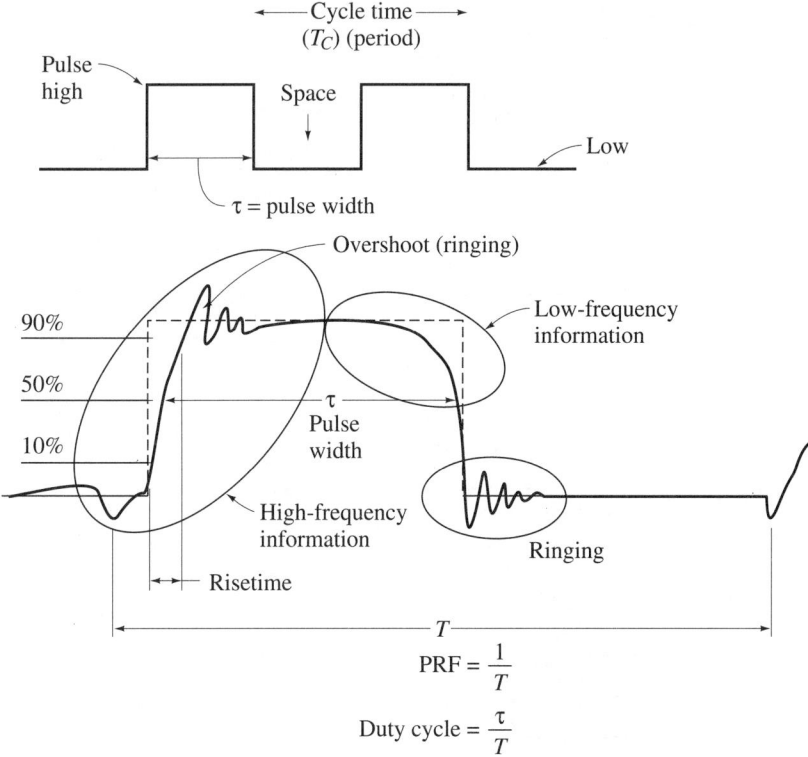

FIGURE 8.24 The output pulse.

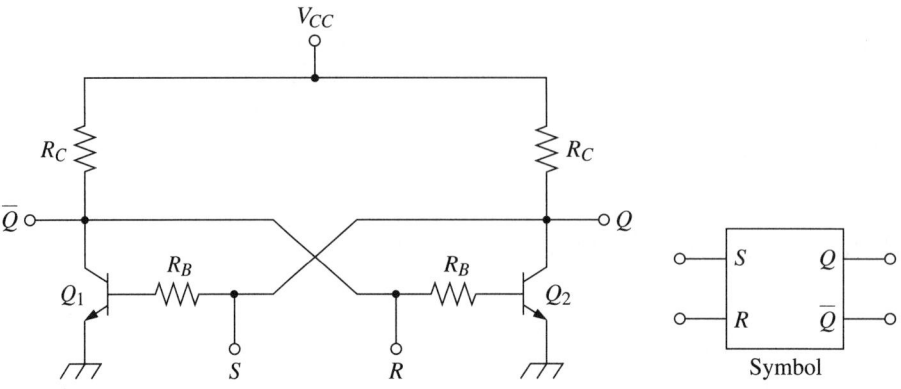

FIGURE 8.25 Flip-flop.

8.22 THE *RS* FLIP-FLOP

TABLE 8.1 Truth Table

S	R	Q	$\overline{Q}$
High	—	High	Low
—	High	Low	High

The discrete *RS* flip-flop circuit is shown in Figure 8.25. It consists of a pair of cross-coupled transistors. When Q_1 is saturated, Q_2 is in cutoff; when Q_1 is in cutoff, Q_2 is saturated. The output, Q, is either high or low. The other output, not Q, is the complementary output. These conditions are summarized in Table 8.1.

Let's add some additional circuitry to the *RS* flip-flop, as in Figure 8.26. If Q is high, Q_3 will saturate and the capacitor, C, discharges, placing the noninverting terminal of the

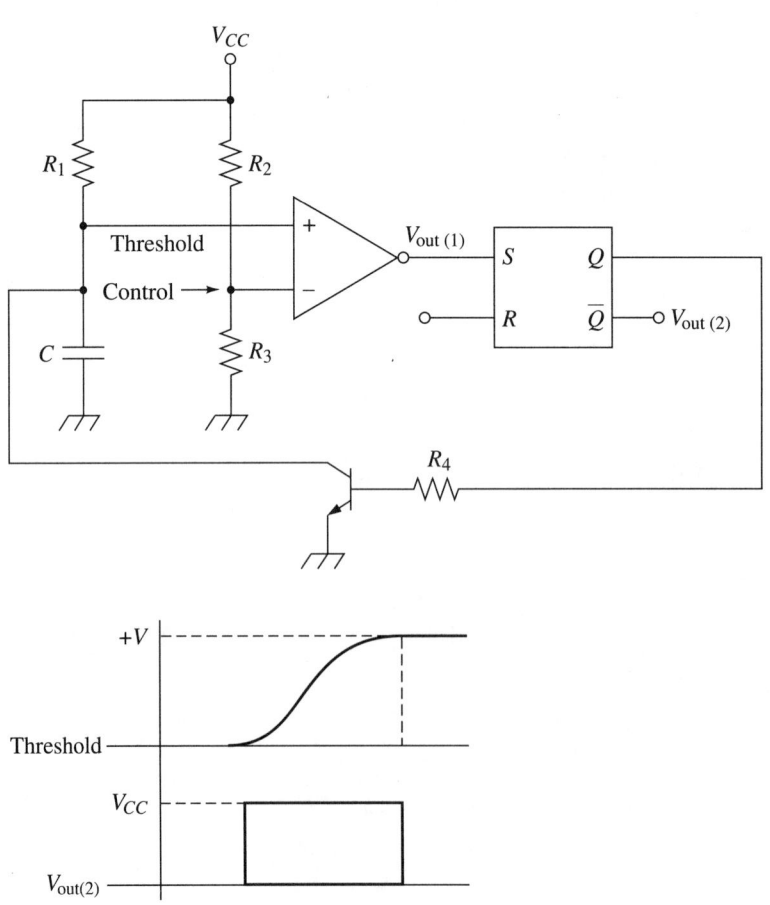

FIGURE 8.26 Adding control circuitry to the *RS* flip-flop.

comparator at ground. If a high voltage is now applied to the reset terminal, R, the output, Q, goes low and Q_3 turns off. The capacitor will charge until its charge equals the control voltage level on the inverting terminal. The comparator output goes high and sets the flip-flop. Q goes high again, and the process continues.

The circuit is modified and presented in Figure 8.27. Most of the time the control voltage is held at two-thirds of V_{CC}. When V_C exceeds this level, the high output of comparator A will set the flip-flop, and the output, Q, goes high. The capacitor can be external and variable according to user requirements. When pin 4 is grounded, the flip-flop will not operate.

The voltage on the noninverting terminal of comparator B is held at one-third of V_{CC}. A voltage on pin 2 that is less than this value will force the output of comparator B to go high and reset the flip-flop.

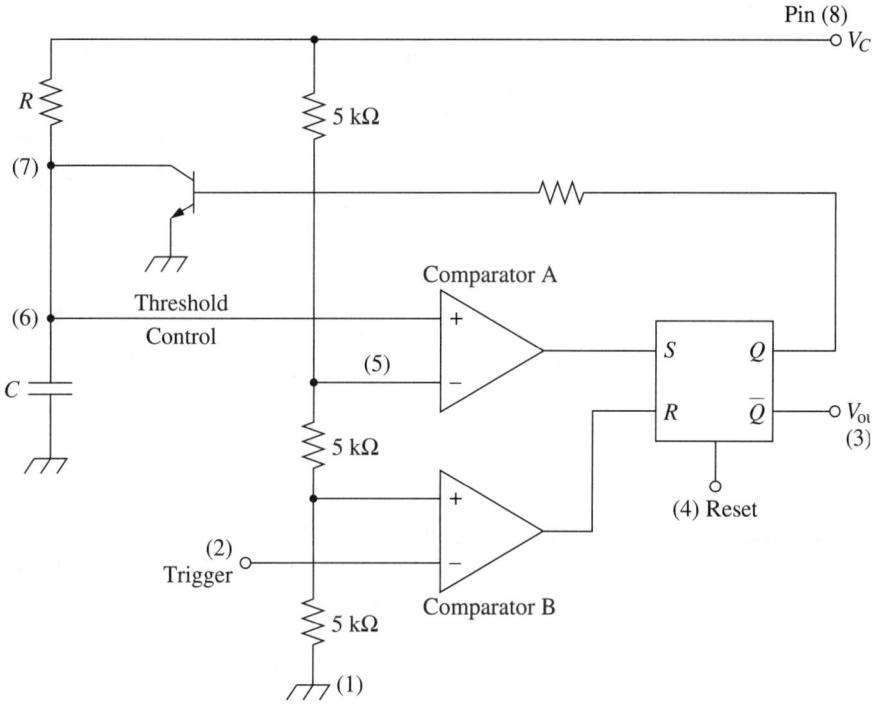

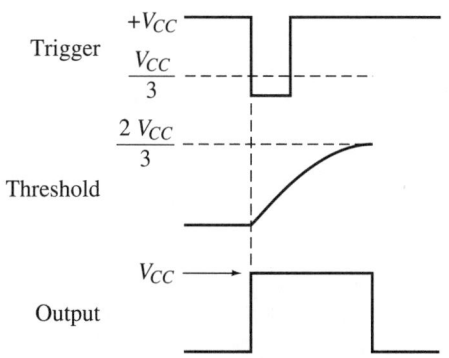

FIGURE 8.27 Complete *RS* flip-flop.

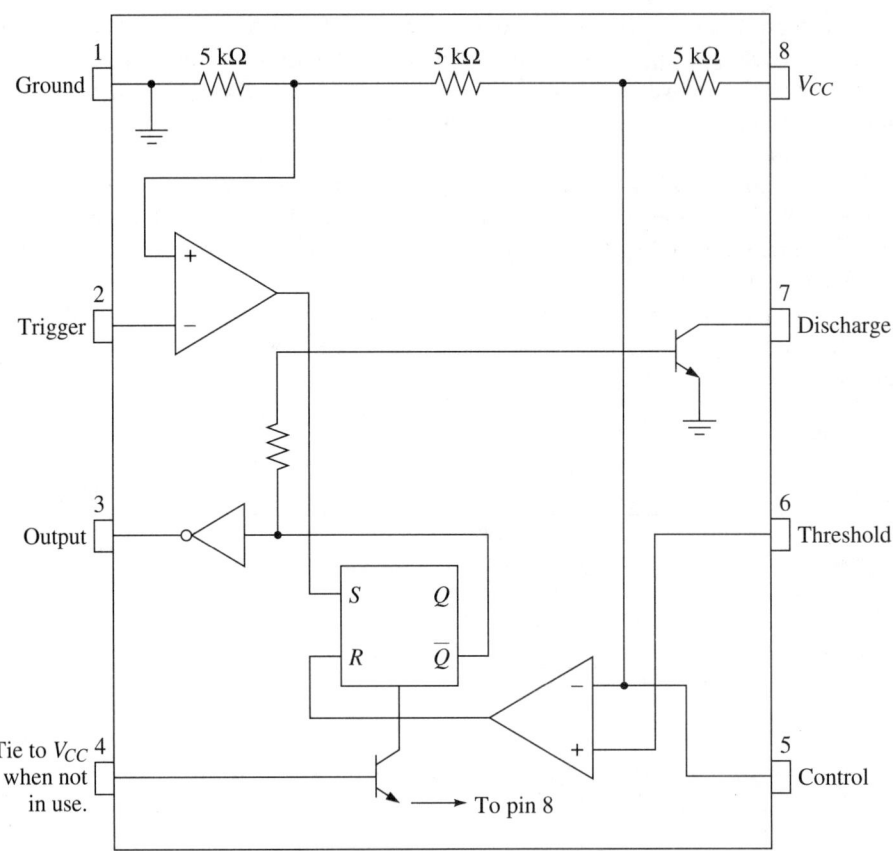

FIGURE 8.28 555 timer.

8.23   THE 555 TIMER

The basic circuitry just described can be found in an IC chip called the 555 timer. This is a very versatile, 8-pin IC switching circuit. The 555 timer is shown in Figure 8.28. The circuit contains two comparators, a flip-flop, a buffer, and several resistors and transistors.

One popular scheme is the monostable timer circuit of Figure 8.29. Another is the astable circuit in Figure 8.30.

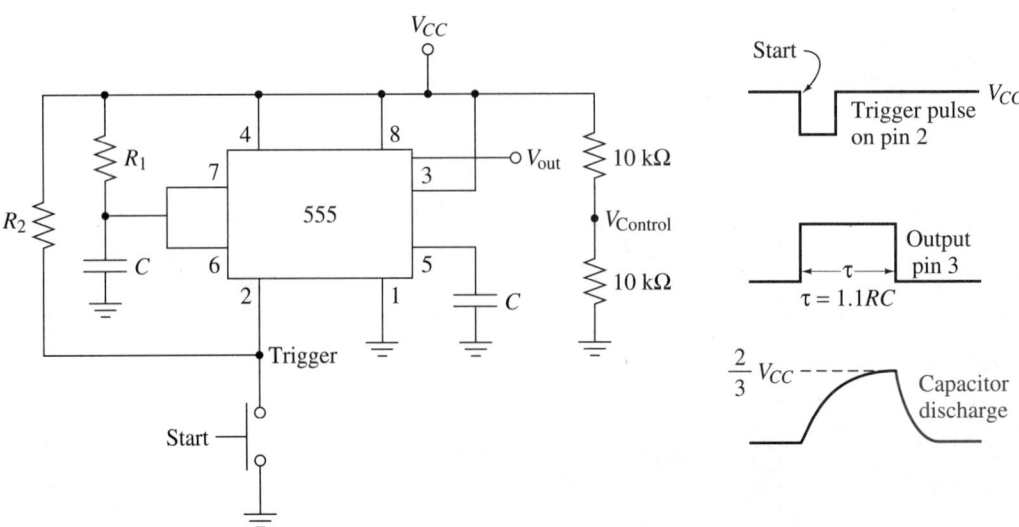

FIGURE 8.29 Monostable circuit.

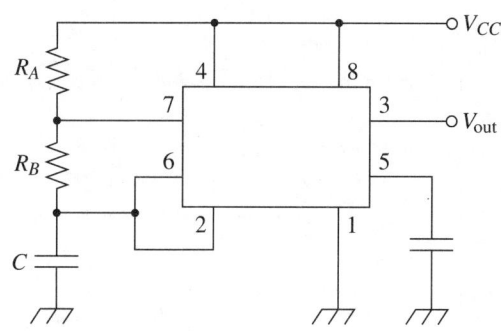

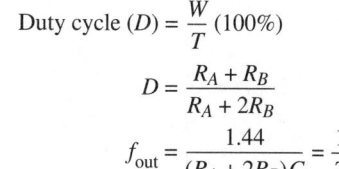

$$\text{Duty cycle } (D) = \frac{W}{T}\,(100\%)$$

$$D = \frac{R_A + R_B}{R_A + 2R_B}$$

$$f_{\text{out}} = \frac{1.44}{(R_A + 2R_B)C} = \frac{1}{T}$$

Charge time: $\tau = (R_A + R_B)C$

Discharge time: $\tau = R_B C$

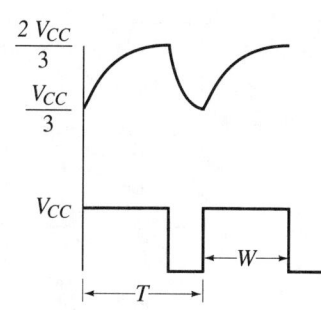

FIGURE 8.30 555 astable circuit.

8.24 TROUBLESHOOTING THE 555 TIMER CIRCUIT

1. Check the V_{CC} and ground connections.
2. Make sure that the reset pin (4) is inactive (high).
3. Check for a valid trigger pulse of pin 2.
   a. If pin 5 is not used, the trigger must be less than one-third of V_{CC}.
   b. If pin 5 is tied to a voltage divider, the trigger must be less than one-half of the control voltage.
4. Check the resistance and capacitance values.
5. For intermittent problems, check the trigger signal. The problem may be due to the frequency of operation. Check this by placing a capacitor from pin 8 to ground.
6. If R_A of the astable is shorted, the 555 will short circuit the supply voltage when pin 7 goes low, and the 555 will be destroyed.
7. If R_B of the astable is shorted, the discharge time constant goes to 0 and the output will not be a pulse. It will be a level dc voltage.

Things to check when troubleshooting the 555 timer circuit

8.25 PROBLEMS

1. Find V_{CEQ} and I_{CQ} for the class C amplifier in Figure 8.31.
2. Can you identify the fault in Figure 8.32?

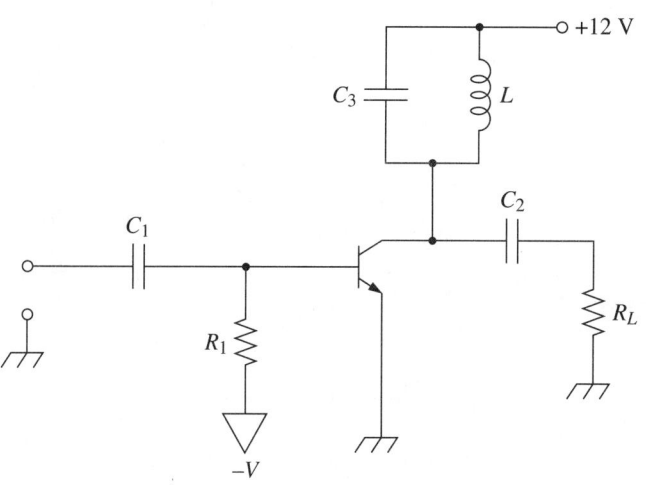

FIGURE 8.31 Problem 1.

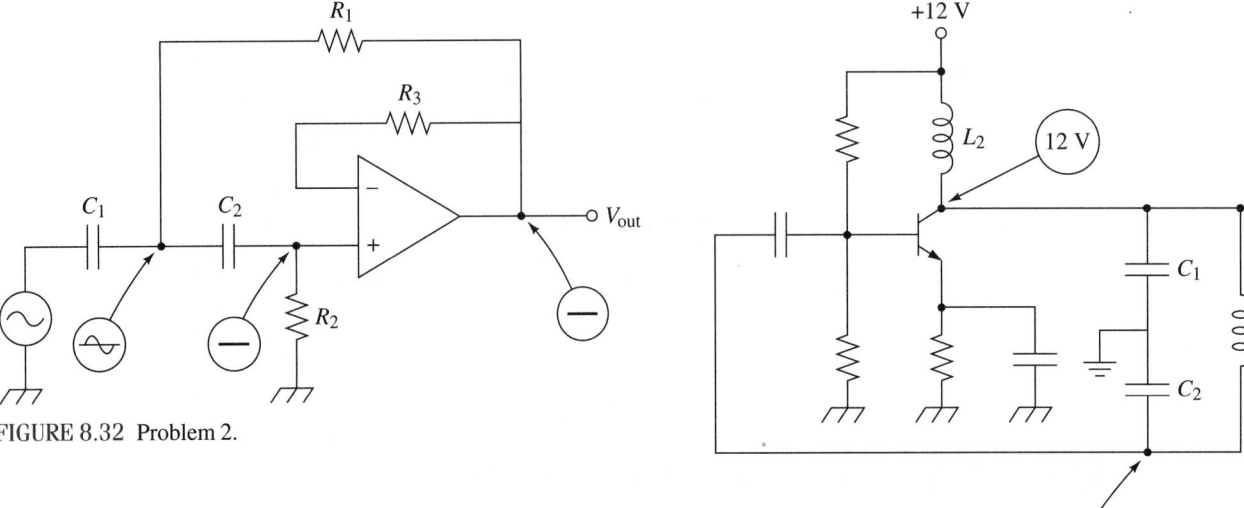

FIGURE 8.32 Problem 2.

FIGURE 8.33 Problem 3.

3. The oscillator in Figure 8.33 does not work. What is wrong with the circuit?

4. Find V_{out} for Figure 8.34.

5. Calculate the BW and Q for the frequency response curve in Figure 8.35.

6. What causes the crossover distortion shown in Figure 8.36?

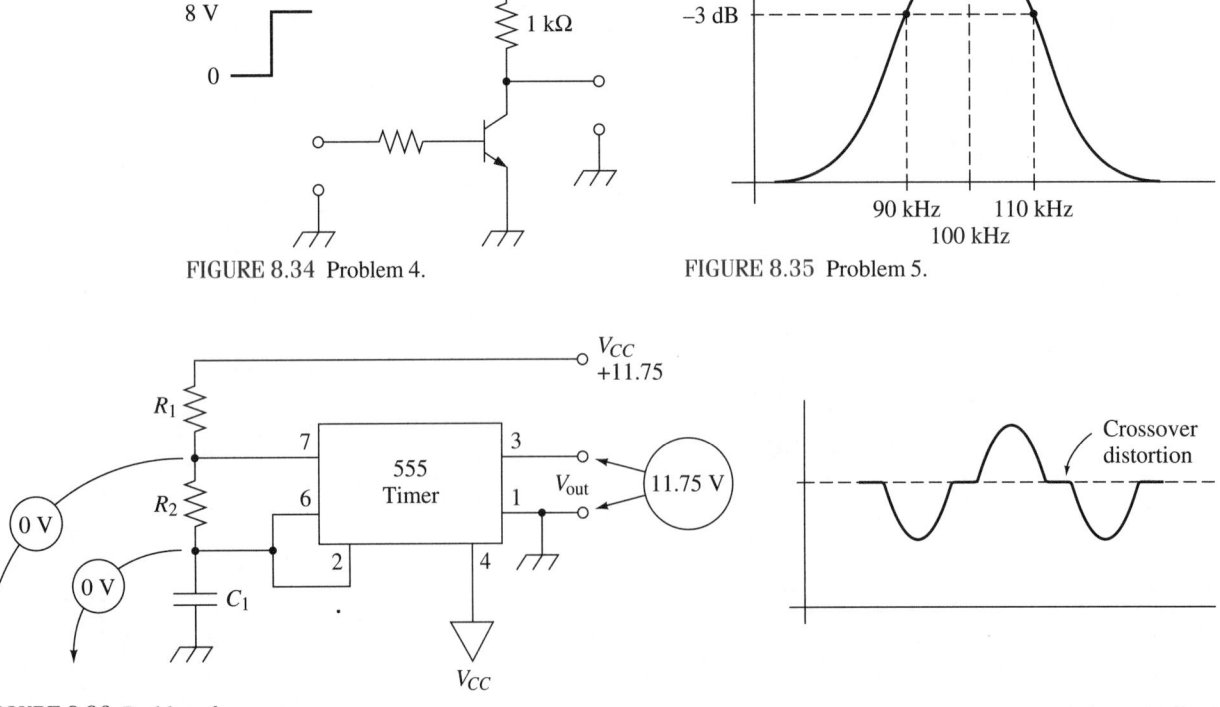

FIGURE 8.34 Problem 4.

FIGURE 8.35 Problem 5.

FIGURE 8.36 Problem 6.

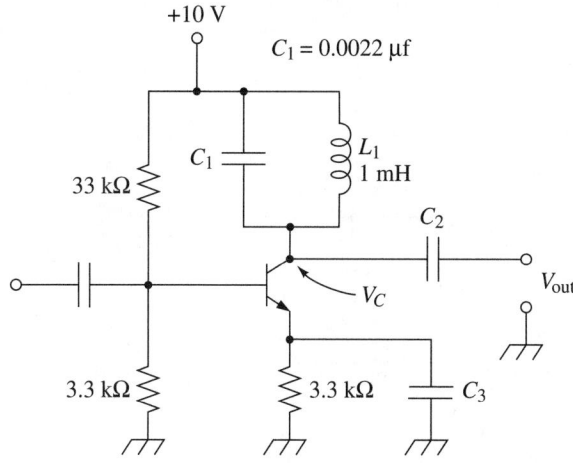

FIGURE 8.37 Problems 7–10.

7. Determine the center frequency of the tuned amplifier in Figure 8.37.

8. Is the circuit in Figure 8.37 class A, B, or C?

9. What happens to the V_{out} in the circuit of Figure 8.37 when C_3 is shorted?

10. Refer to Figure 8.37. The value of V_C is found to be equal to $+10$ V, and V_{out} measures 0 V. What is the most likely fault?

8.26 TROUBLESHOOTING EXAMPLE 1

The circuit was constructed as shown in Figure 8.38. The center frequency was calculated as $f_R = \frac{1}{2} \pi \sqrt{LC}$ and found to be 107 kHz. When V_{CC} was applied, the output was found to be zero at 107 kHz. What happened? V_6 was measured and found to equal 0.90 V. The ac input voltage was determined to be acceptable. R_L was found to be good. V_E was checked and passed. This leads to the tank circuit. Component values were checked and C was found to be 0.0022 µH instead of 0.0022 µF. The correct capacitor solved the problem.

Troubleshooting Example 1

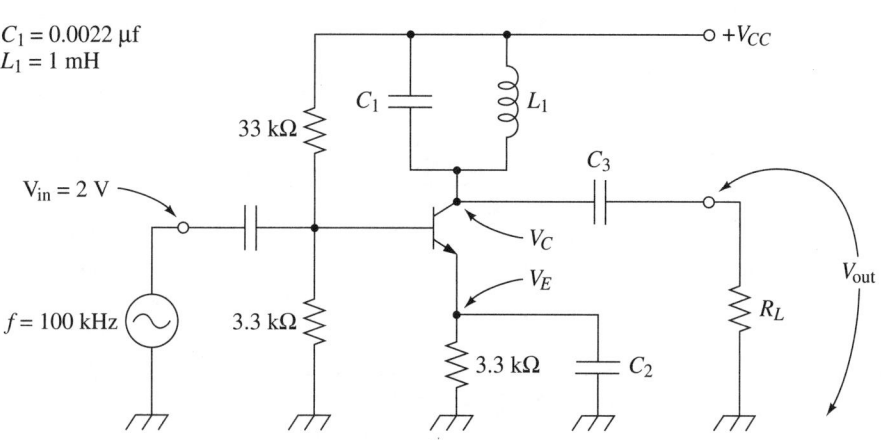

FIGURE 8.38 Troubleshooting Example 1.

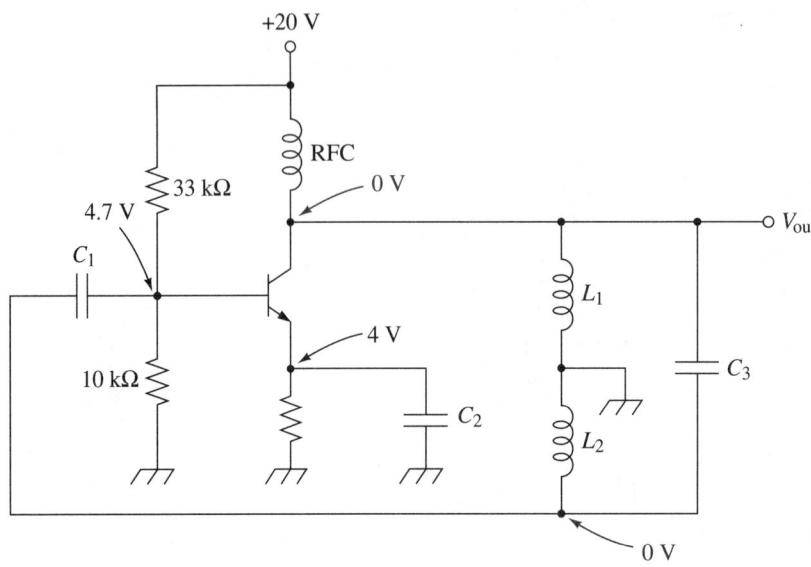

FIGURE 8.39 Troubleshooting
Example 2.

8.27 TROUBLESHOOTING EXAMPLE 2

Troubleshooting Example 2

The circuit shown in Figure 8.39 does not oscillate. The dc voltage was found to be as shown. A check for shorted components proved negative. The RF choke was checked and found to be open. Replacing the RF choke solved the problem.

8.28 TROUBLESHOOTING EXAMPLE 3

Troubleshooting Example 3

The circuit in Figure 8.40 is faulty. Logic probe measures are shown. Q should be low. The possible faults are

- D flip-flop is internally stuck at $+5$ V.
- C is internally stuck at $+5$ V.
- Pin 5 is shorted to $+5$ V.
- Q is stuck at ground.
- D flip-flop is bad.

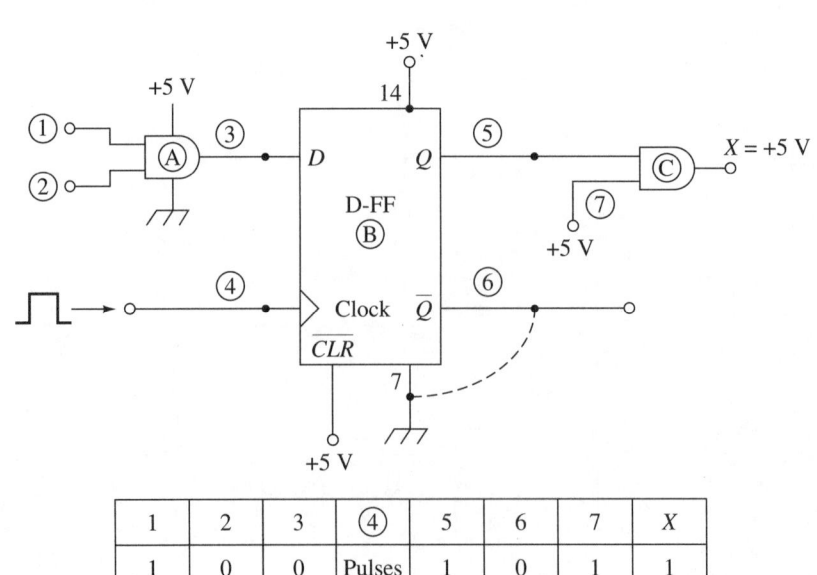

1	2	3	④	5	6	7	X
1	0	0	Pulses	1	0	1	1

FIGURE 8.40 Troubleshooting
Example 3.

Procedure:

1. Probe pins 14 and 7 to test voltage load and ground. They are acceptable.

2. Check the clock signal. It is within specifications.

3. Probe Q (pin 6). It is stuck low.

4. Replace the flip-flop. The problem still exists.

5. Perform an ohmic check on pin 6. It is found to be extremely shorted to ground.

6. Locate the short and repair it.

9

TROUBLESHOOTING RADIO FREQUENCY CIRCUITS

9.1 INTRODUCTION

It is customary to refer to signal frequencies above 20 kHz as radio frequencies (RF). Table 9.1 lists the general breakdown of the RF spectrum. All the components that we have been discussing operate normally at low frequencies; however, as the signal frequency goes up, the behavior of components starts to change. In fact, at very high frequencies, they don't act the same at all. A resistor can take on the personality of an inductor or capacitor. An inductor will have capacitance between its windings, and a capacitor offers some inductance to the circuit. To make the situation worse, even the component leads, circuit wiring, and printed circuit board traces can become tiny antennas and radiate signals that interfere with the circuit performance. This means more headaches to you when troubleshooting an RF circuit. See Figure 9.1.

Special tools are required for testing RF circuits. Even touching the circuit can cause frequency changes.

1. Metal screwdrivers will upset the circuit. Therefore, special plastic alignment tools are used.

2. It is absolutely necessary to use RF probes with your DMM or VOM. These probes will rectify the RF information into a dc voltage. *Do not use meters without the RF probe.*

3. Oscilloscopes can be used for signal tracing. Even if the frequency exceeds the vertical response, the signal amplitude will decrease, but you will still be able to detect

How to test RF circuits

CAUTION
Do not use meters without the RF probe.

TABLE 9.1 RF Spectrum

30 Hz	300 Hz	3 kHz	30 kHz	300 kHz	3 MHz	30 MHz	300 MHz	3 GHz
Extremely low frequency	Voice frequency	Very-low frequency	Low frequency	Medium frequency	High frequency	Very-high frequency	Ultra-high frequency	Super-high frequency

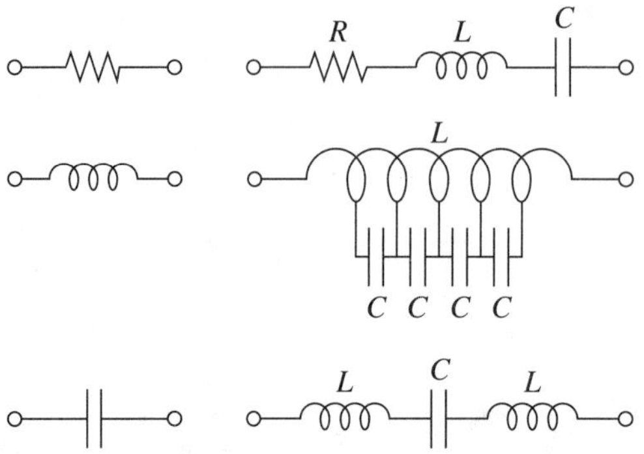

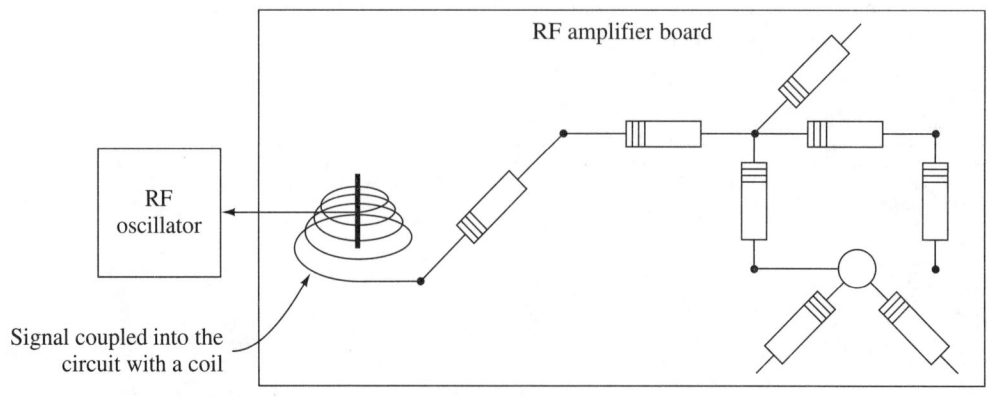

FIGURE 9.1 RF equivalent of discrete components.

the presence or absence of the signal. Be sure to check the compensation of the oscilloscope probes. It is best to use 10× probes rather than 1×. The 1× probe will pass on the scope's input capacitance to ground to the circuit under test. This capacitance will probably be in the picofarads range and may cause significant problems.

4. Signal generators can be used to inject a signal for tracing. Interface the generator with the circuit under test through a 52-Ω coaxial cable to prevent the signal from picking up any stray noise. It is also possible to couple the RF signal into the circuit inductively by using a coil, as shown in Figure 9.2.

Why do we need these radio frequencies? It is true that most of our communication signals originate as low-frequency or audio frequencies. The problem is that it becomes impossible to transmit these signal frequencies over distance for a number of reasons. One very important factor is the size of the antenna that would be required to do this.

Therefore, we have to find a way to use high frequencies for the transmission of the lower-frequency information. Then smaller antennas are practical.

The basic mathematical representation of a radio-frequency sine wave is

$$e_c = E_{max}\sin(\omega t + \theta) \tag{9.1}$$

where

e_c = the instantaneous voltage of the high-frequency carrier signal

E_{max} = the peak value of the signal

$\omega = 2 \times 3.14 \times f$

θ = the phase of the signal

FIGURE 9.2 Inductive coupling.

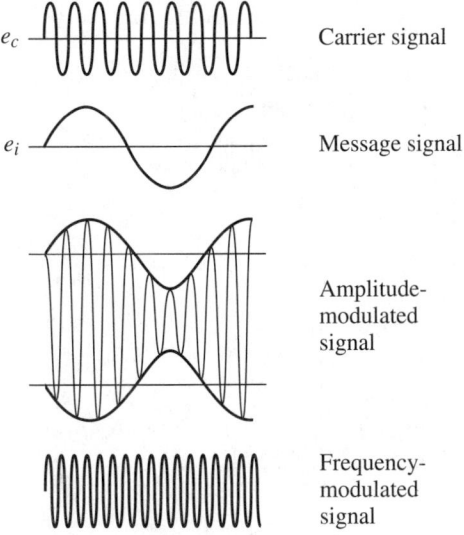

e_c — Carrier signal

e_i — Message signal

Amplitude-modulated signal

Frequency-modulated signal

FIGURE 9.3 Time-domain signals.

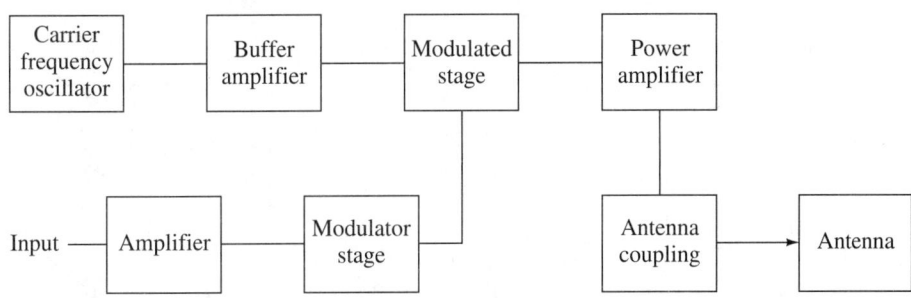

FIGURE 9.4 Typical transmitter block diagram.

How can this carrier signal be modified to indicate an information signal? There are three variable parameters in Equation (9.1): amplitude, frequency, and phase. By varying the amplitude of the carrier signal, we have amplitude modulation. By varying the frequency, we end up with frequency modulation; varying the last parameter, phase, gives phase modulation. Figure 9.3 is a time domain plot of the AM and FM wave depicting the carrier signal, message signal, and modulated carrier signals. Although transmitter circuits are not discussed in this text, block diagrams of typical AM and FM transmitters are shown in Figure 9.4.

9.2 THE AM MODULATION PROCESS

How do we impress a low-frequency signal onto the high-frequency signal? The two signals can be mixed linearly, as in Figure 9.5. However, the resultant signal is merely the algebraic summation of the two voltages over time. If you look at the individual components in the frequency domain, you will see that you really have not accomplished anything.

We can also use a nonlinear device to "mix" the two signals. As shown in Figure 9.6, the resultant equation produces many frequency components. The final waveform will have the following:

1. A dc voltage
2. A voltage at the original information frequency

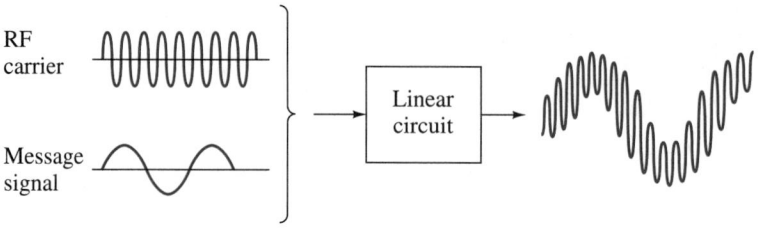

FIGURE 9.5 Linear mixing.

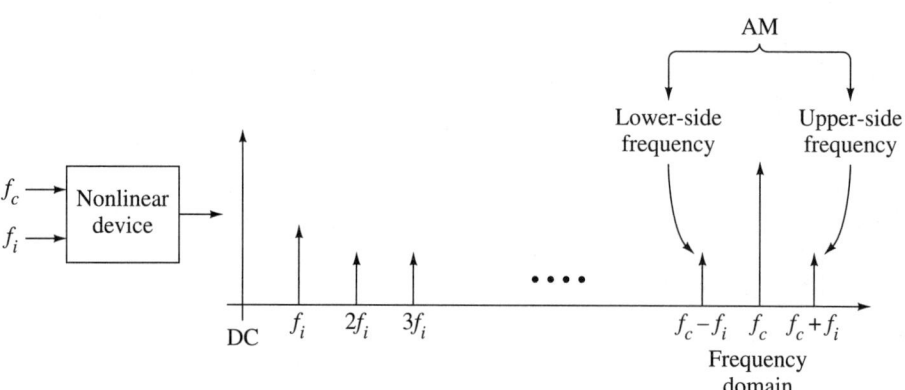

FIGURE 9.6 Creating the AM signal.

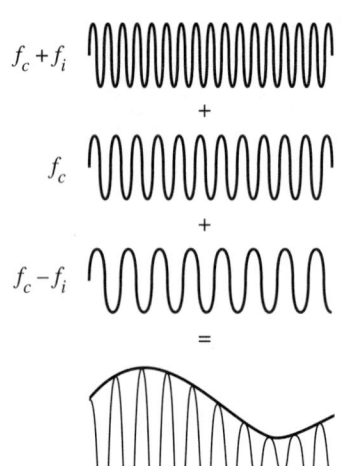

FIGURE 9.7 The transmitted signal.

3. Components of voltage at the harmonics of the information frequency
4. The carrier frequency
5. Components at harmonics of the carrier frequency
6. A voltage at a frequency equal to the carrier frequency minus the information frequency
7. A voltage at a frequency equal to the carrier frequency plus the information frequency

The components of frequency that become important to us are the ones listed in 4, 6, and 7. So, we will filter out the ones we do not want and transmit the ones we do want.

The transmitted signal, then, is the vector sum of all these voltages, as in Figure 9.7.

An important measurement of the amount of modulation is known as the percent modulation, or the modulation index, given by

$$M = \frac{E_i}{E_c} \tag{9.2}$$

If the level of the information signal is too strong, overmodulation will occur; the resultant AM signal will go to 0, causing distortion.

Various circuits are used as modulators. Figure 9.8 shows some simple ones.

The dual-trace oscilloscope can be used to display a pattern known as the trapezoidal

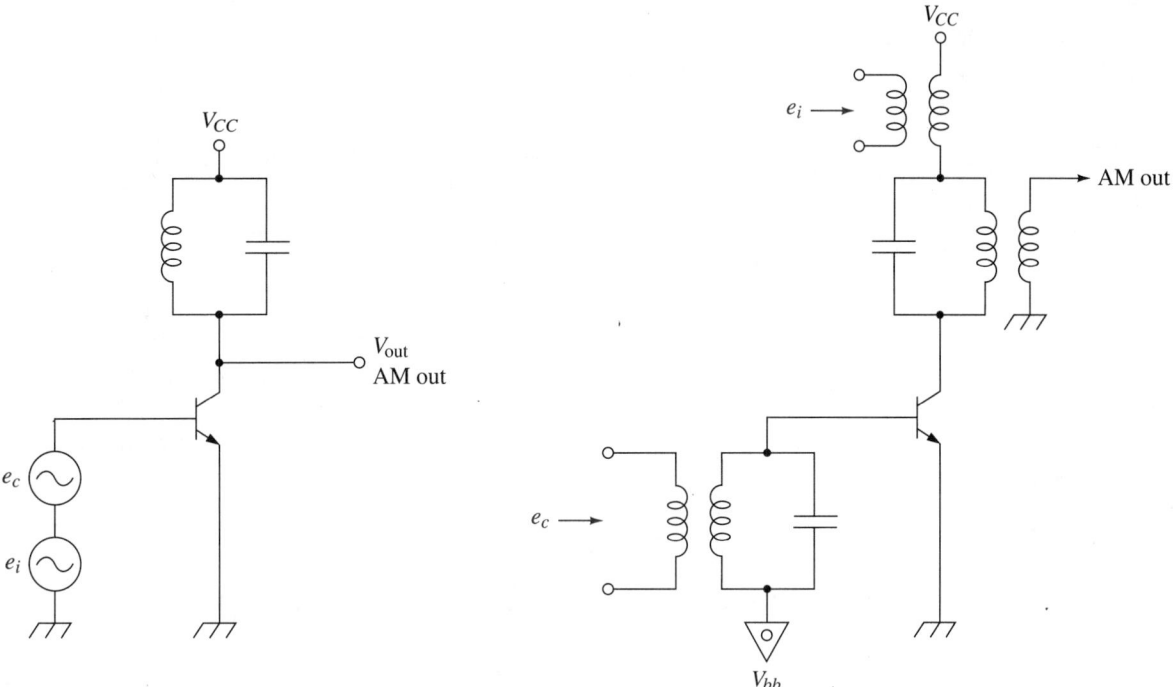

FIGURE 9.8 Typical modulators.

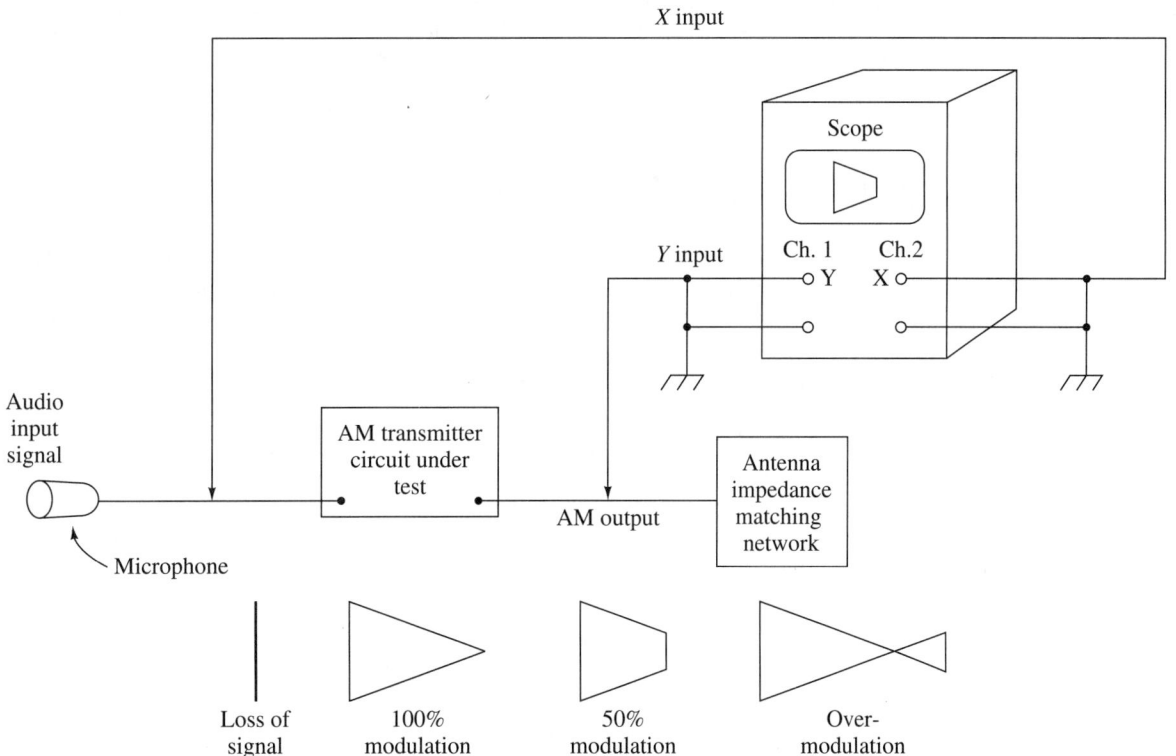

FIGURE 9.9 Trapezoidal pattern test setup.

pattern of the AM envelope. Figure 9.9 shows the proper connections of the oscilloscope needed to obtain the pattern.

The final package of signals is transmitted through space, intercepted by the receiver's antenna, and processed.

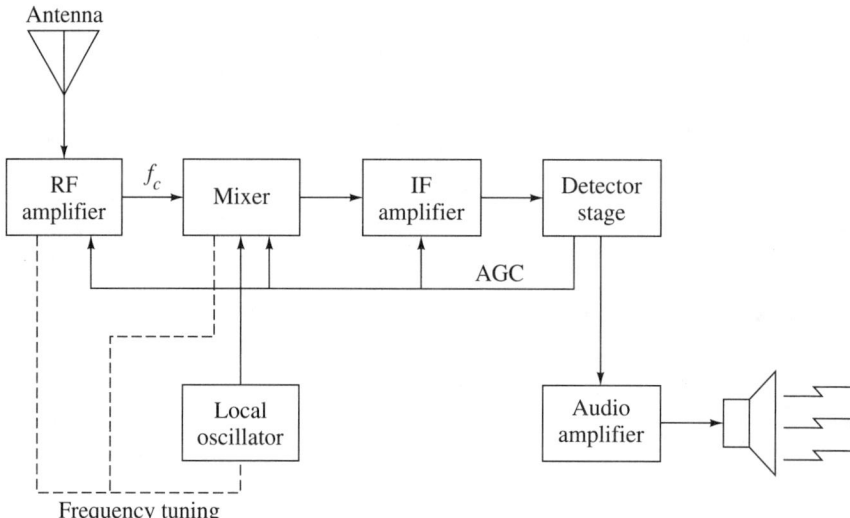

FIGURE 9.10 Typical superheterodyne receiver.

9.3 THE SUPERHETERODYNE AM RECEIVER

A block diagram of a typical AM superheterodyne receiver is shown in Figure 9.10. The electromagnetic wave traveling through space cuts the antenna and induces very small currents in it. The resulting minute voltage is amplified by the RF amplifier and fed to the mixer stage, where it combines with a stable signal from the local oscillator. The output of the mixer stage is filtered and sent to the IF amplifier. The IF amplifier sends an amplified signal to the detector stage, where the audio frequency is separated from the high frequency. Finally, the small audio signal is amplified and converted to sound by the speaker.

9.4 THE RF AMPLIFIER

The RF amplifier is usually an FET or MOSFET circuit like those in Figure 9.11. At this point in the receiver, the signal level is extremely small, and the amplifier must not inject

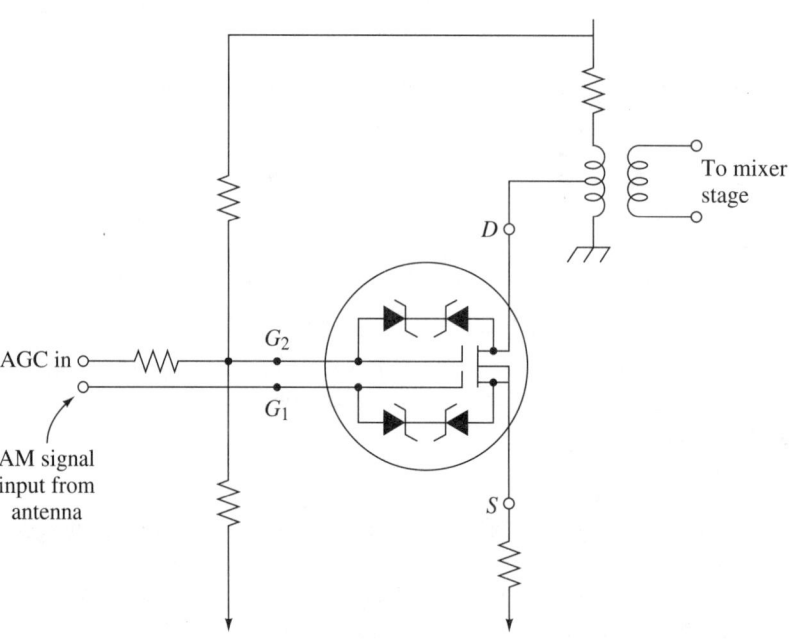

FIGURE 9.11 RF amplifier.

any noise. Some receivers have more than one RF stage. Usually the RF amplifier is a tuned-input, tuned-output amplifier with a voltage gain of about +10 dB. Transformer coupled and, in general, tuned circuits are not used in low-frequency amplifiers. It is important for you to be able to look at a circuit and identify the type of stage being used. In this case, due to the use of a ferrite rod antenna, you can assume small-signal operation. We know that the frequency range of the RF amplifier is 540 to 1650 kHz. The dc bias arrangement used is a combination of collector feedback and voltage-divider bias. Typically, this is a class A amplifier. If voltages are listed on the schematic, these are dc operating voltages with no ac signal applied. You can estimate the dc operating levels by drawing the dc equivalent circuit in Figure 9.12. To approximate the value of the gain of the amplifier, use the ac equivalent circuit shown in Figure 9.13. The gain is found by the equation

$$A = \frac{r_c}{r'_e + r_e} \tag{9.3}$$

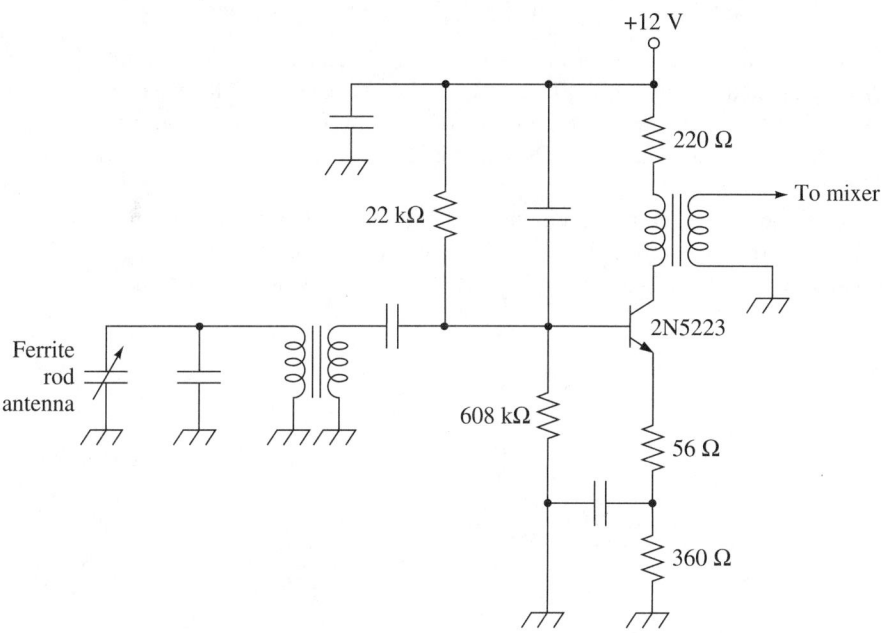

DC equivalent circuit

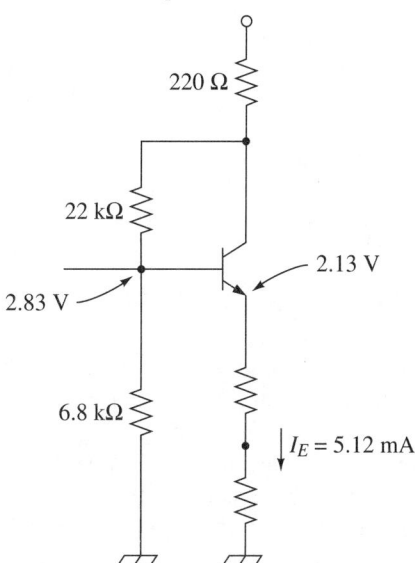

FIGURE 9.12 DC equivalent circuit.

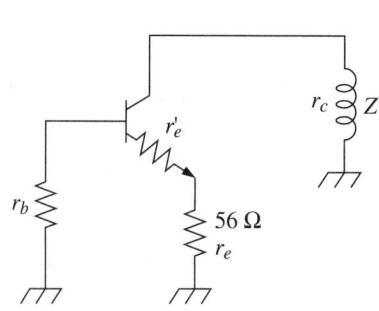

FIGURE 9.13 AC equivalent circuit.

9.5 TROUBLESHOOTING RF AMPLIFIERS

How to troubleshoot RF amplifiers

CAUTION

Do not operate a transformer-coupled stage with the secondary circuit open.

The ac resistance of the collector is determined by the reflected impedance into the primary of the transformer. This value will, therefore, determine the gain of the stage. The dc resistance of the primary is very small. Thus, the dc collector voltage should be about equal to the supply voltage level. This means that the ac output will actually rise above the supply voltage level. Do not operate a transformer-coupled stage with the secondary circuit open (Figure 9.14).

1. If the load is removed and the secondary circuit is open, the reflected impedance and the circuit gain will rise dramatically. The ac collector signal will, therefore, be very large. The large voltages can destroy the transformer and the transistor.

2. If the secondary circuit is shorted, the ac collector signal will go to 0.

3. You will find the dc voltage at the collector to be 0 V when the primary is opened.

4. The ac collector voltage will go to 0 if the primary is shorted.

Because the tuned transformer circuit is a parallel resonant tank circuit, its impedance is high at the resonant frequency. The incoming ac signal will have many frequencies varying around the resonant frequency of the tank circuit. Therefore, the amplifier's gain will also vary up and down. The result is an amplifier that amplifies a narrow band of frequencies around the resonant point. This band is known as the *bandwidth* of the circuit. The higher the Q of the circuit, the narrower the bandwidth. See Figure 9.15.

It is practical to use your oscilloscope to signal-trace through the receiver. However, be sure to use a low-capacitance probe with the scope to avoid circuit loading.

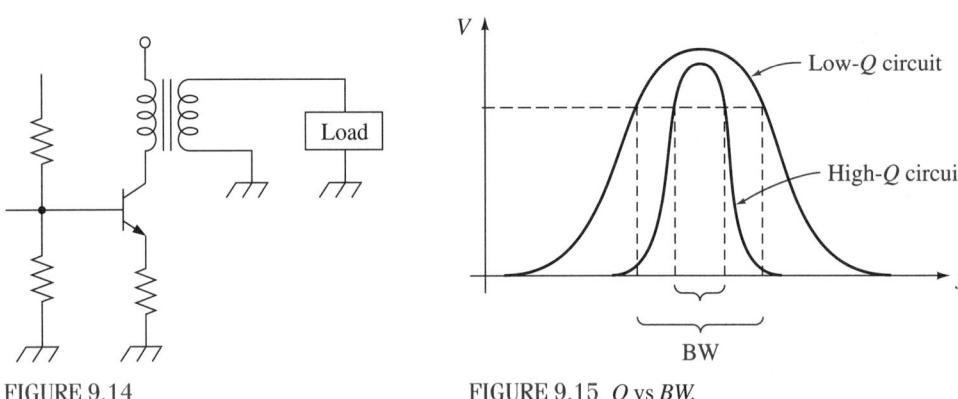

FIGURE 9.14   FIGURE 9.15 Q vs BW.

9.6 THE MIXER STAGE

The intermediate stages of the AM receiver use a frequency of 455 kHz. Therefore, the output of the RF amplifier must be mixed with another signal called the *local oscillator signal*. This mixing is done by combining the two in a nonlinear device that produces many sum and difference frequencies, as shown in Figure 9.16. Some typical mixer circuits are shown in Figure 9.17 and 9.18. These circuits are normally class C and deliver pulses of current to the tank circuit. The circuit in Figure 9.17 is normally found in conjunction with RF amplifier stages. The signal from the local oscillator is fed to the emitter of the mixer, whereas the output of the RF amplifier is applied to the base. Figure 9.17 also shows the dc and ac equivalent circuits. If no RF amplifier is used, the typical mixer stage has a circuit like Figure 9.18.

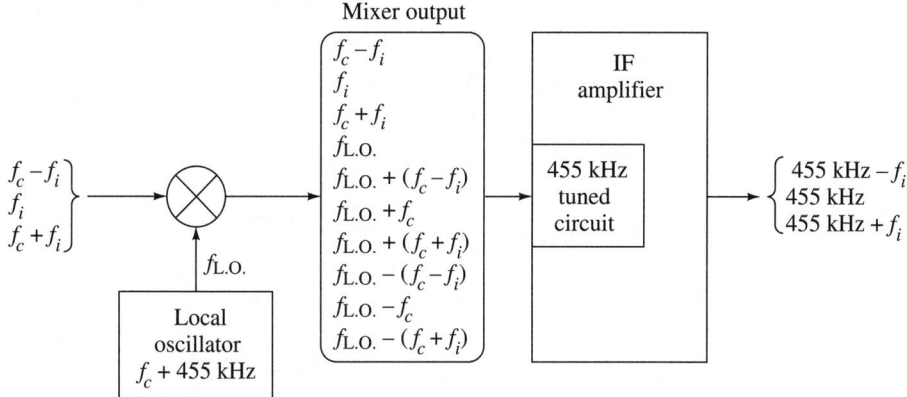

FIGURE 9.16 Mixer stage.

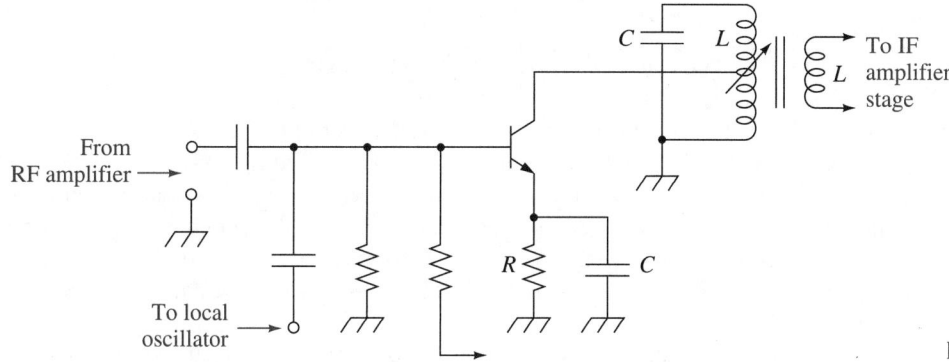

FIGURE 9.17 Typical mixer circuit.

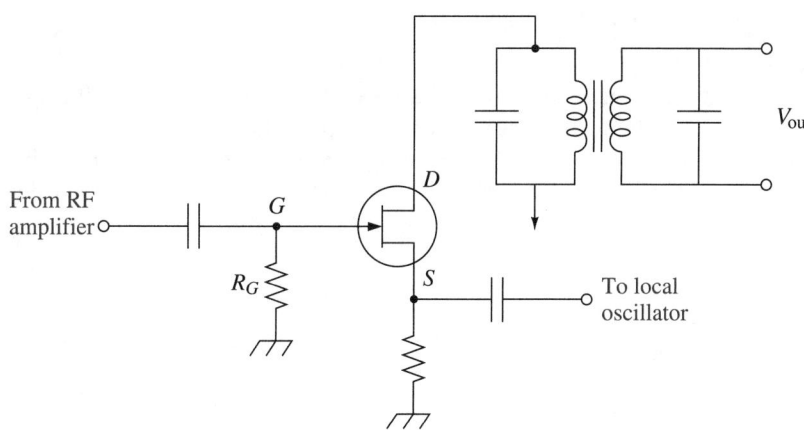

FIGURE 9.18 Typical mixer circuit.

9.7 TROUBLESHOOTING THE MIXER STAGE

If the mixer fails, there will be no audio and no input to the IF stage. Use your oscilloscope to

1. Check the input to the mixer stage.
2. Make sure that the local oscillator signal is present.
3. Check the dc bias voltages.
4. Recheck the resonant frequency of the tank circuit.
5. Perform static tests on the various circuit components.

Steps to take in troubleshooting
the mixer stage

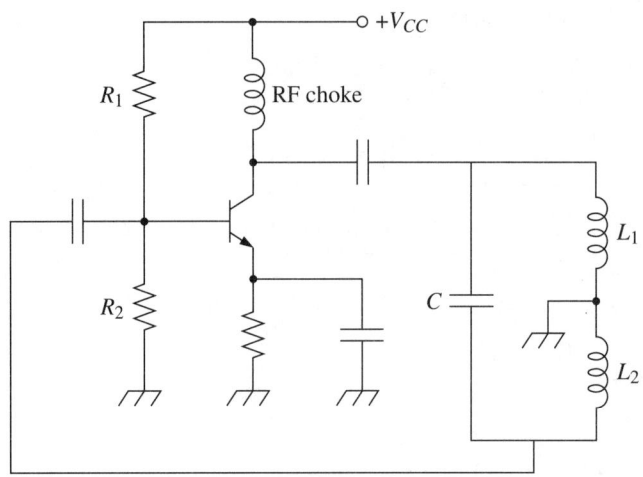

FIGURE 9.19 Local oscillator.

9.8 THE LOCAL OSCILLATOR STAGE

Generally the local oscillator stage is a sine-wave oscillator like the one in Figure 9.19. It must operate over the frequency range of 540 kHz + 455 kHz to 1650 kHz + 455 kHz. The Hartley oscillator is the most common. It usually operates class A. If the frequency is within the bandwidth of the oscilloscope, you can observe the output of the oscillator. Otherwise, you must measure the dc voltage level to determine if the circuit is operational. By using an RF probe with your oscilloscope, you should be able to monitor the RF signal at the coil winding.

If there is no signal, then try adjusting the oscillator and watch for any signs of oscillations.

9.9 THE IF AMPLIFIER STAGE

The frequency spectrum of the signal leaving the mixer stage is shown in Figure 9.20. As you can see, the front end of the IF stage must be a filter tuned to 455 kHz. The IF amplifier has a dual role. It must amplify the 455-kHz signal to a level suitable for detection and reject any other frequencies. This circuit must be very stable and have a level frequency

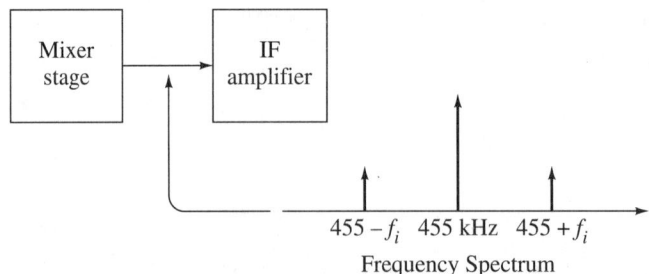

FIGURE 9.20 Frequency spectrum.

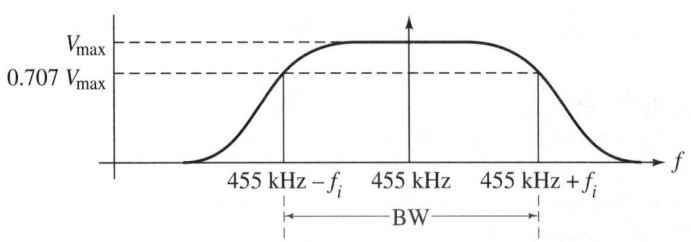

FIGURE 9.21 IF frequency response.

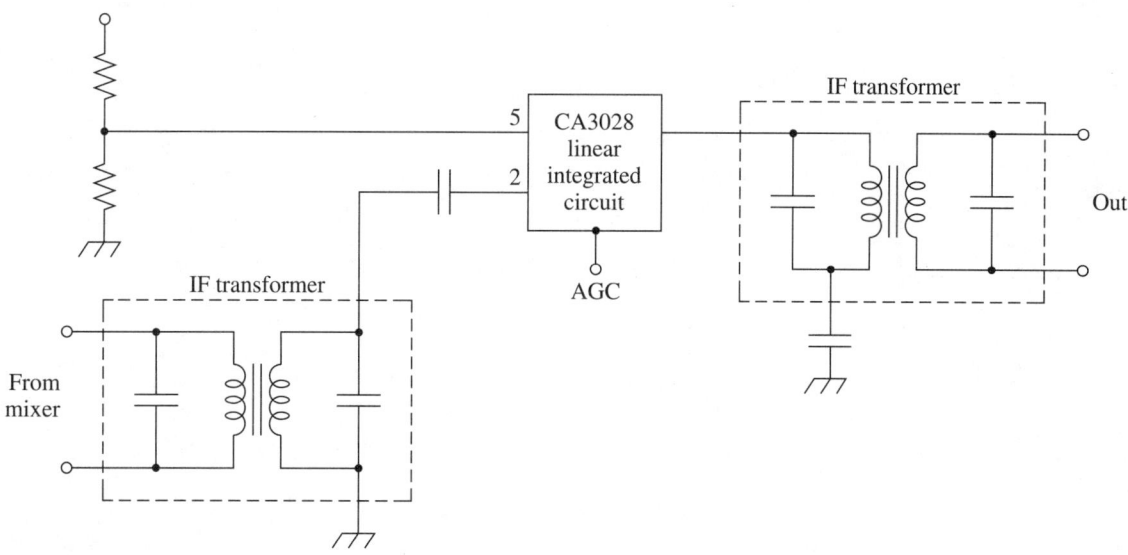

FIGURE 9.22 Typical IF amplifier.

response. The incoming IF signal contains sidebands that are 5 kHz above the carrier frequency and 5 kHz below it. These sidebands contain the information. A typical frequency response curve is shown in Figure 9.21. The IF amplifier is essentially the same as the RF amplifier except at a lower frequency. Figure 9.22 shows a typical IF amplifier circuit.

9.10 TROUBLESHOOTING THE IF AMPLIFIER STAGE

The basic procedures used for RF amplifiers apply to IF amplifiers. The typical problems might be no output, low output, distorted output, or an output at the wrong frequency.

Troubleshooting IF amplifiers

1. Check the output with your oscilloscope for indications of the problem.
2. Check the output of the mixer stage.
3. Perform static voltage tests on the circuit.
4. Check the frequency response using the procedures already covered.
5. Static-test the circuit components.

Distortion can be analyzed by waveform analysis. An overload condition will normally cause clipping of the signal. In this case the automatic gain control circuits are suspect.

9.11 THE AM DETECTOR STAGE

The detector (also called demodulator) is an application of the basic diode clipper circuit, as in Figure 9.23. It is essentially a half-wave rectifier followed by a low-pass filter. Since

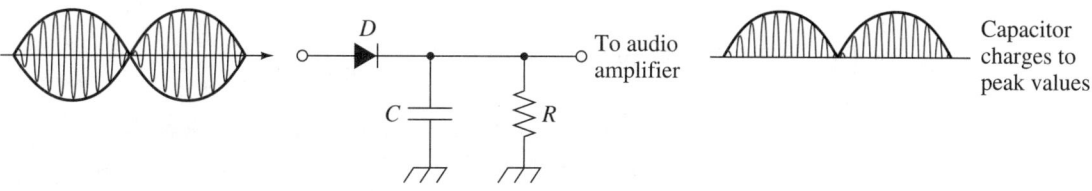

FIGURE 9.23 Detector.

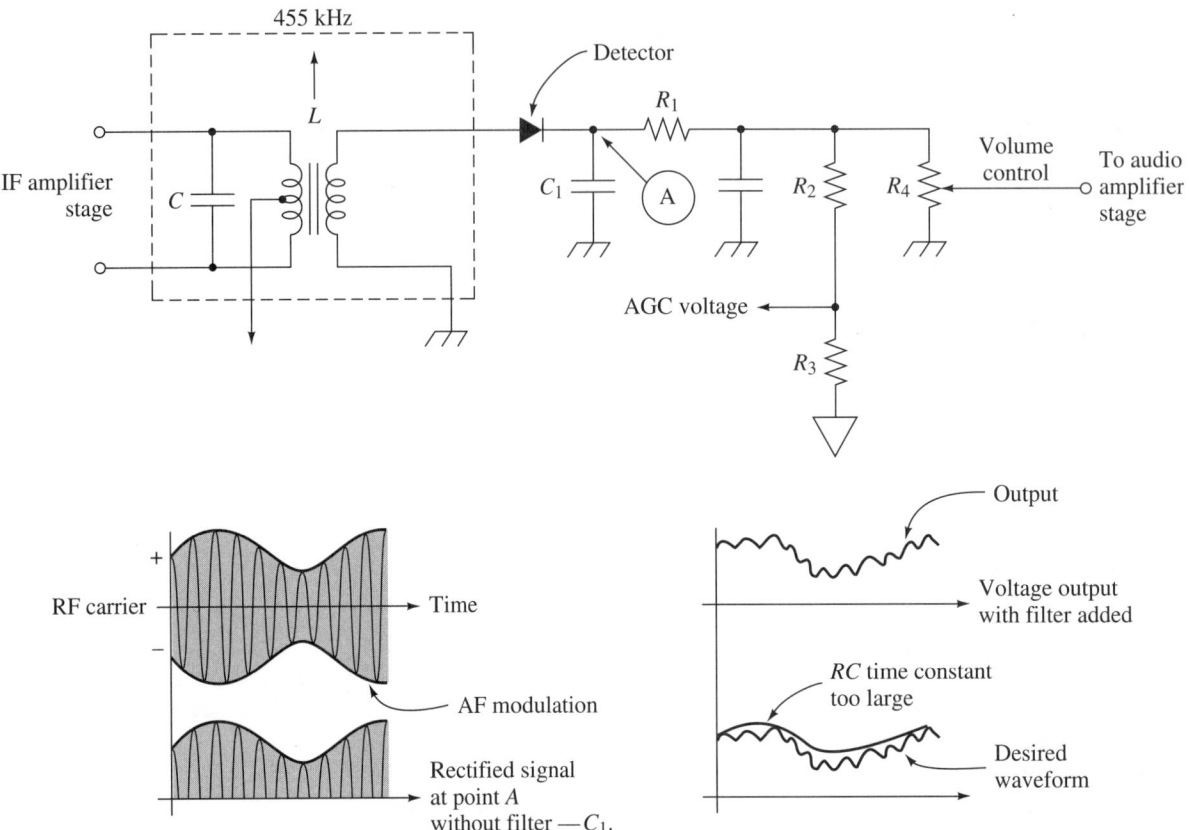

FIGURE 9.24 Effect of a time constant that is too long.

the signal from the IF stage is symmetrical about its axis, the diode can point either way. The filter removes the higher frequency and leaves the original audio signal, which is then sent to the audio amplifier. A dc voltage is picked from this circuit and sent back to the IF amplifiers to perform automatic gain control (AGC). With reference to Figure 9.24, a loss of high frequencies will result if the circuit time constant is too long compared to the signal fluctuations. This will happen if the values or R and C are too large or if an open-circuit condition exists. Be sure to note whether or not the diode is biased slightly on. If it is, then the proper polarity of the diode placement is critical. If the diode were placed in the circuit backward, clipping would occur, causing distortion.

9.12 AM RECEIVER TESTING

The most important specifications for the receiver are

1. Proper alignment of the RF and the IF stages
2. The sensitivity of the receiver
3. The selectivity of the receiver
4. The receiver's image rejection
5. The receiver's noise figure

9.13 ALIGNMENT

IF alignment is done by tuning the interstage coupling transformers for a maximum signal at the proper frequency. The following is a typical procedure:

1. Remove the local oscillator from the circuit in Figure 9.25.

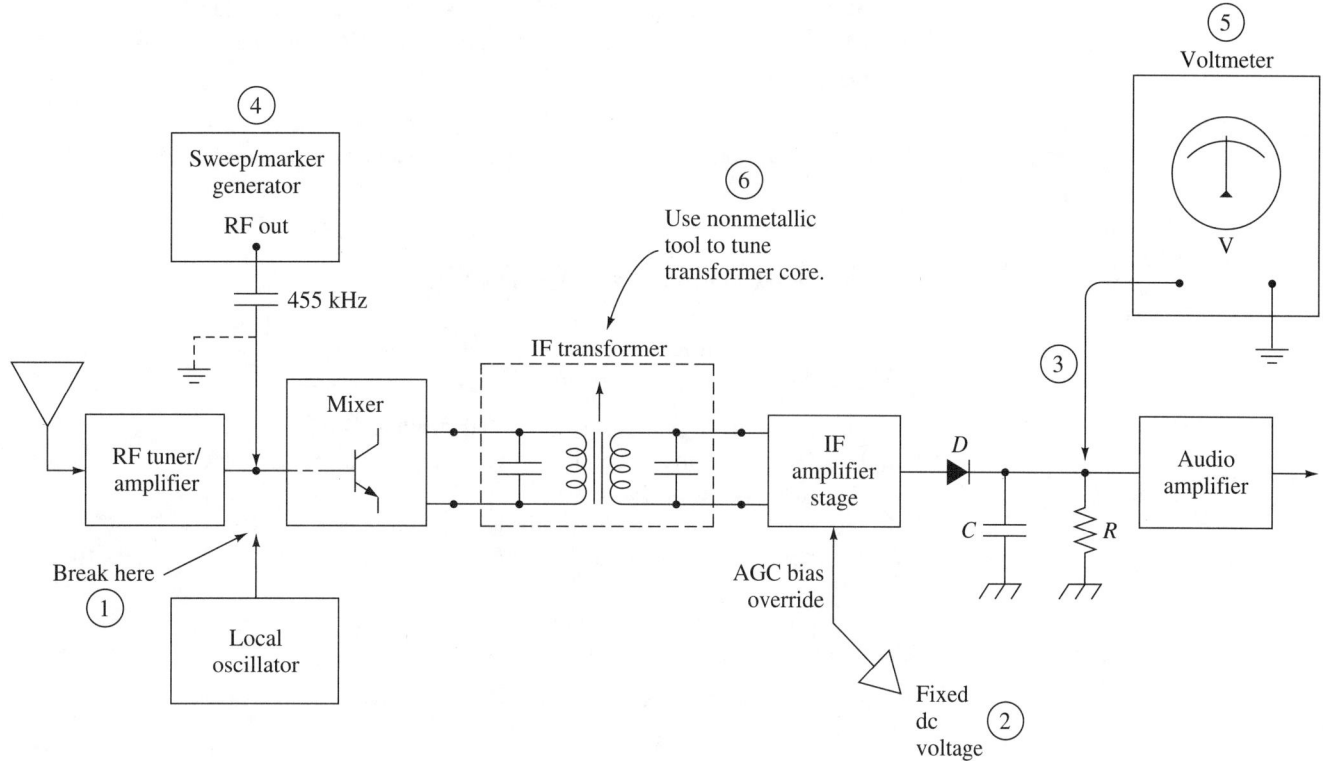

FIGURE 9.25 IF alignment.

2. Tie the AGC path to a fixed dc voltage.
3. Place your dc voltmeter across a load resistor on the output of the detector.
4. Apply a 455-kHz RF signal through a capacitor to the base of the mixer transistor.
5. Increase the output of the generator until the voltmeter indicates about 1.0 V.
6. Now, working back from the detector, use a nonmetallic tool and tune each transformer to obtain a maximum level on the dc voltmeter.
7. Enable the local oscillator and the AGC.

9.14 RECEIVER SENSITIVITY

The equipment required is shown in Figure 9.26. The sensitivity of a receiver defines the minimum input signal that will result in a signal-to-noise ratio of 20 dB at the detector. A typical technique is to use the quieting method of measurement.

1. Replace the speaker with a resistive load.
2. With the generator at 0, turn the volume controls to maximum.

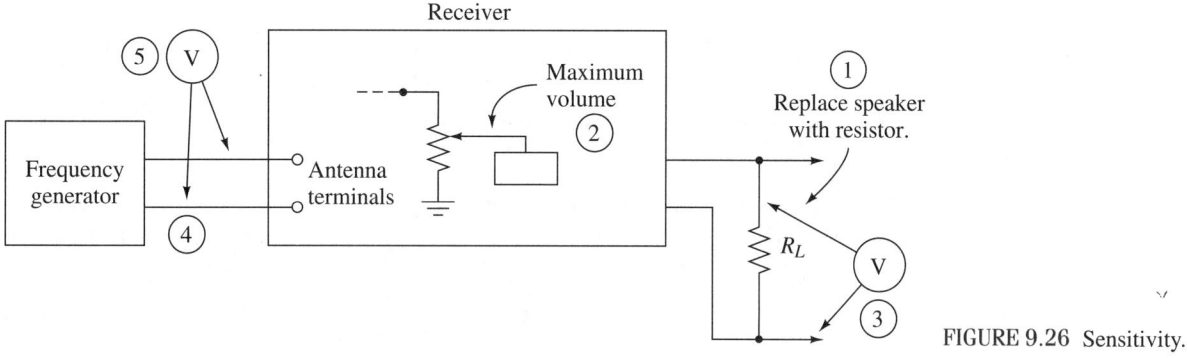

FIGURE 9.26 Sensitivity.

3. Measure the voltage across the load. This will be due to noise levels.
4. Now increase the output of the generator until the load voltage drops 20 dB. Use the equation

$$dB = 20 \log_{10}\left(\frac{V_{initial}}{V_{final}} \right) \qquad (9.4)$$

5. Read the output level of the generator. This value indicates the sensitivity of the receiver.

9.15 RECEIVER SELECTIVITY

The selectivity of a receiver defines its ability to reject any unwanted signal frequencies. The amplifier's IF stages contribute primarily to the selectivity.

1. Place your dc voltmeter across the detector load resistor, as shown in Figure 9.27.
2. Inject a signal at the IF frequency to the output of the mixer stage.
3. Disable the local oscillator.
4. Measure the voltage and label this the 0-dB reference.
5. Vary the frequency above and below the IF while noting the output voltage for each frequency.
6. Plot your data and find the upper and lower -6-dB points.
7. The frequency difference between each -6-dB point is the receiver bandwidth.

As the selectivity of the receiver increases, the bandwidth will decrease.

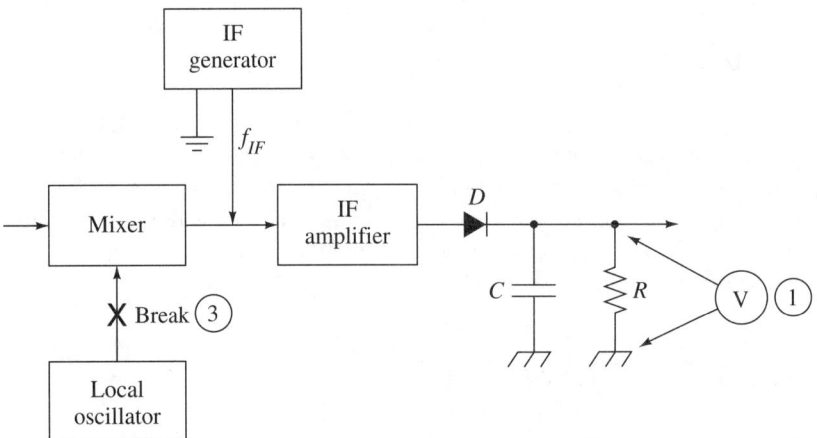

FIGURE 9.27 Selectivity.

9.16 RECEIVER IMAGE RESPONSE

For a given signal frequency, there will always be another that, when added to the local oscillator, gives a difference frequency equal to the IF frequency. The response can be measured as follows:

1. Set the signal generator and the receiver at a given frequency. Read the voltage across the detector load resistor. The generator output is V_1.

2. Tune the generator above and below the frequency given in step 1 until you hear the signal at the speaker.

3. Adjust the generator for the same voltage level across the detector resistor, as measured in step 1. The generator output is V_2. Calculate the image rejection in dB as follows:

$$dB = 20 \log_{10}\left(\frac{V_2}{V_1} \right) \qquad (9.5)$$

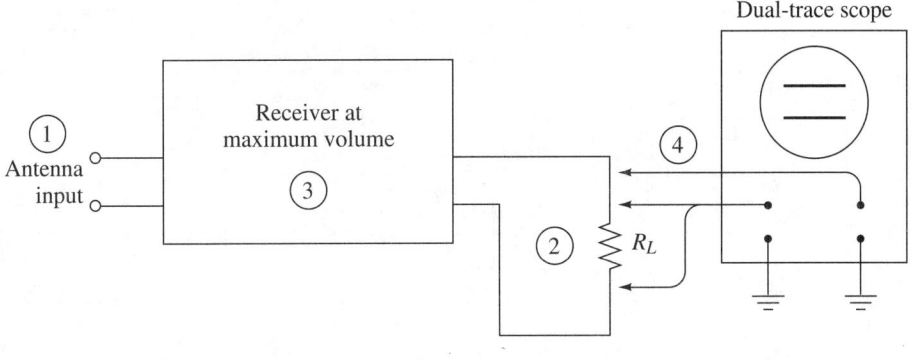

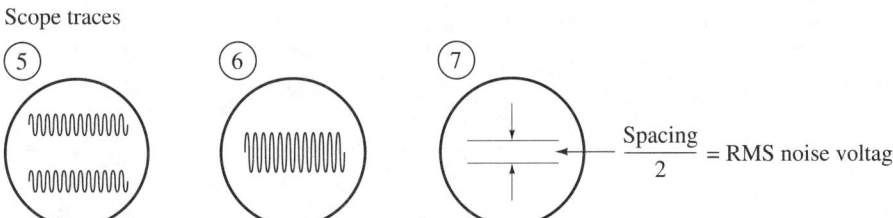

FIGURE 9.28 Noise figure.

9.17 RECEIVER NOISE FIGURE

By using a dual-trace oscilloscope, you can make a good measurement of the noise voltage being generated within the receiver, as shown in Figure 9.28.

1. Disable the antenna input.
2. Replace the speaker with a resistor.
3. Turn on the receiver at maximum volume.
4. Connect both channels across the load resistor.
5. With the scope input coupling on ac, view the noise voltage on each trace.
6. Now, move both traces together until they just begin to overlap.
7. Set the input coupling of the scope to ground and measure the spacing between the two ground traces.
8. The RMS noise voltage is equal to one-half of the value determined in step 7.

9.18 AM RECEIVER TROUBLESHOOTING

Figure 9.29 is a basic typical AM receiver schematic. Let's assume that there is no output. However, there is noise. Where would you start to troubleshoot this circuit?

Steps to take to troubleshoot an AM receiver

1. Inject an audio signal at the output of the detector. This effectively divides the circuit between the RF path and the audio path. Inject at test point *A*.

2. If you hear a tone, you have eliminated the audio section as a source of problems.

3. If you do not hear tone, you should troubleshoot the audio amplifier section, as we have already discussed.

4. Next split the RF and IF sections by injecting a 455-kHz modulated signal at the input to the first IF stage at test point *B*.

5. If you hear a tone, you have cleared everything from the first IF to the speaker.

6. If you do not hear tone, check the output of the first IF with your oscilloscope.

    a. A bad output means you must troubleshoot the first IF stage.

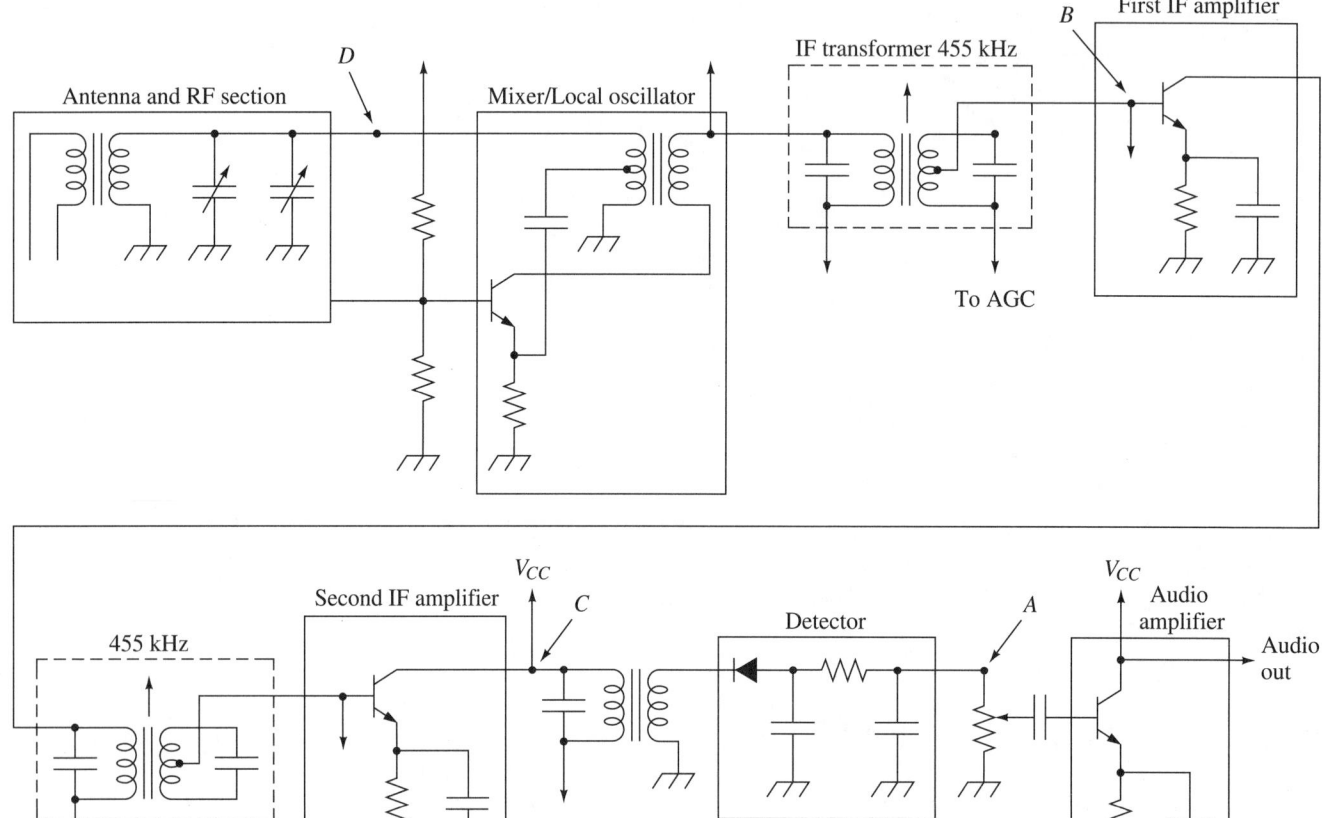

FIGURE 9.29 AM receiver partial schematic.

b. If the output is good, check the output of the second IF at point *C*.

c. If there is no output from the second IF, you must troubleshoot the stage.

7. If you have cleared the circuit from the first IF stage to the speaker, then you need to inject a modulated RF signal at the output of the RF amplifier. Test point *D*.

a. If you do not hear tone, check the mixer stage and the local oscillator stage.

b. A tone at the speaker means that the RF amplifier stage is faulty.

To summarize, signal injection and tracing consists of splitting the circuit in sections (as few as possible). We checked the audio sections first, then proceeded to the IF sections, and, finally, checked the RF section. By checking out the circuit in parts, you can rule out several stages at a time. Once you have localized the trouble to a stage, the next step is to isolate the stage and perform static tests to identify the faulty component or components.

9.19 THE FM MODULATION PROCESS

In FM modulation, the information signal is used to vary the frequency of the carrier oscillator. The voltage level of the intelligence determines how much the carrier frequency will vary, whereas the frequency of the intelligence signal determines how often the changes occur. Figure 9.30 is a time-domain display of the audio signal and associated FM signal. When no information signal is present, the carrier oscillates at its normal (*rest*) frequency.

The process of changing the carrier frequency back and forth around the rest frequency creates many sidebands around the carrier, as shown in Figure 9.31. The total of the voltages in the carrier frequency component and those in the sideband components will equal the voltage level of the carrier when it is at rest.

The amount of frequency change is known as the *deviation*. The amount of maximum deviation is set by the FCC and is ± 75 kHz for broadcast FM. This is referred to as 100% modulation. The percent modulation can be calculated as shown in Figure 9.32. The FM radio frequency band covers the frequencies from 88.0 MHz to 108.0 MHz. At 100% modulation, the carrier deviates ± 75 kHz, for a total swing of 150 kHz. The FCC allows guard bands of 25 kHz, so the total bandwidth of an FM channel is 200 kHz.

9.20   THE FM RECEIVER

The block diagram of a typical FM receiver in Figure 9.33 is very much like the AM receiver circuit discussed earlier. When you are looking at a receiver circuit to identify its purpose, the first place to look is the detector stage. The second is the IF stage. FM IF stages typically operate at 10.7 MHz.

The RF and IF stages of the FM circuit are very much like those in the AM circuit. Notice the limiter stage. This is found only in the FM receiver. Its purpose is to clip excessive levels and, therefore, cut off any noise spikes. After the limiter stage, you will find the detector stage, which is known as the *discriminator*, *ratio detector*, or *phase-*

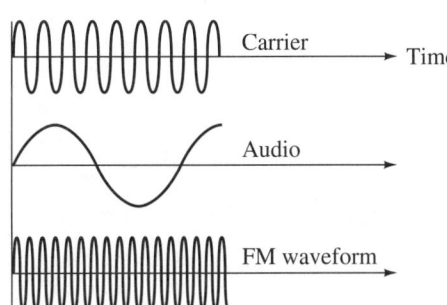

FIGURE 9.30   Time-domain display.

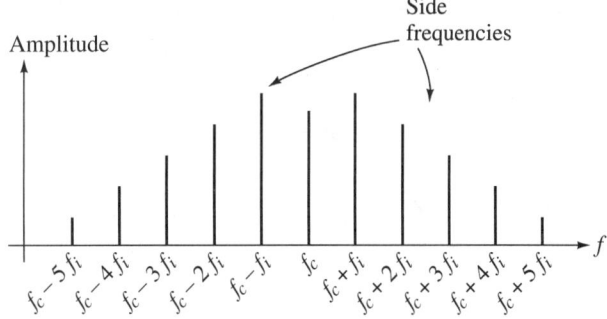

FIGURE 9.31   FM spectrum display.

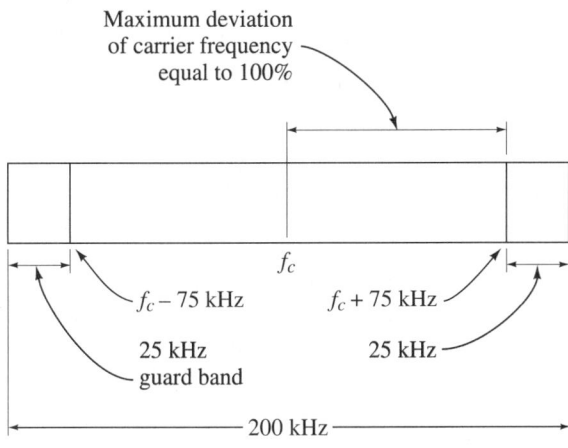

Actual deviation of $f_c = \Delta f$

$$\% \text{ modulation} = \frac{\Delta f}{75} (100)$$

FIGURE 9.32   FM channel.

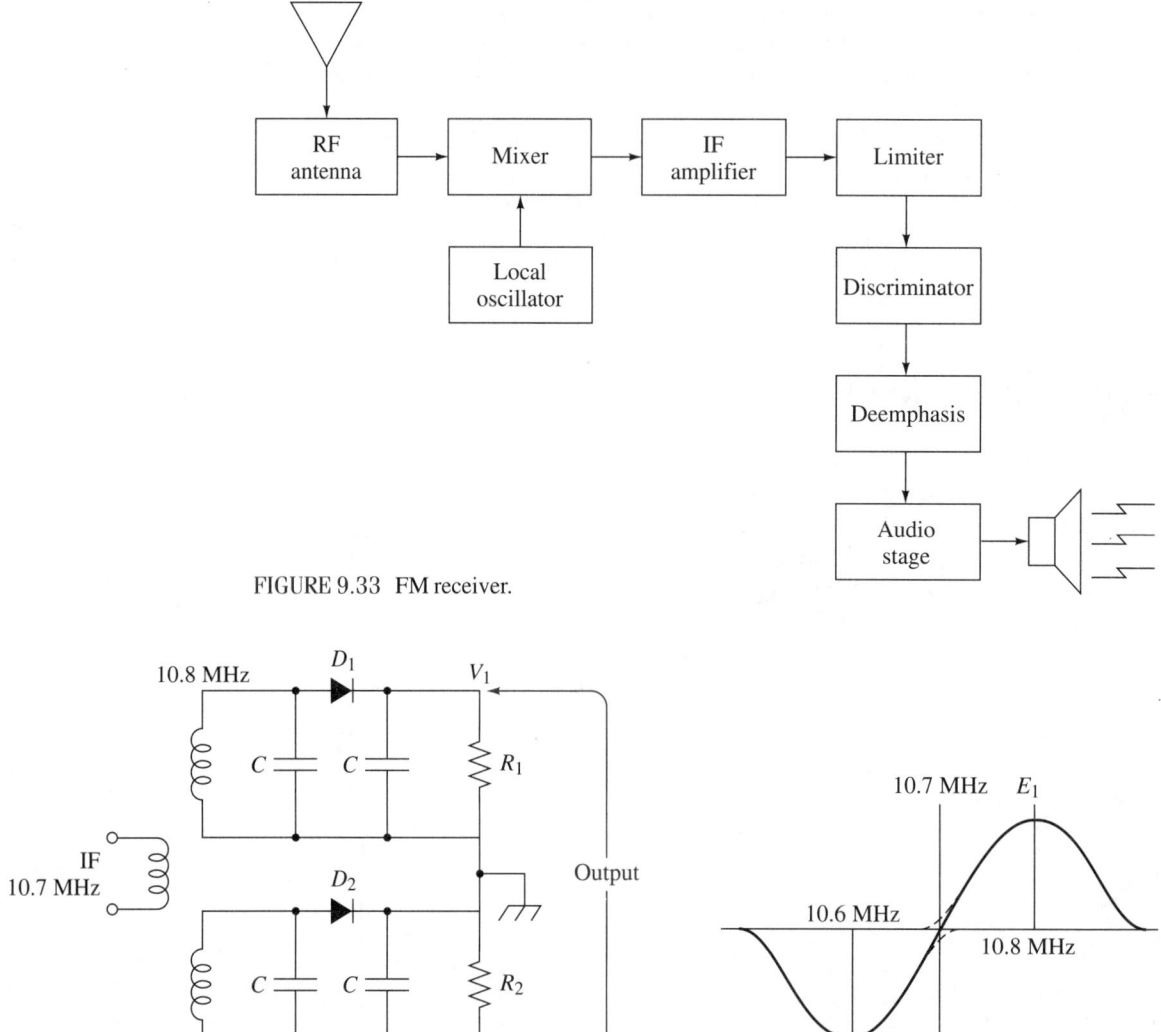

FIGURE 9.33 FM receiver.

FIGURE 9.34 Dual-slope detector.

FIGURE 9.35 Total response curve.

locked loop. Once the signal has been converted to audio, the audio amplifiers will perform the same functions as in the AM receiver.

9.21   THE DUAL-SLOPE DETECTOR

The circuit in Figure 9.34 is a combination of two slope detectors. Since the IF frequency is 10.7 MHz, each detector is tuned to the frequencies shown. When the carrier signal is at the rest frequency of 10.7 MHz, the output will be 0 because equal and opposite voltages will be developed across R_1 and R_2. The total response curve is shown in 9.35. Any positive shift in frequency causes E_1 to increase, and the output will ride up the curve. Any decrease in frequency causes a corresponding change down the S-curve. You can see why the curve is referred to as the S-curve.

9.22   THE DISCRIMINATOR

The voltages in Figure 9.36 can be expressed in vector form. The voltage across the diodes is the vector sum of V_1 and V_3, and V_2 and V_3. At the carrier rest frequency of

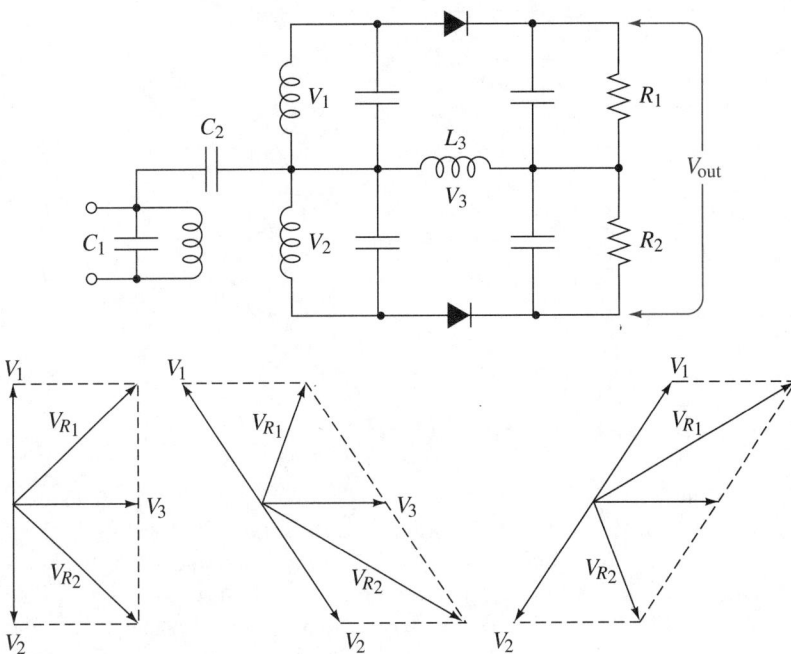

FIGURE 9.36 Discriminator.

10.7 MHz, the resultant voltage $V_0 = 0$. For frequencies above the carrier rest frequency, the voltage V_1 leads V_3 by an angle greater than 90°, and the total output voltage changes. The changing frequency produces a changing dc voltage output.

Note that a frequency change causes a corresponding change in the sum of the voltages and the voltage ratio. Note also that the diodes point in the same direction.

Another important point is that the discriminator requires a limiter stage to reject any AM due to noise.

9.23   THE RATIO DETECTOR

The ratio detector circuit of Figure 9.37 uses C_3 to perform the task of limiting the input signal. Therefore, no separate limiter stage is needed.

Note the direction of the diodes in the circuit. Since one of the diodes is reversed, the output is one-half the value of the output from the discriminator. The vector sum of the voltages remains constant: however, the ratio of the voltages will change with frequency shifts.

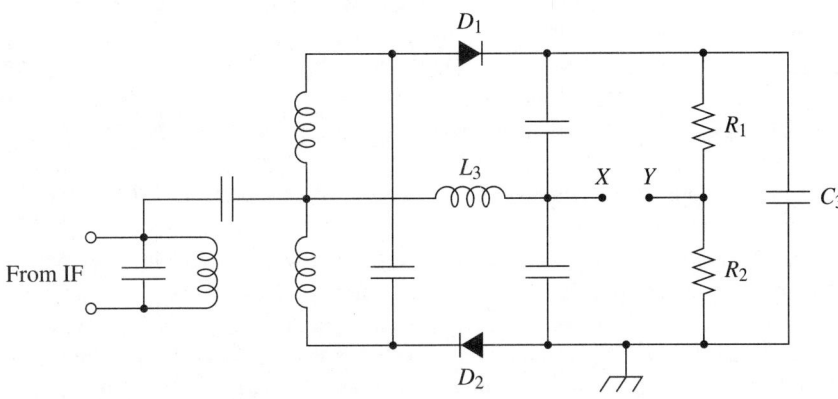

FIGURE 9.37 Ratio detector.

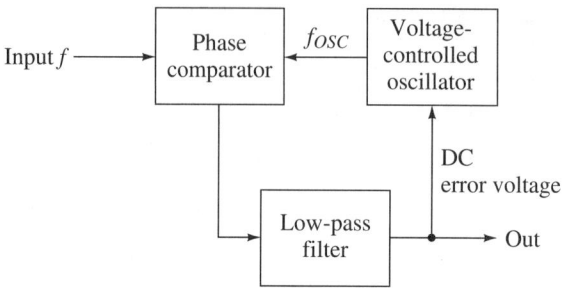

FIGURE 9.38 PLL block diagram.

9.24 THE PHASE-LOCKED LOOP

The typical block diagram of a phase-locked loop is given in Figure 9.38. You can see that it is a feedback system made up of comparator, oscillator, and filter circuits. The free-running frequency of the VCO is determined by external *RC* components. At the comparator, the frequency of the VCO is compared with that of the incoming signal. This incoming signal in the FM receiver is the 10.7-MHz signal from the IF stage. When the two inputs to the comparator are equal, the output is a dc voltage that will be $V_{CC}/2$ for a single-supply circuit or 0 V for a dual-supply circuit. This voltage is passed through the low-pass filter to remove the IF frequency. The resulting dc voltage is fed back to the VCO as a control signal. Any shift in the carrier frequency will produce a corresponding shift in this dc level, and the VCO output frequency will stabilize to lock with the input signal. Since a positive carrier shift results in a positive change to the dc voltage and a negative carrier shift gives a corresponding negative dc voltage, you can see that the dc voltage actually represents the information signal that originally created the frequency shifts in the transmitter.

When the VCO adjusts its frequency to be equal to the input frequency, the two are synchronized, or locked. As the input carrier frequency changes, the VCO responds accordingly. The VCO will remain locked (*lock range*) over a range of frequencies and then lose lock and return to its free-running frequency. After this occurs, the two frequencies must be brought close enough together to lock again. How close they need to be for lock to occur is known as the *capture range*.

9.25 FM RECEIVER ALIGNMENT

When the detector is a phase-locked loop circuit, there is no need for alignment.

Figure 9.39 shows how to connect a sweep generator to your receiver and use your oscilloscope to generate the S-curve. The generator must be capable of FM modulation and must also provide markers at 10.6, 10.7, and 10.8 MHz. Inject the generator output to the RF amplifier output terminals through a suitable capacitor, such as a 0.001-μF model. Connect the detector output to the vertical amplifier of the oscilloscope. Allow the FM generator to sweep ± 150 kHz around the center frequency of 10.7 MHz. Next, connect the sweep output of the generator to the horizontal input of the oscilloscope. The oscilloscope should be set in the X-Y mode. The familiar S-curve should be displayed on the oscilloscope. Finally, tune the detector's transformer coils to center the 10.7-MHz marker and obtain a linear curve.

If you do not have a sweep generator, you can still perform a good alignment with a simple RF generator and a voltmeter. Connect the output of your generator to the antenna terminal of the receiver and monitor the dc voltage at the output of the detector. With the generator set to the IF frequency, adjust the detector transformer for 0 V on the meter (Figure 9.40). RF and IF amplifier alignments are essentially the same as with the AM receiver.

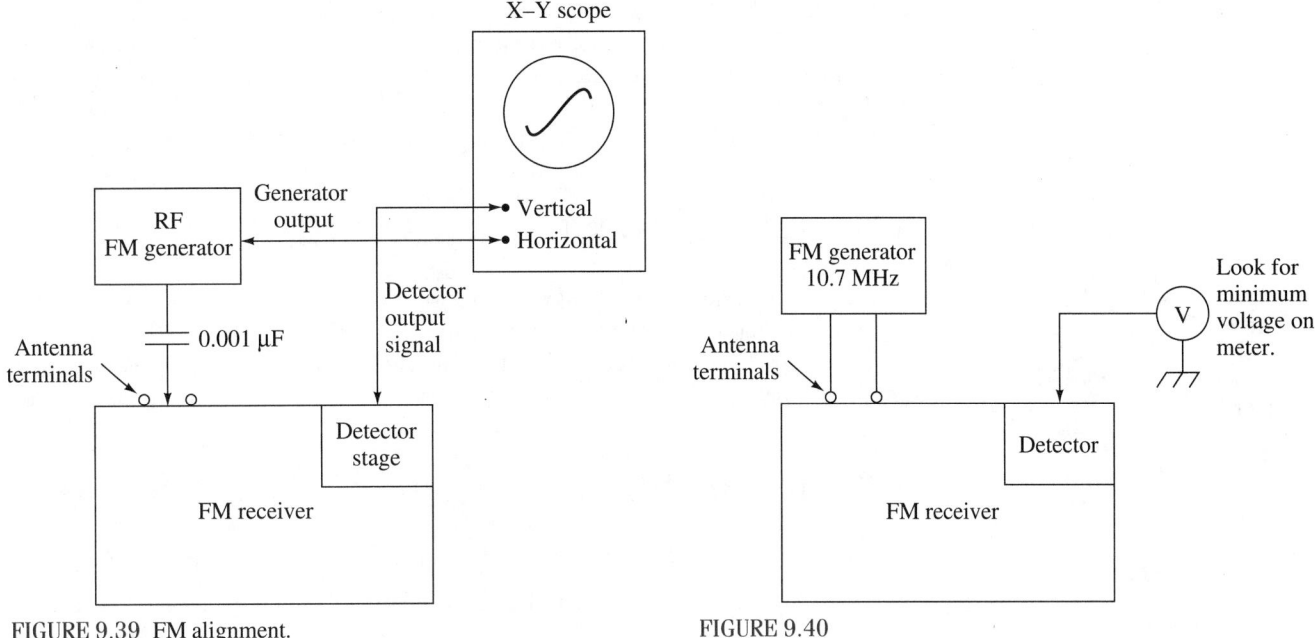

FIGURE 9.39 FM alignment.

FIGURE 9.40

9.26 FM RECEIVER SELECTIVITY

Use the same procedure as for the AM receiver. However, remember to use the correct dummy antenna. The FM receiver input impedance is either 300 Ω or 75 Ω.

9.27 FM RECEIVER SENSITIVITY

A quick, but not the most exact, method to check FM receiver sensitivity is the 20 dB of quieting method. As shown in Figure 9.41, an RF unmodulated signal is applied through an attenuator and dummy antenna to the antenna terminals of the receiver. Replace the speaker with its equivalent resistance. This will be your output location for the voltmeter. With no RF input, read the noise level indicated on the meter when the receiver is at max-

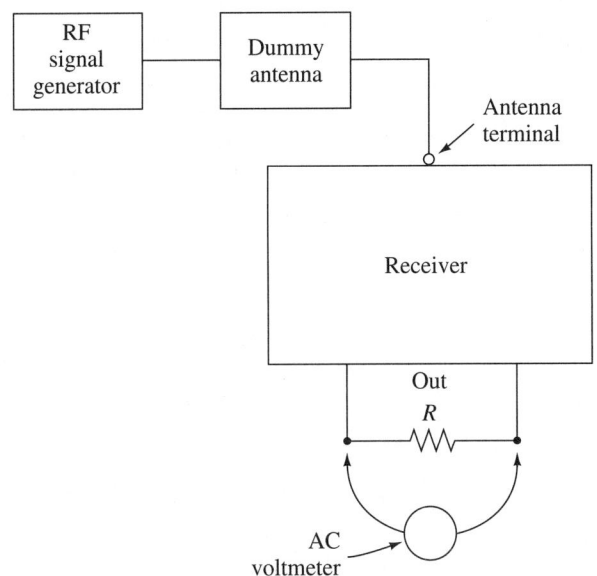

FIGURE 9.41 Receiver sensitivity.

imum volume. Now turn on the RF generator and tune the receiver to match its frequency. Increase the output of the generator until the meter indication is 20 dB down (one-tenth) from the first reading. Measure the RF signal level at the receiver terminals. This is the 20-dB quieting threshold value. This represents the minimum RF input signal required for the AGC to reduce the gain of the IF stages for noise reduction.

9.28 FM RECEIVER TROUBLESHOOTING

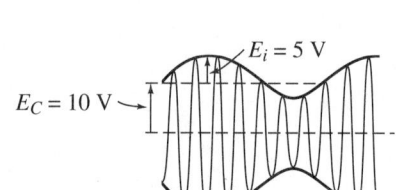

Basic methods used for FM receivers

The basic troubleshooting methods used for AM receivers are also used for FM receivers. However, the important differences to remember are the limiter stage, the detector stage, the input impedance, and the fact that the input capacitance of your oscilloscope probe can detune the circuits. Using an RF probe, you can trace a signal through the receiver.

9.29 PROBLEMS

1. What sine-wave parameters can be varied to represent intelligence?

2. Why is mixing two signals in a nonlinear device required for modulation?

3. What is the percent modulation of Figure 9.42?

4. What is the IF value for AM circuits?

5. You are troubleshooting your AM receiver and localize the trouble in the mixer stage. What are your next steps?

6. Explain the function of the AM detector circuit.

7. What is meant by receiver sensitivity and selectivity?

8. What is the S-curve?

9. Explain the operation of the PLL circuit.

10. Where will you find a ratio detector circuit?

11. A 1-MHz sine wave is amplitude modulated by a 4-kHz tune.

    a. Sketch the result and waveform in the time domain.
    b. Sketch the frequency domain display.

12. Calculate the percent modulation from the trapezoidal pattern in Figure 9.43.

13. What can you tell from the trapezoidal waveform in Figure 9.44?

14. Define the operation of a phase-locked-loop circuit.

15. Given the spectrum analyzer display in Figure 9.45, what is the modulating frequency?

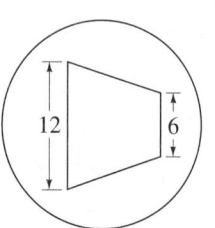

FIGURE 9.43 Problem 12.

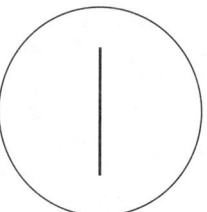

FIGURE 9.44 Problem 13.

FIGURE 9.42 Problem 3.

$E_i = 5$ V

$E_C = 10$ V

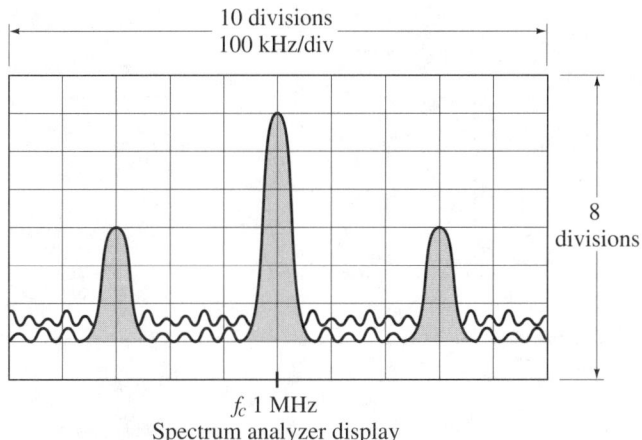

Spectrum analyzer display

FIGURE 9.45 Problem 15.

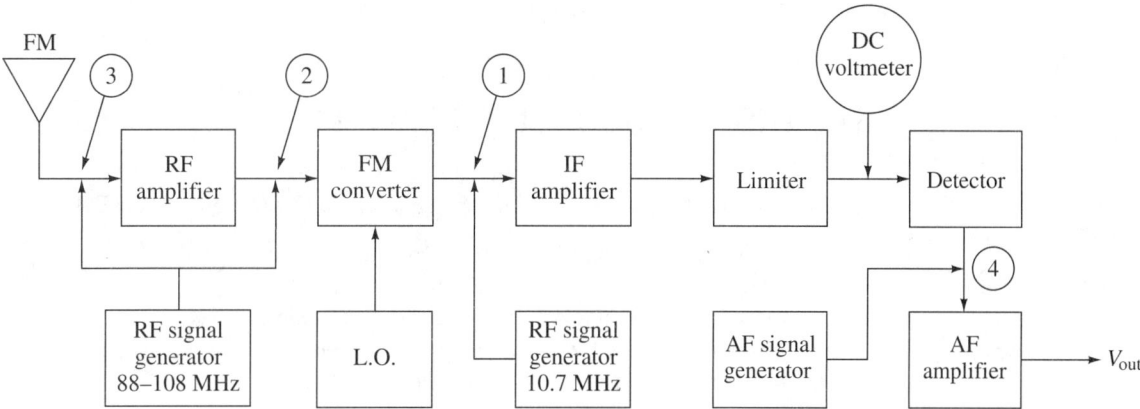

FIGURE 9.46 Troubleshooting Example 1.

9.30 TROUBLESHOOTING EXAMPLE 1

The FM receiver in Figure 9.46 has no output. Signed injection at points 1, 2, and 3 prove that the RF amplifier, the FM convention, and the IF amplifier are acceptable. Injection of an audio frequency signal at point 4 shows that the AF amplifier is good. It is determined that the detection stage is bad. The detector is a ratio detector, as shown in Figure 9.47.

Troubleshooting an FM receiver

The voltmeter is connected across the ratio-detector capacitor. Apply an IF signal at the limiter stage and look for a dc voltage. Vary the signal and watch for a corresponding positive and negative swing to the dc voltage. If this happens, the detector is all right.

1. A zero dc voltage across C indicates

    a. A bad transformer.
    b. A bad D_1 or D_2.
    c. C_1 is shorted.

Conduct ohmmeter checks to verify.

2. If the dc voltage does not swing through 0, this indicates

    a. A bad D_1 or D_2.
    b. A leaky C_1.
    c. Secondary Q transformer not aligned properly.

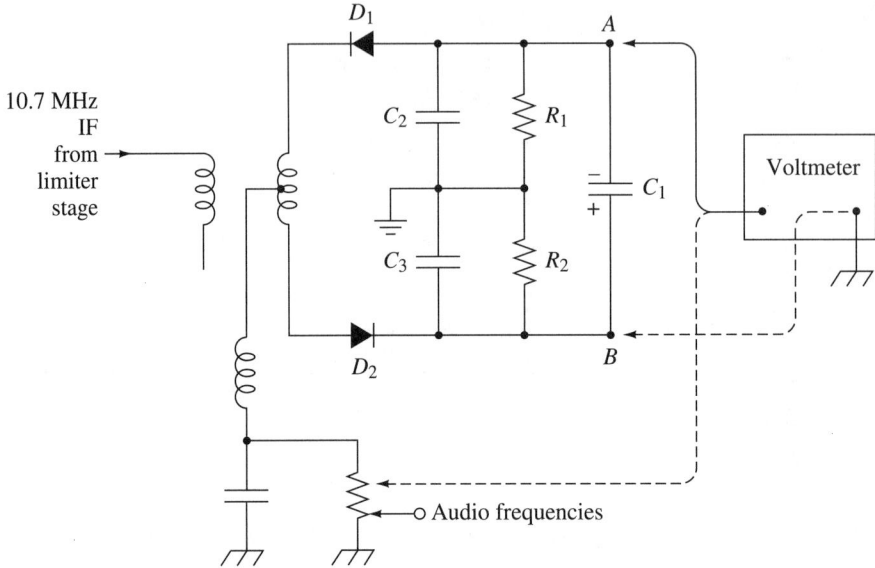

FIGURE 9.47 Troubleshooting Example 1.

9.31 TROUBLESHOOTING EXAMPLE 2

1. What kinds of signals would you inject at points 1 through 5 in Figure 9.48?

    a. Test point 1: RF amplitude-modulated signal in the range of 550 kHz to 1705 kHz.

    b. Test point 2: Same as A except larger in amplitude.

    c. Test point 3: 455-kHz signal amplitude modulated.

    d. Test point 4: Same as C except larger in amplitude.

    e. Test point 5: Audio frequency signal.

    2. Let's assume that no stations can be heard. There is, however, noise when the volume control is turned to maximum. Where will you begin troubleshooting?

    Step 1. Inject an audio signal at test point 5. This divides the receiver into RF and AF sections.

Steps to take when troubleshooting

    a. If no signal is heard at the output, the audio stages must be checked.

    b. If your signal is detected at the output, the audio circuits are all right. Of course, you probably guessed this because of the noise you heard in the speaker.

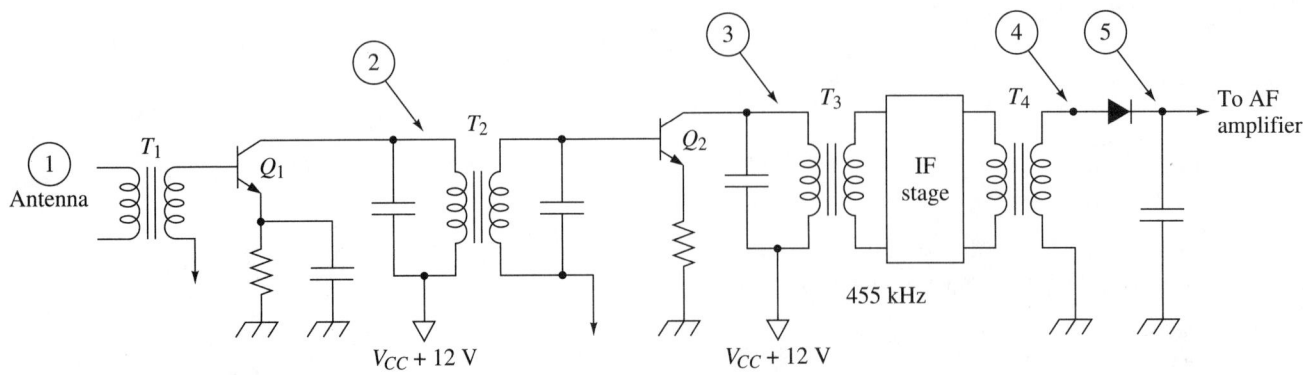

FIGURE 9.48 Troubleshooting Example 2.

Step 2.   Next, inject a 455-kHz modulated signal at test point 3. This splits the circuit in half.

  a. If you hear no signal, you have isolated the trouble to the IF stage.
  b. If you hear the signal, the fault is in the first two stages of the receiver.

Step 3.   Now, inject an RF modulated signal at test point 2. Be sure to tune the receiver to the test frequency.

  a. A good response indicates a faulty RF amplifier stage.
  b. No response means the trouble is in the converter stage of Q_2.

Step 4.   Now, perform voltage checks in the Q_2 circuitry. Be sure to use an RF probe with your scope and meter. If everything checks out, replace Q_2. Perform ohmic checks on all resistors. This should isolate the faulty component.

10

INTRODUCTION TO DIGITAL CIRCUIT TROUBLESHOOTING

10.1 INTRODUCTION

When information is transmitted in analog form, all the frequencies within the signal combine to produce a nonsinusoidal analog signal, as shown in Figure 10.1. These signals can be sent over telephone lines or radio links.

The same information can be transmitted in a digital system; however, the complex signal must be changed to some form of digital code because only two voltage levels are used in the transmission of digital signals. The digital signal in Figure 10.2 consists of only two logic levels, 0 and 1. Specific voltages are used to represent the 0 and the 1, depending on the digital device used.

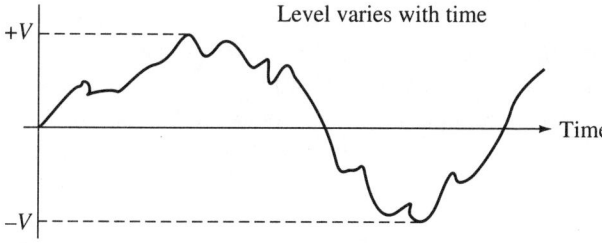

FIGURE 10.1 Analog signal.

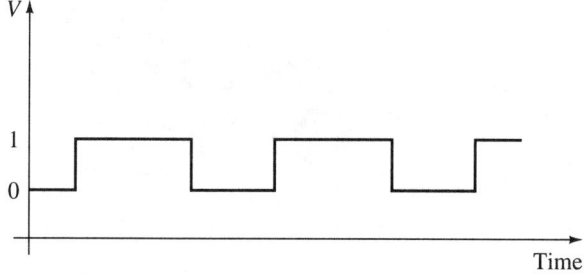

FIGURE 10.2 Digital signal.

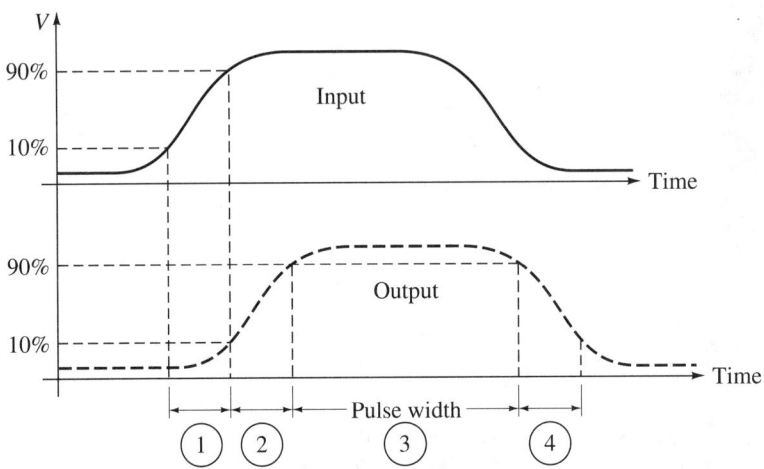

① Propagation delay: Time between the start of the input pulse and the start of the output pulse

② Rise time: Time between the 10% and 90% levels of the pulse

③ Pulse width

④ Fall time: Time between the 90% and the 10% levels

FIGURE 10.3 Pulse parameters.

There are some important parameters about the digital pulse that we need to review. When a pulse is transmitted through cable or other circuits, the output will be delayed in time, as shown in Figure 10.3. The total delay time can be critical if the frequency is high enough. This is known as the *propagation delay*.

When a digital circuit switches from a low (0) to a high (1) output, the transition is not immediate. The *rise time* is defined as the time interval between the 10% level and the 90% level. On the other end of the pulse, the time taken for the level to drop from 90% to the 10% value is called the *fall time*. The time interval between the two 90% points is defined as the *pulse width*.

10.2 DIGITAL INTEGRATED CIRCUIT CHARACTERISTICS

A digital *chip* is a piece of silicon (most common) material referred to as the *substrate,* on which resistors, diodes, and transistors have been fabricated. The chip is then enclosed in a plastic or ceramic package. Figure 10.4 shows the most common of these, the dual-in-line package (DIP). There is an identifying dot or notch to indicate the location of pin 1,

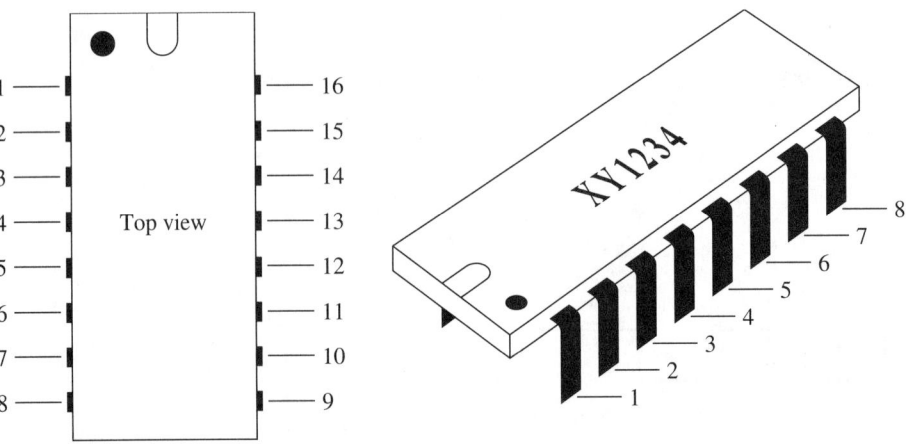

FIGURE 10.4 Dual-in-line package (DIP).

Pin assignments

and the remaining pins are numbered counterclockwise from this point. DIPs come in 14-, 16-, 20-, 24-, 28-, 40-, and 64-pin packages. The type of chip is determined by the number of logic gates on the substrate. The standard categories are summarized in Table 10.1.

The most popular types of ICs used in small-scale and medium-scale integration are made with bipolar junction transistors. These are known as TTL devices. Becoming increasingly more popular are CMOS (complementary metal-oxide semiconductor) devices. The older type, ECL (emitter-coupled logic), is not in common use today except in very high-frequency applications.

Each type (family) uses specific voltage levels to represent logic 1 and logic 0. These can be positive, negative, or both. Table 10.2 summarizes these levels for TTL, ECL, and CMOS. Notice that TTL (transistor-transistor logic) uses positive voltages only to represent the 1 and 0. There are various series within the TTL family. These are summarized in Table 10.3. Refer to Appendix C for more complete data. CMOS voltage levels depend on the power-supply voltage and have thresholds that are usually one-half of V_{CC}. The CMOS family also contains several series, which are summarized in Table 10.4. More complete information is found in Appendix D.

TABLE 10.1 Categories of DIPs

Number of Gates	Category
0–12	SSI — small-scale integration
12–99	MSI — medium-scale integration
100–9,999	LSI — large-scale integration
10,000–99,999	VLSI — very large-scale integration
>100,000	ULSI — ultra large-scale integration

TABLE 10.2 Device Voltage Levels

Device Family	Logic 1	Logic 0
TTL	Input >2 V Output >2.4 V	< 0.4 V
ECL	> 0.2 Typical: 0.4	< 0.2 Typical: −0.4
CMOS $\left(\begin{array}{c}\text{Assume}\\ V_{CC} = +5\text{ V}\end{array}\right)$	$5 < V < 3.5$ $\geq 70\%$ of V_{CC}	$2.5 < V < 0$ $\leq 30\%$ of V_{CC}

TABLE 10.3

TTL Logic Family	
Standard	74
High-speed	74H
Low-power	74L
Schottky	74S
Low-power Schottky	74LS
Advanced Schottky	74AS
Advanced low-power Schottky	74ALS

TABLE 10.4

CMOS Family	
Metal-gate	40 or 140
Metal-gate	74C
Silicon-gate high-speed	74HC
Silicon-gate high speed—TTL compatible	74HCT
The 74HC and 74HCT are pin-for-pin compatible with corresponding TTL 7400 series.	

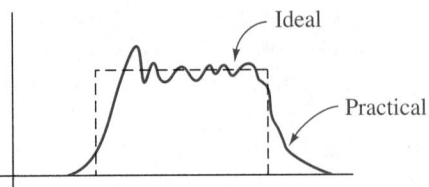

FIGURE 10.5 Digital waveform.

Digital signals are not usually nice square waves but normally have some ringing due to reactive elements or noise. A practical waveform is shown in Figure 10.5. The noise spikes can cause erroneous readings. The circuits could consider large transitions due to noise spikes to mean a logic 1.

If a TTL input is left unconnected (*floating*), it will act like a logic 1. The input may be susceptible to the noise voltages and should never be left unconnected. The floating dc level would be about 1.5 V.

The CMOS inputs should never be left floating because the device may overheat and be destroyed.

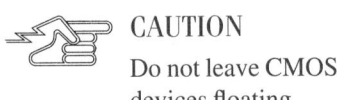 CAUTION
Do not leave CMOS devices floating.

10.3 FUNDAMENTAL OPERATIONS

Most logic functions are a combination of four basic operations: inverter, AND, OR, and exclusive-OR. Throughout this chapter the traditional symbology will be used. However, the new rectangular symbols introduced in 1984 under the IEEE/ANSI STD. 91–1984 are also included. A complete comparison can be found in Appendix A. In Figure 10.6, the small right triangle used with the rectangle symbol replaces the small bubble used with the older symbol to indicate a logic-level inversion. Additionally, the presence or absence of the rectangle signifies active-low or active-high inputs or outputs. The (1) inside the box designates one input. The presence of the right triangle indicates that the output will be active low when the input is active high.

The & symbol inside the AND box means that all inputs must be active high to give an active-high output.

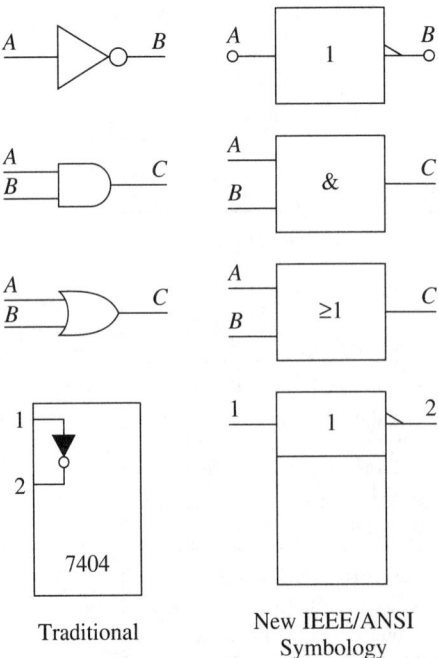

Traditional   New IEEE/ANSI Symbology

FIGURE 10.6 New IEEE/ANSI symbology.

The $>=$ symbol means that the output will be active high when one or more of the inputs are active high.

This new symbology is rapidly becoming more common. Manufacturers are now generating schematics with them and military contracts are requiring them. The new standard uses rectangular symbols for all devices. Integrated circuits that contain a large number of logic gates are represented by rectangular symbols, as shown in Figure 10.6.

10.4 INVERTER

The basic common-emitter transistor circuit in Figure 10.7 operates as an inverter. When the base is low, the collector is high. Therefore, if the base switches the transistor between saturation and cutoff, the output voltage will be either 0 or V_{CC}. The output is inverted (180° out of phase with the input). Figure 10.8 is the logic symbol and Table 10.5 is the truth table for the inverter. The truth table indicates all variations of input and output. The operation of various inverter circuits is summarized in Figures 10.9, 10.10, and 10.11.

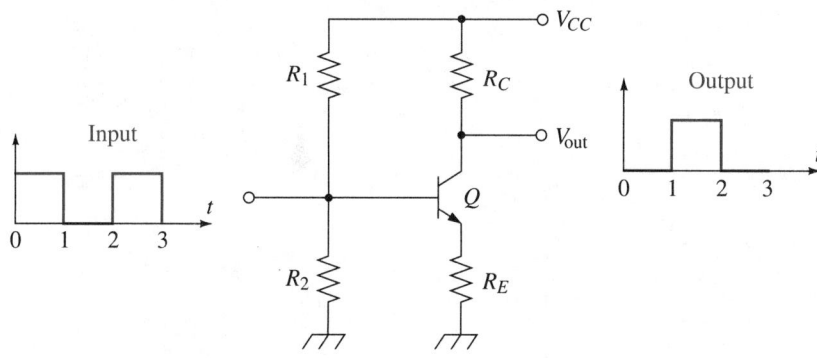

FIGURE 10.7

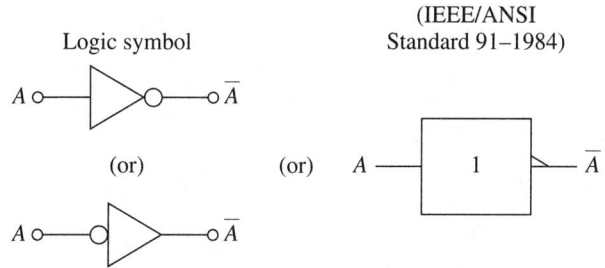

Logic symbol

(IEEE/ANSI Standard 91–1984)

TABLE 10.5 **Truth Table**

In	Out
0	1
1	0

Boolean expression

$A = \overline{A}$

Electronic schematic

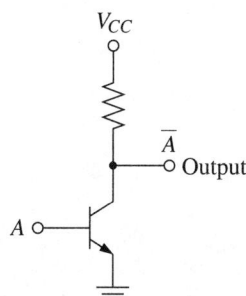

FIGURE 10.8 Inverter gate.

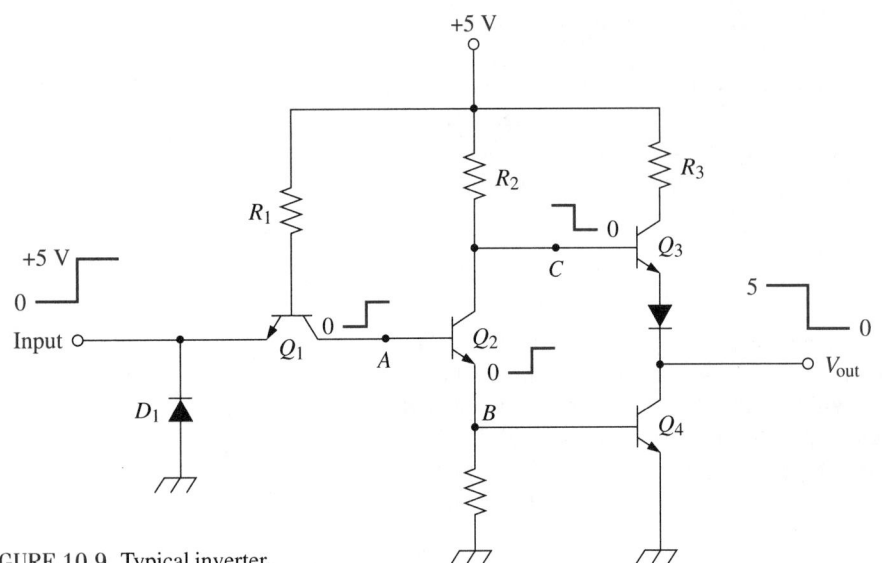

FIGURE 10.9 Typical inverter.

Totem-pole output Q_3 and Q_4. Q_2 is the driver circuit. The input circuit must be able to sink the emitter current of Q_1.

When the emitter of Q_1 is grounded, Q_1 is forward-biased. The load on the collector of Q_1 is R_2 and the base-collector junction of Q_2.

Point A goes low, cutting off Q_2. Point B goes low and C goes high. Q_3 turns on; Q_4 turns off. V_{out} equals +5 V.

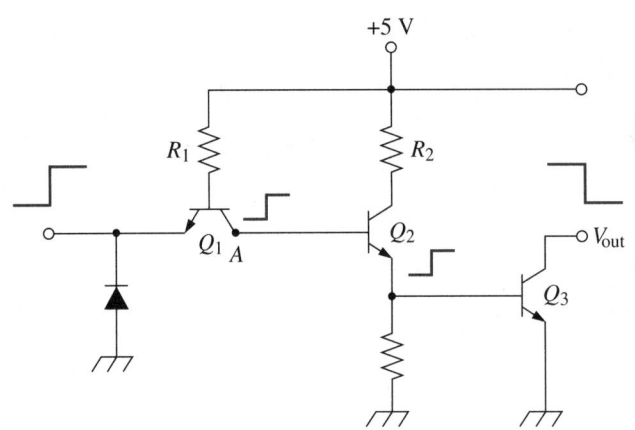

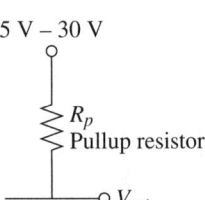

FIGURE 10.10 TTL inverter.

A pullup resistor is required. This allows direct connection with lamps, LEDs, diodes, relay coils, and non-TTL circuits.

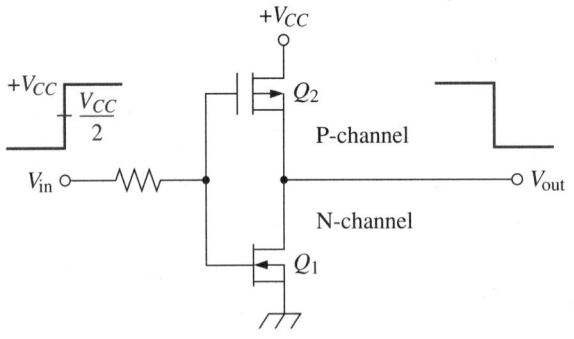

FIGURE 10.11 CMOS inverter.

V_{CC} range can be 4.5 V to 18 V.

When V_{in} is low: Q_1 is off. Q_2 is on. V_{out} is high.

Device will not change state until V_{in} passes $\dfrac{V_{CC}}{2}$.

10.5 THE AND GATE

The discrete-circuit AND gate, logic symbol, and truth table are shown in Figure 10.12 and Table 10.6. You can see that the output will be high only when all the inputs are high. When the AND gate has any 0s at the inputs and a 1 on the output terminal, the gate becomes a NAND gate, as shown in Figure 10.13 and Table 10.7.

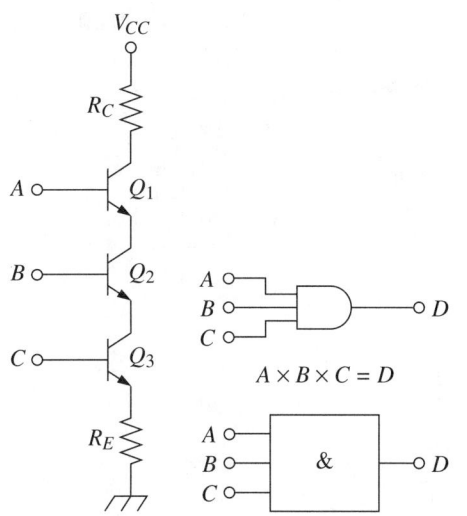

TABLE 10.6 Truth Table

A	B	C	D
0	0	0	0
0	0	1	0
0	1	0	0
0	1	1	0
1	0	0	0
1	0	1	0
1	1	0	0
1	1	1	1

Discrete circuit   Logic symbols

FIGURE 10.12 AND gate.

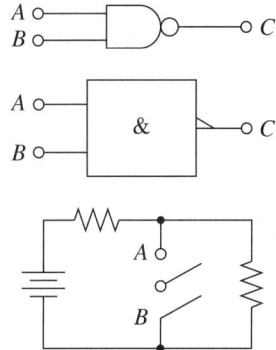

TABLE 10.7 Truth Table

A	B	C
0	0	1
0	1	1
1	0	1
1	1	0

FIGURE 10.13 NAND gate.

Logic symbol

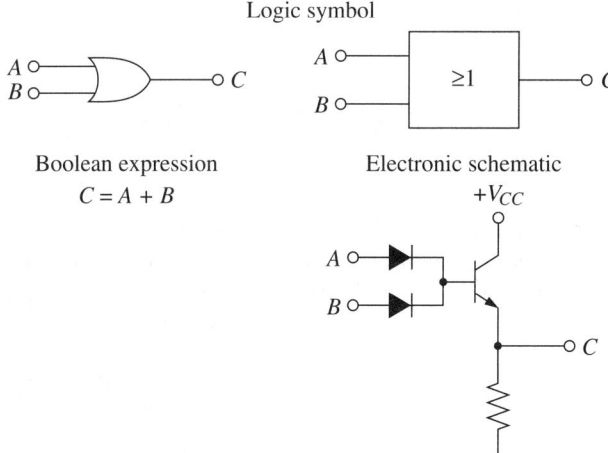

Boolean expression
$C = A + B$

Electronic schematic

TABLE 10.8 Truth Table

A	B	C
0	0	0
0	1	1
1	0	1
1	1	1

FIGURE 10.14 OR gate.

10.6 THE OR GATE

The OR gate (Figure 10.14) has a high output when any one of the inputs is high. See Table 10.8. The corresponding NOR circuit is shown in Figure 10.15. See Table 10.9.

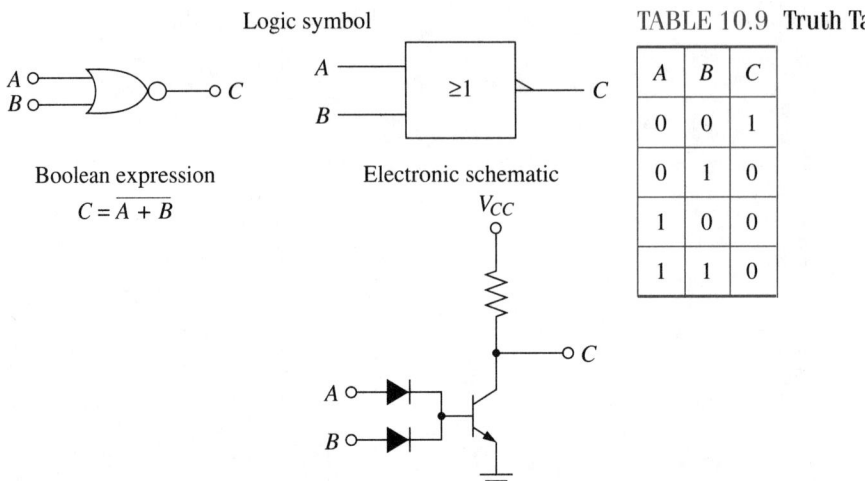

Logic symbol

Boolean expression
$$C = \overline{A + B}$$

Electronic schematic

FIGURE 10.15 NOR gate.

TABLE 10.9 Truth Table

A	B	C
0	0	1
0	1	0
1	0	0
1	1	0

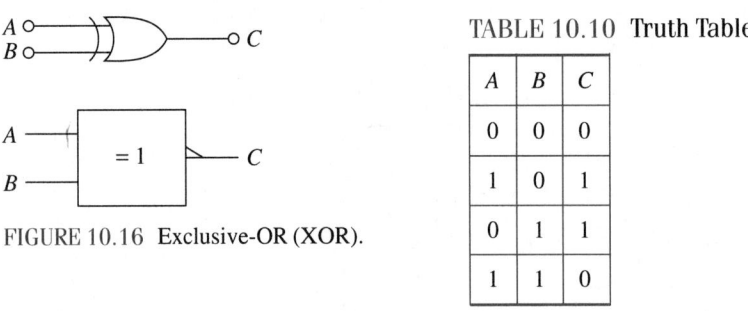

FIGURE 10.16 Exclusive-OR (XOR).

TABLE 10.10 Truth Table

A	B	C
0	0	0
1	0	1
0	1	1
1	1	0

10.7 THE EXCLUSIVE-OR GATE

The exclusive-OR gate is shown in Figure 10.16. This circuit has a high output only when the inputs are different. The truth table is given in Table 10.10.

10.8 STORAGE DEVICES—ASYNCHRONOUS *RS* FLIP-FLOP

The discrete-device circuit is shown in Figure 10.17. Here is how it works.

The circuit stores a 0 on the Q output and a 1 on the NOT-Q output. The logic symbol and truth table are given in Figure 10.18 and Table 10.11. The circuit stores a single binary digit. An input with a bar over it is active only when the input is low. This is termed

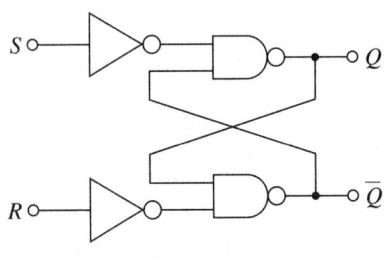

FIGURE 10.17 Asynchronous flip-flop.

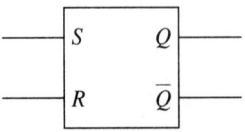

Logic symbol
FIGURE 10.18 Logic symbol.

TABLE 10.11 Truth Table

In		Out	
S	R	Q	$\overline{Q}$
0	0	No change	
1	0	1	0
0	1	0	1
1	1	Not allowed	

active low. See Figure 10.19 and Table 10.12. For an active-high device, as in Figure 10.20, if S is high, the Q output stores a high (1) level, and if R is high, the Q output goes low. The NOT-Q terminal will always be the complement of the Q terminal. Assuming no propagation delay, the output responds immediately when the input conditions are changed.

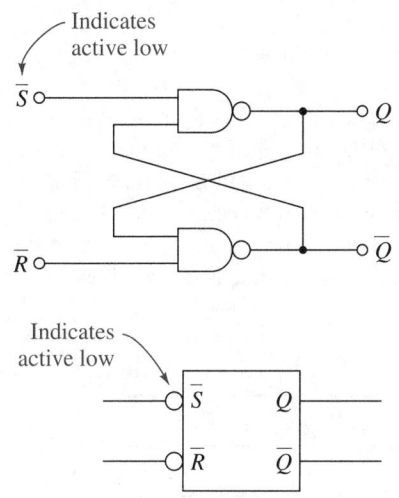

FIGURE 10.19 Active-low inputs.

TABLE 10.12 Truth Table

In		Out	
$\overline{S}$	$\overline{R}$	Q	$\overline{Q}$
0	0	No change	
0	1	1	0
1	0	0	1
1	1	Not allowed	

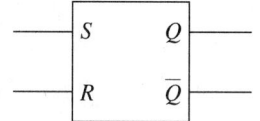

FIGURE 10.20 Active-high device.

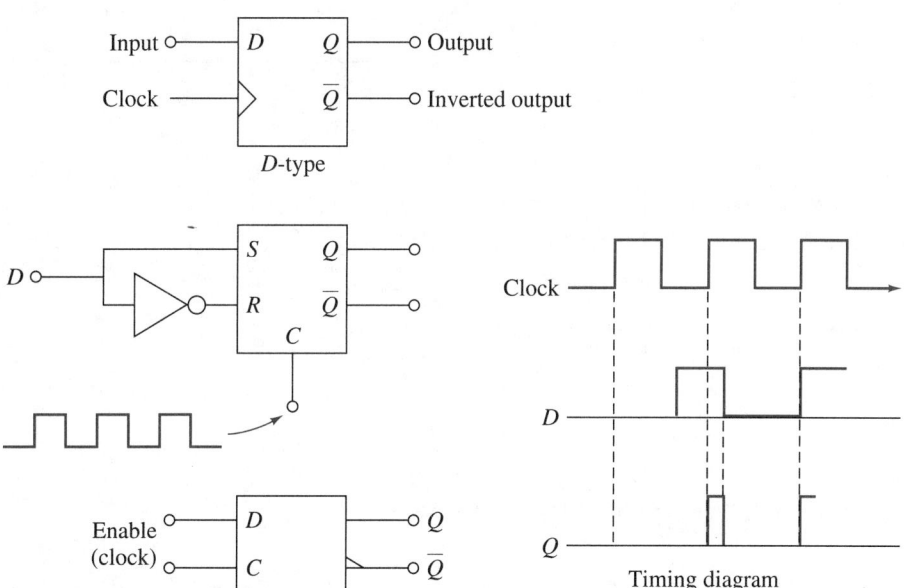

FIGURE 10.21 Synchronous flip-flop.

10.9 SYNCHRONOUS FLIP-FLOP

The synchronous flip-flop, depicted in Figure 10.21, has a separate control terminal called the *strobe*, *clock*, or *latch* to determine when the output condition should change. The input condition must exist or be stable for a given period of time before the control signal appears for switching to occur.

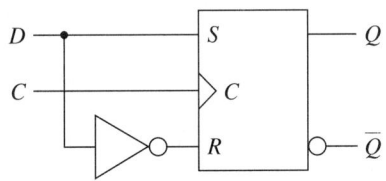

FIGURE 10.22 A leading edge-triggered *D* flip-flop.

The *D*-Type Flip-Flop

The condition on the *D* terminal may change, but the *Q* output will not change until the clock condition goes from low to high. Then the *Q* terminal will go to the value of the *D* terminal. The circuit shown in Figure 10.22 is a leading edge–triggered *D*-type flip-flop. If there is a 0 before the clock terminal, this indicates that the circuit is initiated by the trailing or falling edge of the clock pulse.

The Clocked *J-K* Flip-Flop

The clocked *J-K* flip-flop in Figure 10.23 is triggered by the positive-going edge of the clock pulse. The *J* and *K* inputs control the state of the flip-flop. The truth table (Table 10.13) summarizes the operation of the flip-flop. Notice that if *J* and *K* are both high, the device will go into the toggle mode and change state with each succeeding clock pulse. It is important to note that the flip-flop is not affected by the negative-going edge of the clock pulses. Also, the *J* and *K* inputs cannot affect the output state by themselves.

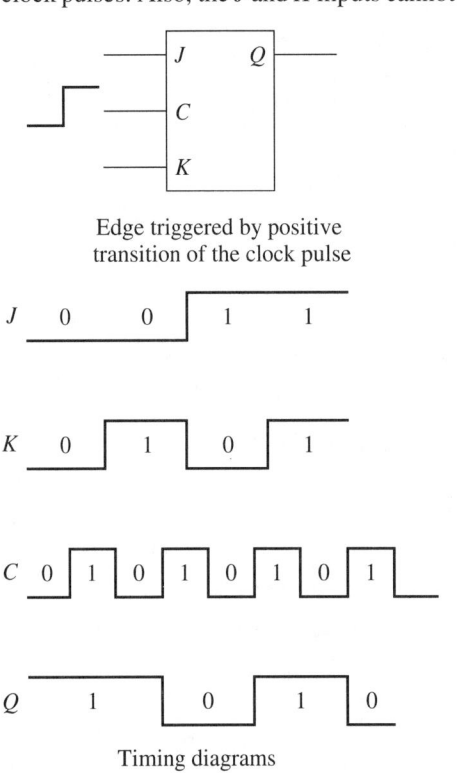

Edge triggered by positive transition of the clock pulse

FIGURE 10.23 *J-K* flip-flop.

Timing diagrams

TABLE 10.13 Truth Table

Inputs			Output
J	*K*	*C*	*Q*
0	0	↑	No change
0	1	↑	0 reset
1	0	↑	1 set
1	1	↑	Toggle

10.10   DATA TRANSMISSION

After the analog information has been converted to a digital data stream, the digital pulse train must be transmitted through the system from one point to another. The two primary methods of accomplishing this are serial data transmission and parallel data transmission.

Serial Data Transmission

In serial data transmission the digital bit stream is sent over a single wire from point *A* to point *B*. As shown in Figure 10.24, the least significant bit (LSB) is sent first. This bit is designated d_0, and the remaining bits follow. Since the bit stream travels a bit at a time, this method is a very slow one. Thus the time needed to transmit an 8-bit word is eight times as long as the time needed to transmit a single bit.

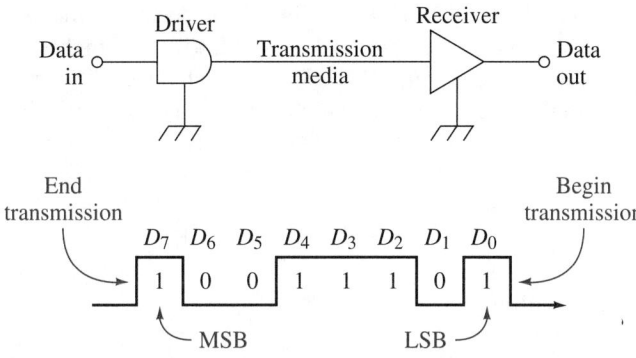

FIGURE 10.24 Serial data transmission.

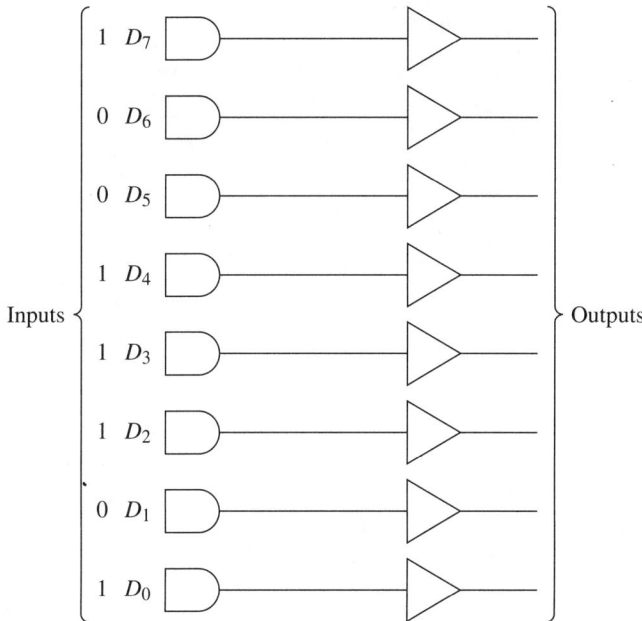

FIGURE 10.25 Parallel data transmission.

Parallel Data Stream

In parallel data transmission, all 8 bits are transmitted simultaneously. Therefore, the interface cable requires one wire for each bit to be transmitted. See Figure 10.25. An 8-bit word needs eight separate lines. Also, each line has to have its own transmitter and receiver. You can see that this method is more expensive. However, it is much faster than the serial method. The speed is limited to the time needed to transmit a single bit.

10.11 DIGITAL CODES AND NUMBERING SYSTEMS

In general, a digital code is a specific combination of bits that represent information according to an assigned reference table. The number of different values that can be represented is determined by the number of bits used in the code. In the code, 1 bit can convey one of two values, such as on or off; yes or no; true or false; or high or low. It follows, then, that 2 bits can convey four different values. See Figure 10.26.

	Value	Condition		
If $d =$	1	Yes	On	High
	0	No	Off	Low

With 2 bits of data we can represent north, south, east, and west as follows:

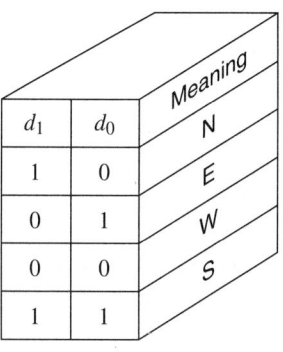

FIGURE 10.26 Digital coding.

The formula 2^n, where n equals the number of bits, can be used to determine the total number of values that can be conveyed. For example, an 8-bit word, with each bit having two possible meanings, can be used to represent 2^8, or 256, different values. The truth table looks like Table 10.14. Let's take a few minutes to review the basic number systems.

TABLE 10.14 Eight-Bit Word Combinations

D_7	D_6	D_5	D_4	D_3	D_2	D_1	D_0	Meaning
0	0	0	0	0	0	0	1	Each combination has a value designated by the user.
0	0	0	0	0	0	1	0	
0	0	0	0	0	0	1	1	
0	0	0	0	0	1	0	0	
0	0	0	0	0	1	0	1	
0	0	0	0	0	1	1	0	
0	0	0	0	0	1	1	1	
0	0	0	0	1	0	0	0	
0	0	0	0	1	0	0	1	
			256 possible combinations					
1	1	1	1	1	1	1	1	

Decimal

1. The decimal system is a base 10 system.

2. It uses the digits 0–9.

3. The value of a number is determined by the digits and their place values.

4. Place values are given in Table 10.15.

5. Each place can be represented as a power of 10, as shown in Table 10.15.

6. The value of a specific number is found by adding the total place values. See Table 10.16.

Binary System

1. The binary system uses only two numbers, 1 and 0. Each one is called a *bit*.

2. It is a base 2 system.

3. The value of a number is determined by the digits and their place values.

4. Place values are given in Table 10.17.

5. Each place represents a power of 2.

6. The specific value of a number is found by adding the total place values. See Table 10.18.

TABLE 10.15 Decimal System

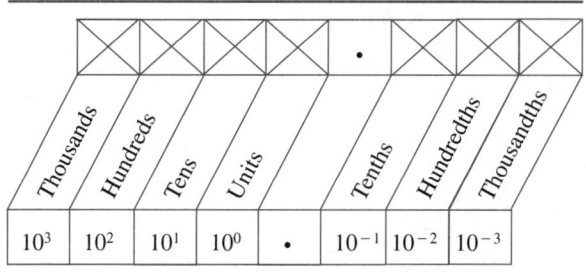

TABLE 10.16 Example

Decimal number	7 9 6 8₁₀			
Place values	1000	100	10	1
Expressed as powers of 10	$7 \times 10^3 + 9 \times 10^2 + 6 \times 10^1 + 8 \times 10^0$			
Simplified	7000 + 900 + 60 + 8			
Sum	7000 900 60 8			
	7968_{10}			

TABLE 10.17 Binary System

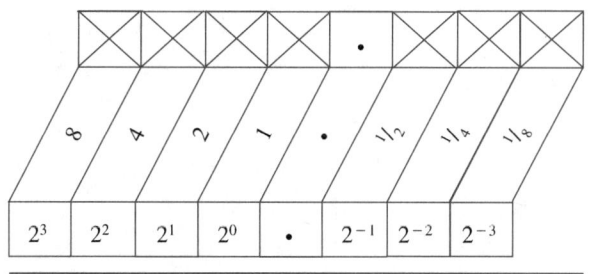

TABLE 10.18 Example

Binary number	(1	1	0	1	1	1	1)₂
Place values	64	32	16	8	4	2	1
Powers of 2	2^6	2^5	2^4	2^3	2^2	2^1	2^0
Value of the digit	64 + 32 + 0 + 8 + 4 + 2 + 1						
Sum	111						
	$(1101111)_2 = (111)_{10}$						

Converting From Decimal to Binary

Figure 10.27 shows the procedure for converting a decimal number to a binary number.

Binary-Coded Decimal—BCD

When dealing with a large decimal number—for example, 7968_{10}—it is easier to use the BCD system. Each decimal number is represented by four binary digits. See Figures 10.28 and 10.29.

Hexadecimal (Hex)

1. The hexadecimal system is a base 16 system.
2. It is used when large numbers are being processed.
3. Place values are as shown in Table 10.19.
4. The truth table is given in Table 10.20.
5. Four binary digits are used to represent a hex number.
6. Figure 10.30 shows the procedure for converting a decimal number to a hex number, hex to decimal, and hex to binary.

To convert decimal 111 (111_{10}) to binary: $1\ 1\ 0\ 1\ 1\ 1\ 1_2$

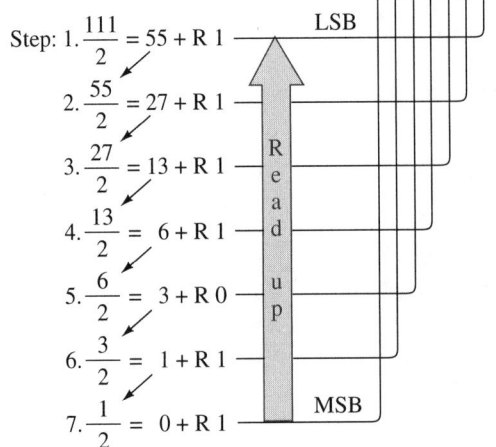

FIGURE 10.27 Decimal-to-binary conversion.

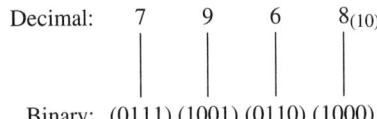

Therefore, $7968_{(10)} = 0111\ 1001\ 0110\ 1000_{(BCD)}$.

FIGURE 10.28 BCD system.

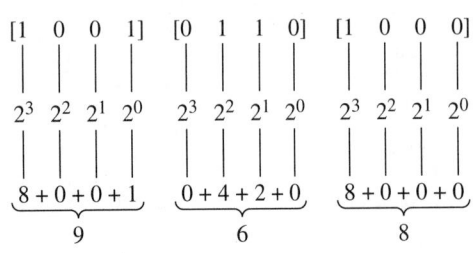

$1001\ 0110\ 1000_{(BCD)} = 968_{(10)}$

FIGURE 10.29 BCD system.

TABLE 10.19 Hex Place Value

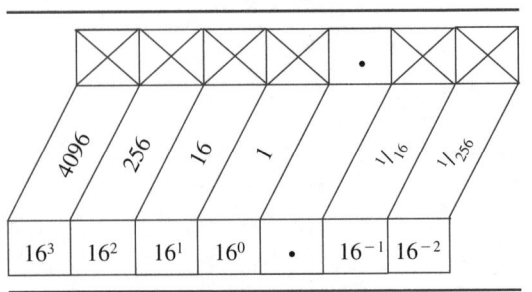

TABLE 10.20 Hex Truth Table

Decimal	Binary	Hex
1	0001	1
2	0010	2
3	0011	3
4	0100	4
5	0101	5
6	0110	6
7	0111	7
8	1000	8
9	1001	9
10	1010	A
11	1011	B
12	1100	C
13	1101	D
14	1110	E
15	1111	F

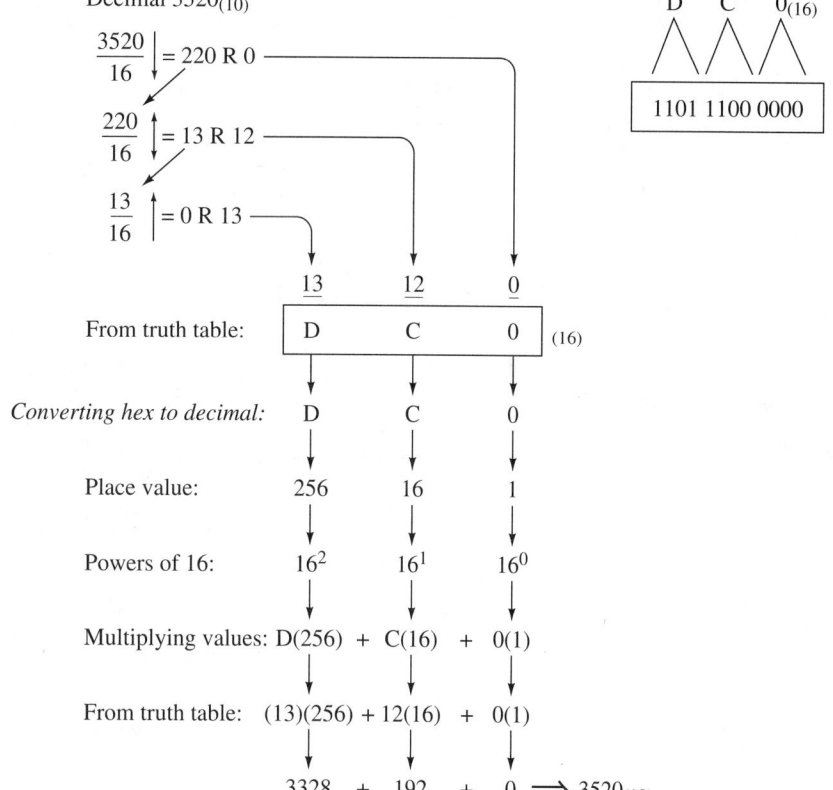

FIGURE 10.30 Converting decimal to hex, hex to decimal, and hex to binary.

TABLE 10.21 Octal System

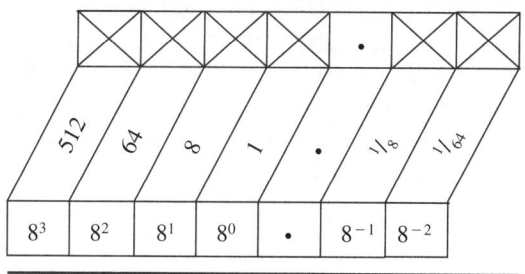

Octal

1. The octal system is a base 8 system.
2. It uses eight digits, 0–7.
3. The place values are shown in Table 10.21.
4. Figure 10.31 shows the procedure for converting a decimal number to its equivalent octal number and back to decimal.
5. The truth table is given in Table 10.22.

Since the octal system uses the numbers 0 through 7, you can see from the truth table that only three binary digits are needed, since 000–111 represent 0–7. Figure 10.32 shows the conversion of octal to binary.

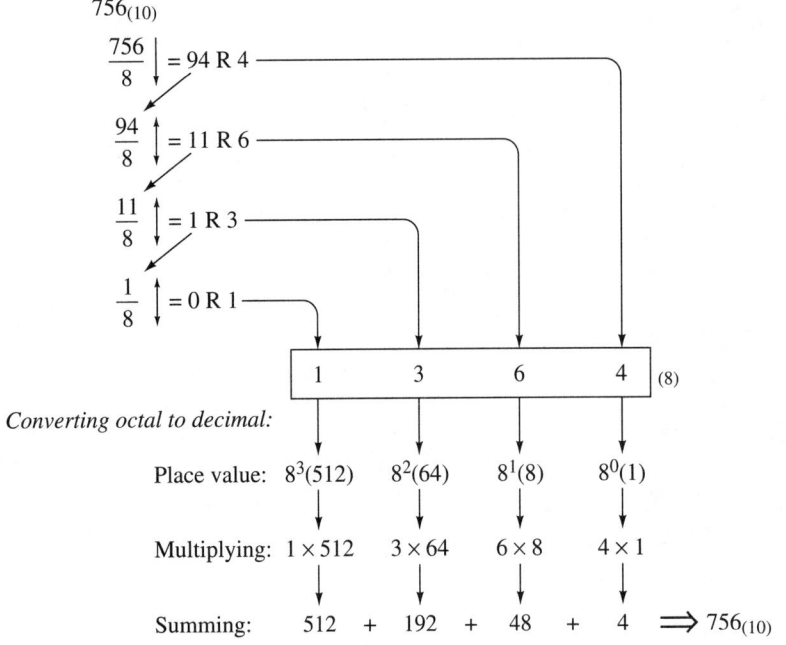

Converting decimal to octal:

$756_{(10)}$

$$\frac{756}{8} \Big| = 94 \text{ R } 4$$

$$\frac{94}{8} \Big| = 11 \text{ R } 6$$

$$\frac{11}{8} \Big| = 1 \text{ R } 3$$

$$\frac{1}{8} \Big| = 0 \text{ R } 1$$

| 1 | 3 | 6 | 4 | $_{(8)}$ |

Converting octal to decimal:

Place value: $8^3(512)$ $8^2(64)$ $8^1(8)$ $8^0(1)$

Multiplying: 1×512 3×64 6×8 4×1

Summing: 512   +   192   +   48   +   4 $\Rightarrow 756_{(10)}$

FIGURE 10.31 Converting decimal to octal and octal to decimal.

TABLE 10.22 Octal Truth Table

Decimal	Binary	Octal
1	0001	1
2	0010	2
3	0011	3
4	0100	4
5	0101	5
6	0110	6
7	0111	7
0	0000	0

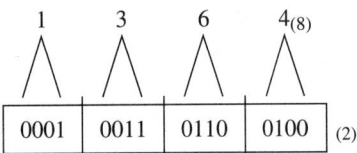

| 0001 | 0011 | 0110 | 0100 | $_{(2)}$ |

FIGURE 10.32 Octal to binary.

10.12 THE ENCODING PROCESS

The process of encoding analog signals to digital codes starts with sampling the analog signal at specific intervals of time. These samples are then quantized to the nearest voltage levels and assigned a binary code. The minimum sampling interval is limited by the Nyquist theorem, which states that a signal can be restored if sampled at a minimum rate of two times the highest-frequency component. If a telephone voice signal whose highest frequency is 4 kHz is sampled at a minimum rate of 8 kHz, it is possible to restore the original signal, as in Figure 10.33.

Circuits that perform this task are called analog-to-digital (A/D) and digital-to-analog (D/A) converters. Their functions are summarized in Figure 10.34.

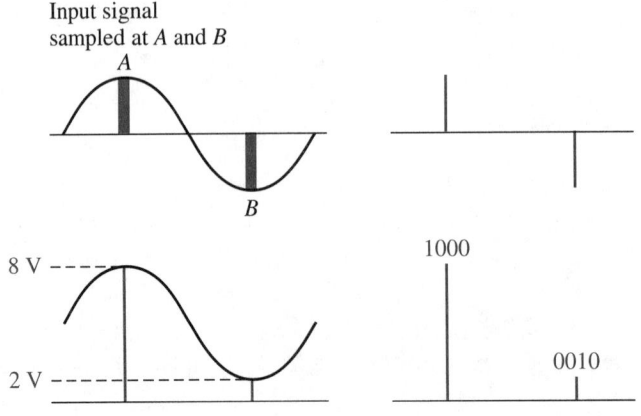

Input signal sampled at *A* and *B*

FIGURE 10.33 Sampling and encoding.

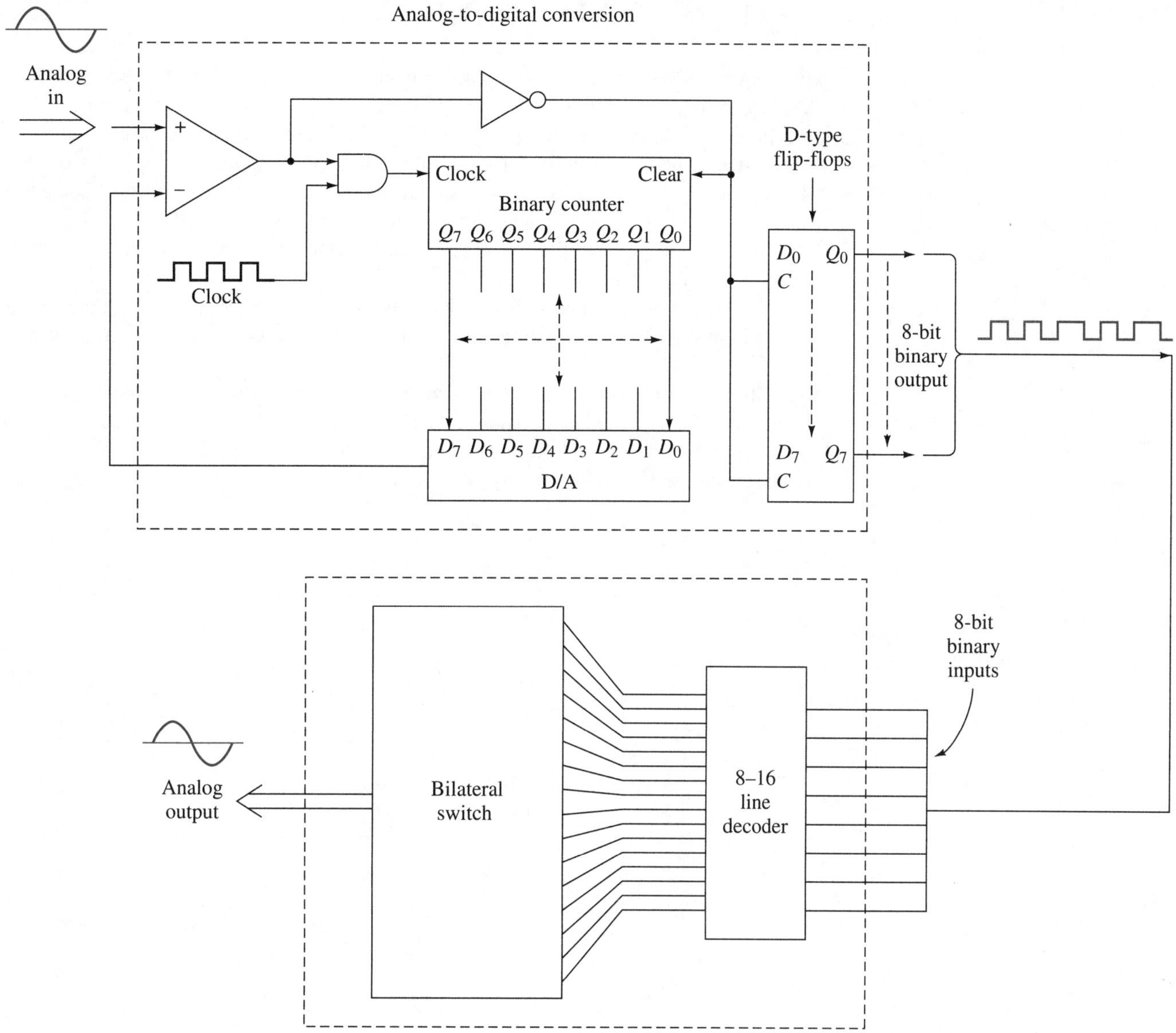

Analog-to-digital conversion

Analog
in

+
−

Clock

Clock   Clear
Binary counter
Q_7 Q_6 Q_5 Q_4 Q_3 Q_2 Q_1 Q_0

D_7 D_6 D_5 D_4 D_3 D_2 D_1 D_0
D/A

D-type
flip-flops

D_0 Q_0
C

8-bit
binary
output

D_7 Q_7
C

Analog
output

Bilateral
switch

8–16
line
decoder

8-bit
binary
inputs

FIGURE 10.34 A/D and D/A conversions.

10.13   DIGITAL SYSTEMS

The most basic digital system is called the *combinational system*. As shown in Figure
10.35, the combinational system does not require a memory section but has an input,
processor, and output section. The input section may be information or data received
from operators, transducers, etc. This section may also include signal-conditioning circuits, storage devices, and gating functions. The processor performs required functions
and then routes data to the outputs. The outputs can be displayed, printed, used to control
process actuators, etc. The combinational system is one that provides an output based on
current inputs and requires no data about past conditions.

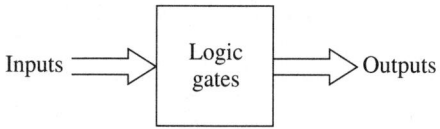

Inputs   Logic
              gates   Outputs

FIGURE 10.35 Combinational system.

10.14 SEQUENTIAL SYSTEMS

In a sequential system, the output is based on previous data in addition to the current data. Therefore, a memory section will be required to retain the past information. It is, in fact, a combinational system with memory, as in Figure 10.36. The memory section consists of registers, which are flip-flops that provide storage of data. One flip-flop will store one data bit. It requires eight flip-flops to store the 8 bits needed for an 8-bit word. The 8 bits are referred to as 1 byte. A typical 3-bit register is shown in Figure 10.37.

For an 8-bit byte (word), each set of eight flip-flops (registers) has a unique address location. See Figure 10.38.

The output of the registers is coupled to the data line with the three-state buffer circuit in Figure 10.39. In the open-circuit condition, the buffer effectively isolates the register from the data bus.

The size of the memory is determined in terms of the number of locations that hold a byte. The reference used is the kilobyte, which is equal to 1024 (2^{10}) locations. The two basic types of memory are RAM and ROM. RAM memory allows both reading and writing capabilities, but ROM can only be read from.

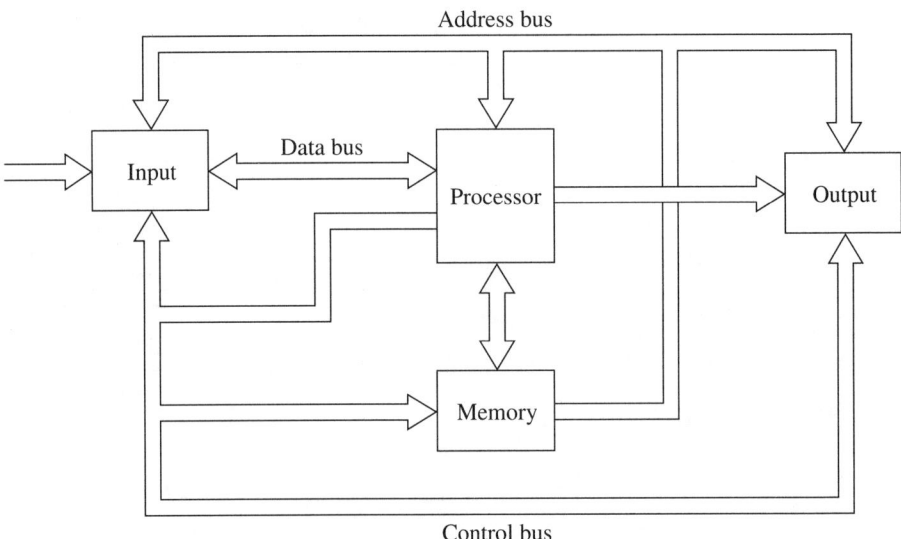

FIGURE 10.36 Sequential system.

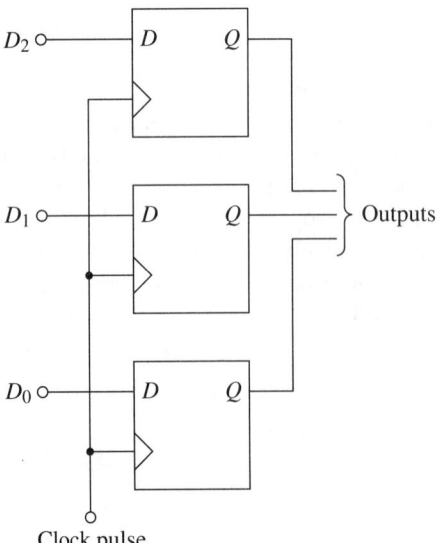

FIGURE 10.37 Three-bit register.

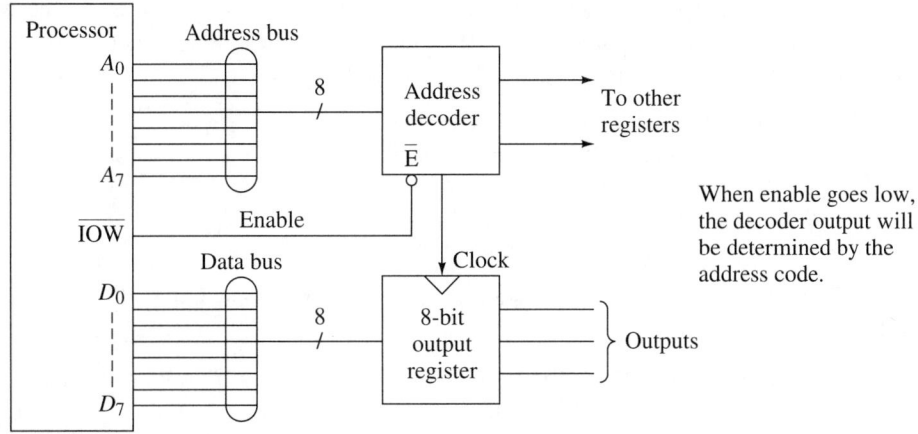

FIGURE 10.38 Eight-bit words.

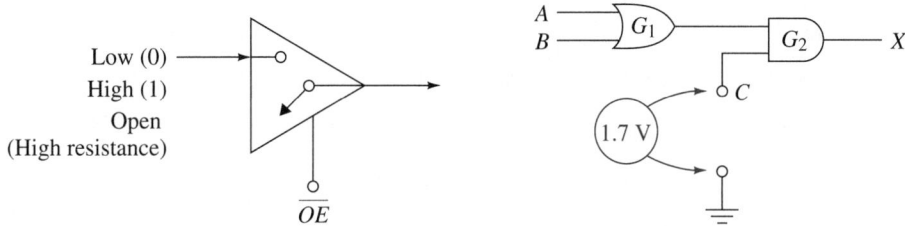

FIGURE 10.39 Three-state buffer.

FIGURE 10.40 Parametric fault.

10.15   FAULTS IN DIGITAL CIRCUITS

The following are some of the most common causes of digital circuit failures:

1. Voltage levels exceed the breakdown ratings of the chips.

2. Circuit requirements exceed the fan-out limits of the chips.

3. Signal levels are too low or too high.

4. Supply voltages are incorrect.

5. Poor power supply regulation causes excessive ripple.

6. Ground loop currents exist.

7. There is too much noise throughout the system. Noise can cause false switching.

8. Shorts between circuit traces occur.

9. Shorts to ground occur.

10. Shorts to V_{CC} occur.

11. There are open circuits.

12. Discrete components fail.

13. Parametric faults, in which some parameter such as gain is bad, exist. The symptoms show up in the outputs as voltages between the normal operating ranges. As Figure 10.40 shows, the 1.7-V read between the TTL input and ground indicates a floating terminal.

14. Logic faults exist when an output terminal stays at an incorrect level. These faults are often referred to as stuck-at-1 or stuck-at-0 faults. The most common causes of logic faults are

    a. Short to ground (voltage is fixed at 0).
    b. Short to power supply (voltage is fixed at 1).
    c. Open circuits
    d. Internal faults
    e. Shorts between leads

15. IC faults can be external or internal. Integrated circuits can fail internally in the following ways:

    a. Bonds within the IC can open. Figure 10.41 shows a TTL IC with an open bond in an internal lead. This leaves a terminal that is floating. In TTL circuits, a floating input will be at about 1.5 V and will look like a logic 1 level.
    b. Bonds within the IC can short to ground. This will effectively hold the pins low or high.
    c. Bonds within the IC can short to V_{CC}.
    d. Shorts can occur between two pins.
    e. Chips within the IC can fail.

External faults are all covered in the preceding text.

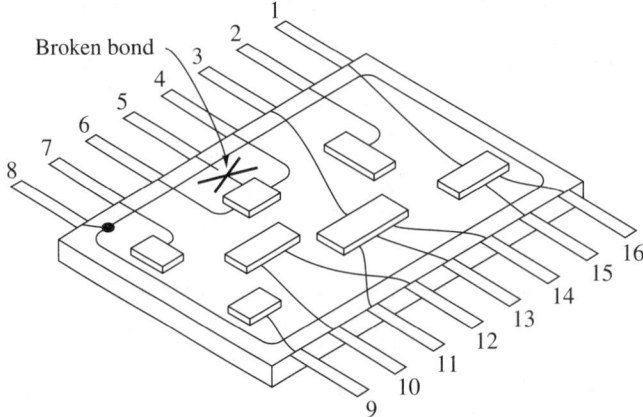

FIGURE 10.41 TTL IC with open bond.

10.16 TEST EQUIPMENT

The basic test instruments that are necessary to troubleshoot a digital circuit include the following:

1. *Multimeter.* General-purpose test instrument to measure voltages, currents, and resistances. You can use either analog or digital. Be sure to remember test probe polarity when making circuit connections. Make certain the multimeter has a large enough input impedance to not load down the circuit under test.

2. *Logic probe.* A device that uses LEDs to indicate the condition of the signal on the line. The LED is on for a logic 1 and off for a logic 0 level (Figure 10.42).

3. *Logic clip.* A clip-on device with 16 contacts to monitor the activity on the 16 pins of an integrated circuit. This way the entire circuit operation can be observed (Figure 10.43).

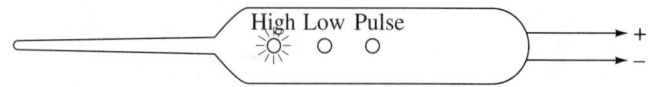

FIGURE 10.42 Logic probe.

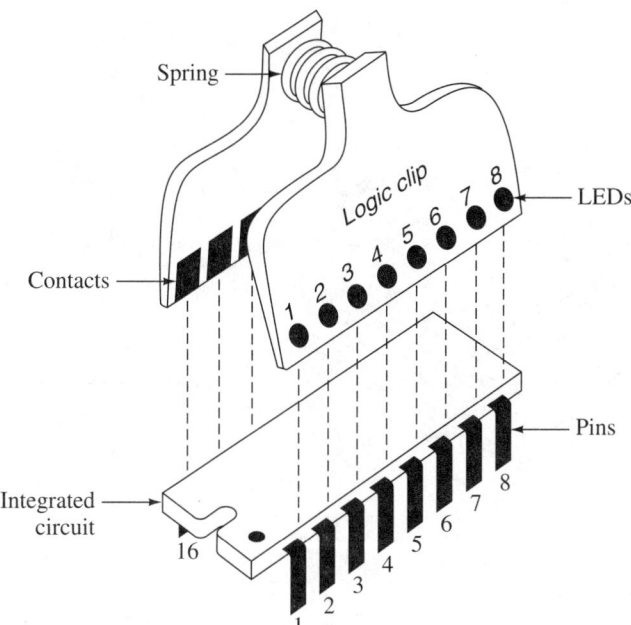

FIGURE 10.43 Logic clip.

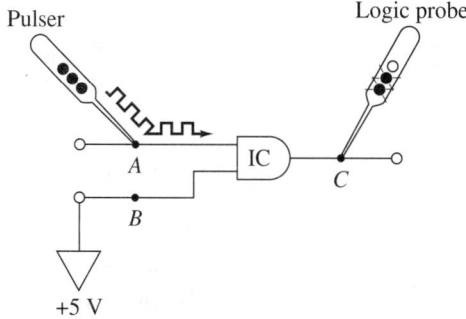

FIGURE 10.44 Tracing a pulse train.

4. *Logic pulser.* Similar to a digital signal generator. The pulser injects a pulse or series of pulses into the circuit at point of contact. The logic probe is then used to trace this pulse or pulse train through the circuit (Figure 10.44).

5. *Logic analyzer.* A device to test the signal and determine whether or not its value is within the limits for logic 1 levels or logic 0 levels. Oscilloscopes are analog instruments and display the voltage level of signals. However, in digital circuits the important thing is whether or not a voltage is present to indicate a 1 or 0. The logic analyzer is able to test many circuit locations simultaneously. Each location value is converted to a 1 or 0 and stored in memory. When needed, the patterns of 1s and 0s can be displayed in timing diagrams. Figures 10.45 and 10.46 show an example of a circuit under test. Table 10.23 shows the associated timing diagrams as seen on a logic analyzer.

6. *Current tracer.* A hand-held probe that senses the magnetic field associated with a current pulse. It can be used to monitor the current level on a line.

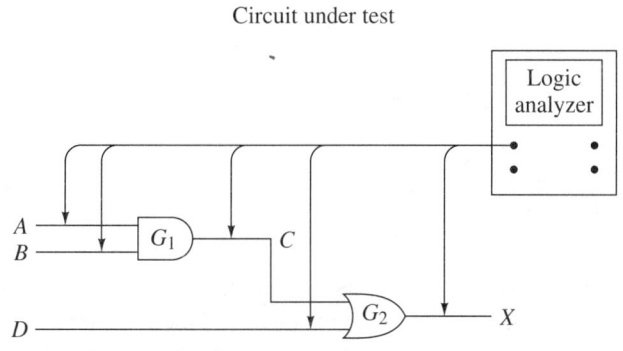

FIGURE 10.45 Use of logic analyzer.

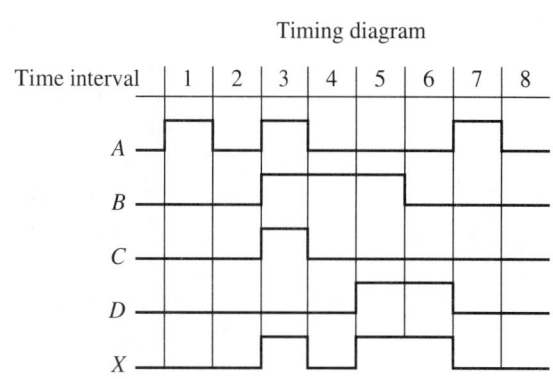

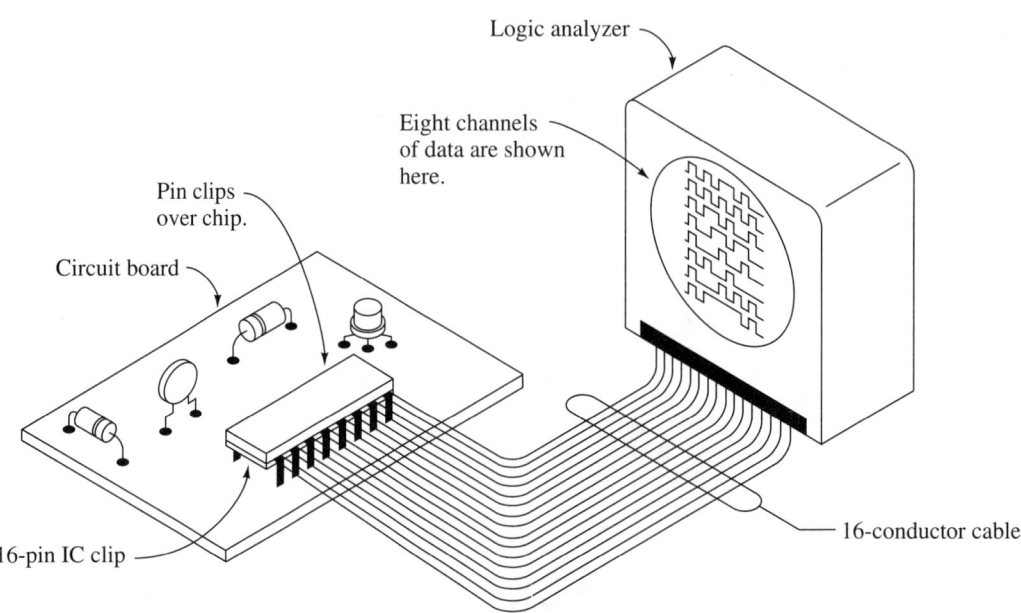

FIGURE 10.46 Logic analyzer.

TABLE 10.23
Timing Diagram

Time Interval	1	2	3	4	5	6	7	8
A	1	0	1	0	0	0	1	0
B	0	0	1	1	1	0	0	0
C	0	0	1	0	0	0	0	0
D	0	0	0	0	1	1	0	0
X	0	0	1	0	1	1	0	0

10.17   TROUBLESHOOTING BASICS

A list of troubleshooting basics

When a fault occurs in a system, the troubleshooter will first try to identify the functional unit that is faulty. A visual inspection of the circuits may accomplish this. But, it usually does not. A study of the system's logic circuit diagram coupled with a complete analysis of the symptoms of the faults may determine the faulty unit. The logic circuit diagram will show you

- All electrical connections
- All pin numbers
- All IC numbers
- Signal identifications
- Supply-voltage levels
- Each gate input and output
- Ground connections

However, in most cases you will have to inject a signal into the system and trace it through until it is lost. By using the logic pulser to inject pulses and the logic probe to check the outputs, you will isolate the area where the trouble is located. Look at Figure 10.47. This figure shows the simple signal-tracing method applied to an AND gate. Note

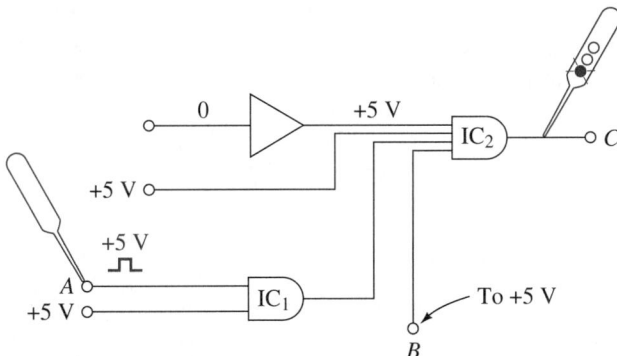

FIGURE 10.47 Pulse tracing.

that, in some cases, you must apply a 1 (+V) to some lines in order to enable the gate and allow the pulse to pass. If you tie point B to +5 V and inject a pulse at point A, the probe should detect the pulse at point C. If not, the gate is bad.

In signal tracing it is important to remember that a degraded semiconductor device will cause significant delays and disrupt the circuit timing. Also, any "pulse jitter" on the scope trace may signify a leaking capacitor.

Complete signal tracing and waveform analysis involves the use of the logic analyzer to observe the signal-timing diagrams at various points in the circuit. By starting at the inputs and observing the waveform timing diagrams, you can isolate the incorrect waveform first and then localize the faulty component. In general, the following procedures are helpful:

1. Isolate the fault to a logic section.

2. Verify that the inputs to the section are correct before beginning.

3. For each gate, observe the timing diagrams for errors.

4. If an output is found to be incorrect, the gate may be faulty. Test the IC out of circuit, if possible. Try replacing the IC if it is found to be faulty. If the gate is all right, check the external circuitry around the IC.

If the circuit timing is off, check the power-supply levels. A supply voltage that is too high can cause digital ICs to speed up. Likewise, a supply voltage that is too low can cause the timing to slow down.

10.18 TROUBLESHOOTING SHORT CIRCUITS IN A LOGIC CIRCUIT

Figure 10.48 is a simple logic circuit with a short to ground. Table 10.24 shows the associated truth tables. The effect is to apply a permanent 0 V on the input to the gate. The out-

How to troubleshoot short circuits in a logic circuit

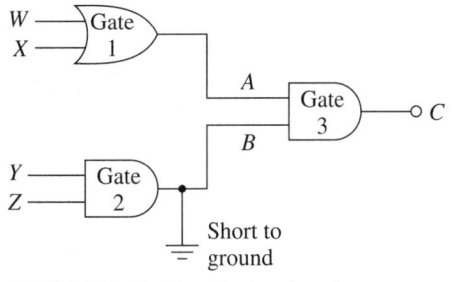

FIGURE 10.48 Troubleshooting shorts.

TABLE 10.24 Truth Tables

W	X	A		Y	Z	B		A	B	C	
0	0	0		0	0	0		0	0	0	
0	1	1		0	1	0		0	1	0	
1	0	1		1	0	0		1	0	0	
1	1	0	(a)	1	1	1	(b)	1	1	1	(c)

put at C will be stuck at 0 V. If both A and B are at logic 1, the output C should be logic 1 also. Both X and Y are normal. If Y or Z were stuck at 1, the other input would have an effect on the output at B. The condition on B would be observed if either Y, Z, or B is shorted to ground. Voltage measurements will isolate the fault to line B.

10.19 TROUBLESHOOTING OPEN CIRCUITS IN A LOGIC CIRCUIT

How to troubleshoot open circuits in a logic circuit

The open in Figure 10.49 disables gate 2 from circuit operation. If Y and Z are brought high (logic 1) and the output of gate 2 is found to be at logic 1, gate 2 is good. However, since the input to gate 3 on line B is stuck low, this means that line B is open somewhere. Voltage measurements will locate the open point.

10.20 INTERNAL FAULTS IN IC LOGIC GATES

Internal failures will cause improper response to inputs. Internal shorts will cause the input or output pins to be stuck in the high or low levels. An internal open in input leads, as in Figure 10.50, will put the corresponding pin in the floating state. This will act as a high-input level for TTL circuits and an unpredictable value for CMOS devices. An open condition in the output lead, as in Figure 10.51, will create a floating condition on input pins to the next IC in line.

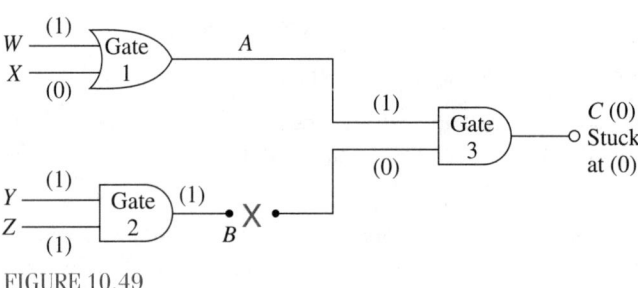

FIGURE 10.49

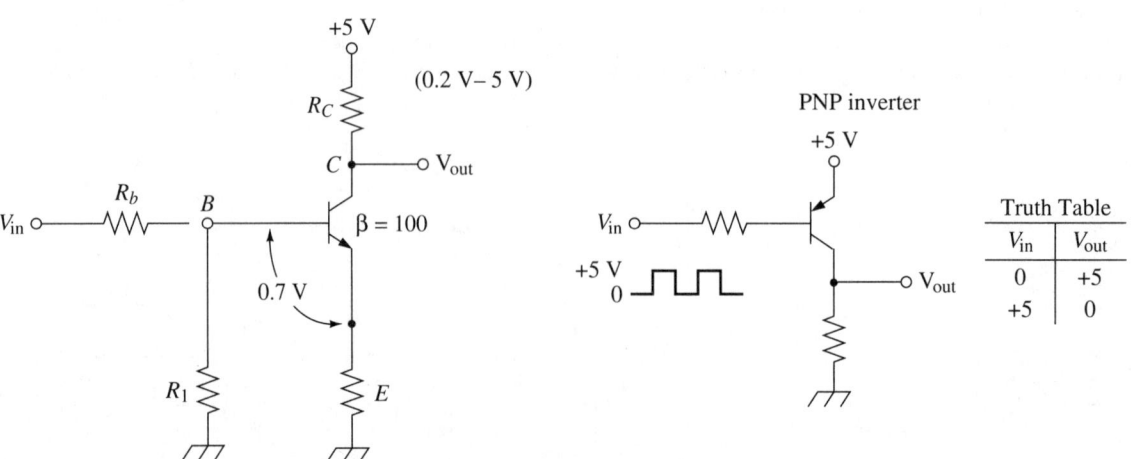

FIGURE 10.50 Inside the simple inverter.

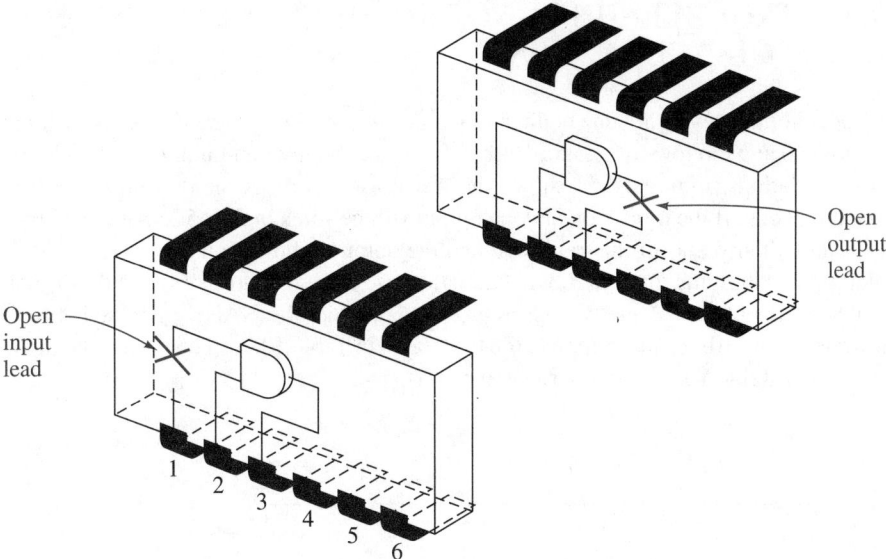

FIGURE 10.51 Internal opens.

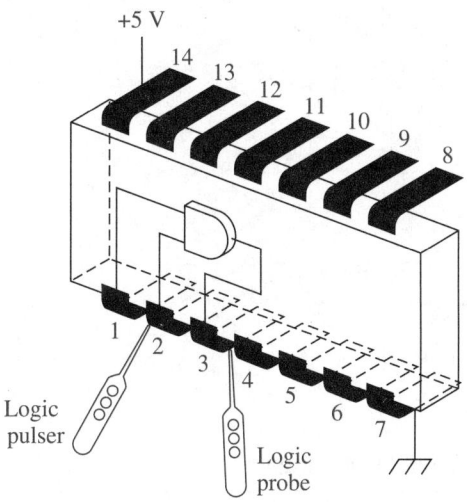

FIGURE 10.52 Troubleshooting a 7408.

These conditions can be tested with a logic pulser and logic probe, as shown in Figure 10.52 for a 7408 AND gate. The typical procedure to follow is

1. Make sure that the V_{CC} and ground pins are connected.

2. Look for a high on pin 3.

3. If no high exists, tie one input (pin 1, for example) to +5.

4. Look for a low on pin 3.

5. If no low exists, pulse the other input (pin 2) with the logic pulser.

6. Observe the output on pin 3 with the logic probe. You should see pulses on pin 3. If not, there has to an open circuit inside the chip between pin 2 and the gate.

7. If pin 2 is all right, pulse pin 1 with the logic pulser while pin 2 is tied to +5.

8. Check for pulses on pin 3. If not, pin 1 is opened.

10.21 TROUBLESHOOTING IMPROPER VOLTAGE LEVELS

Steps to take in troubleshooting an improper voltage level

Improper voltage levels are due neither to an open nor to a short. The voltage is too high for a logic 0 and too low for a logic 1. Let's go inside the inverter circuit again and look at the operations. Figure 10.53 shows the transistor circuit. A short from point B to ground will cut off the transistor, and the output will be stuck high at 5 V. An open in the base lead will also cut off the transistor, and the output will be high at 5 V. However, a leakage path to ground through R_1 will cause the base current to drop. As a result, BI_b will drop, and the output voltage will be improper. Construct the NAND gate latch circuit shown in Figure 10.53, and complete the truth table in Table 10.25. Then insert the faults indicated on Table 10.26 and describe the results.

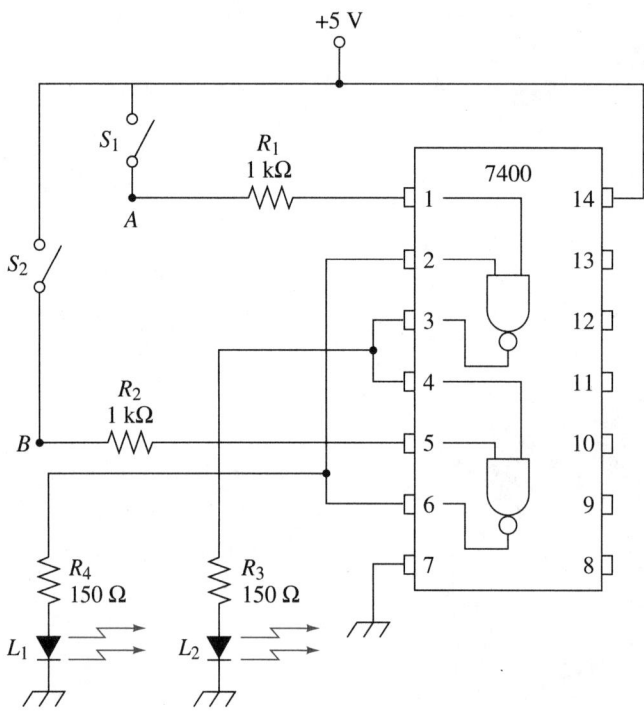

FIGURE 10.53 Digital troubleshooting problem.

TABLE 10.26 Troubleshooting Problem

Fault	Results
Point A shorted to ground.	
Point B stuck high.	
Pin 3 shorted to ground.	
Pin 7 open in ground connection.	
Open to pin 14.	
Open R_2.	
Pin 6 shorted to +5.	
Pin 3 shorted to +5.	
L_2 open.	

TABLE 10.25
Truth Table

A	B	L_1	L_2
0	0		
0	1		
1	0		
1	1		

10.22 TROUBLESHOOTING EXAMPLE 1

The circuit in Figure 10.54 is faulty, and point A is determined to be stuck low by using a logic probe. Point B is checked and determined to have the correct pulse activity. A pulse injected at C does not cause point A to change states. The TSV source was checked and found to be good. R_1 was also determined to be good. The problem was determined to be inside IC_2.

Troubleshooting faulty circuits

10.23 TROUBLESHOOTING EXAMPLE 2

We are given Figure 10.55(a) and the truth tables in Table 10.27. The timing diagrams are shown in Figure 10.55(b).

Examining a timing diagram

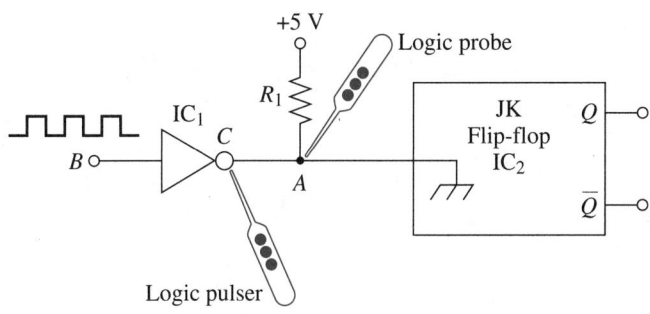

FIGURE 10.54 Troubleshooting Example 1.

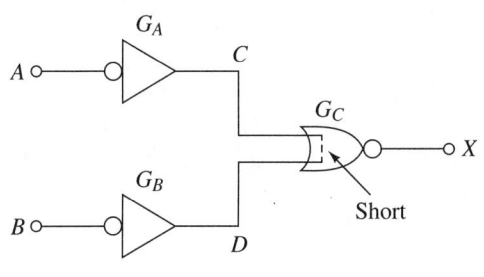

(a) Given circuit

TABLE 10.27 Truth Tables

OR			NOR		
C	D	X	C	D	X
0	0	0	0	0	1
0	1	1	0	1	0
1	0	1	1	0	0
1	1	1	1	1	0
(a)			(b)		

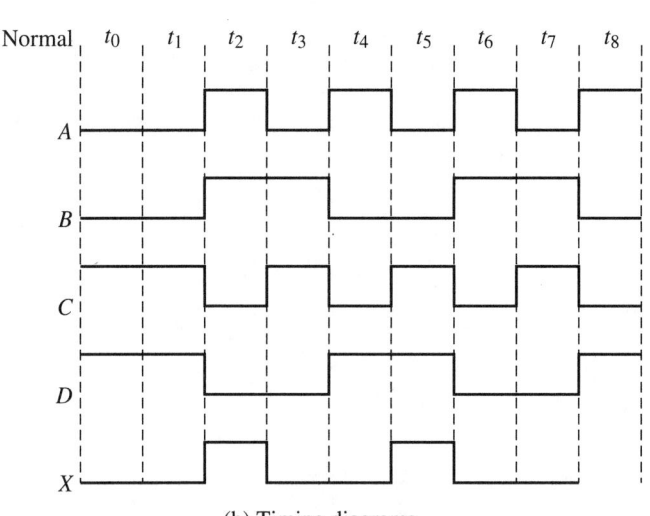

(b) Timing diagrams

FIGURE 10.55 Troubleshooting Example 2.

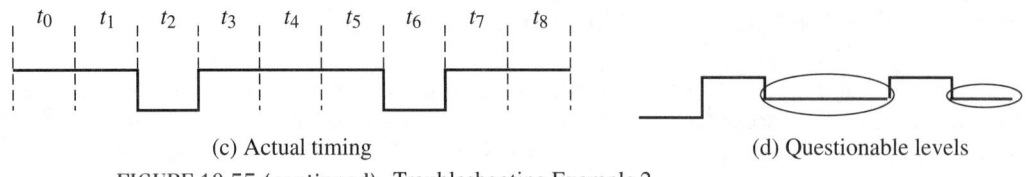

(c) Actual timing   (d) Questionable levels

FIGURE 10.55 (continued)   Troubleshooting Example 2.

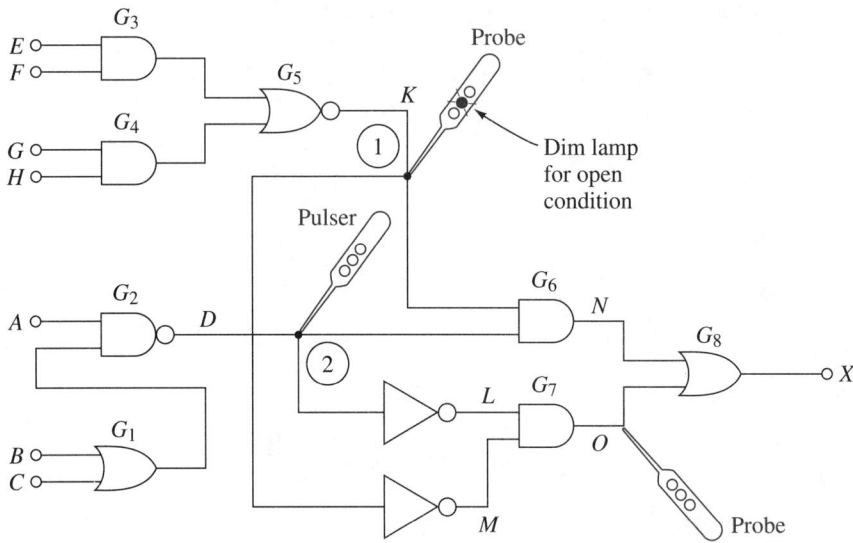

FIGURE 10.56   Troubleshooting Example 3.

Actual timing diagrams appear in Figure 10.55(c). C and D will be equal here due to the short inside G_3. During t_3, t_4, t_5, and t_7, one gate is producing a high output, whereas the other is producing a low output. Therefore, the actual voltage level at C and D will be determined by internal circuitry of logic gate C.

The tri-level output shown in Figure 10.55(d) should suggest a possible short.

10.24   TROUBLESHOOTING EXAMPLE 3

Testing a network

You are testing the network in Figure 10.56. The correct timing diagrams are shown in Figure 10.57.

A shorted condition at G_2 exists. The node probably will be stuck low. Check with the pulser and logic probe, as shown in Figure 10.58.

10.25   PROBLEMS

1. Give the truth table for a three-input NAND gate.

2. A TTL logic-high level is a voltage in a range from _____ to _____ volts.

3. Convert the following binary numbers to decimals:

   a. 0011_2
   b. 1100_2
   c. 11001_2
   d. 1011001010_2

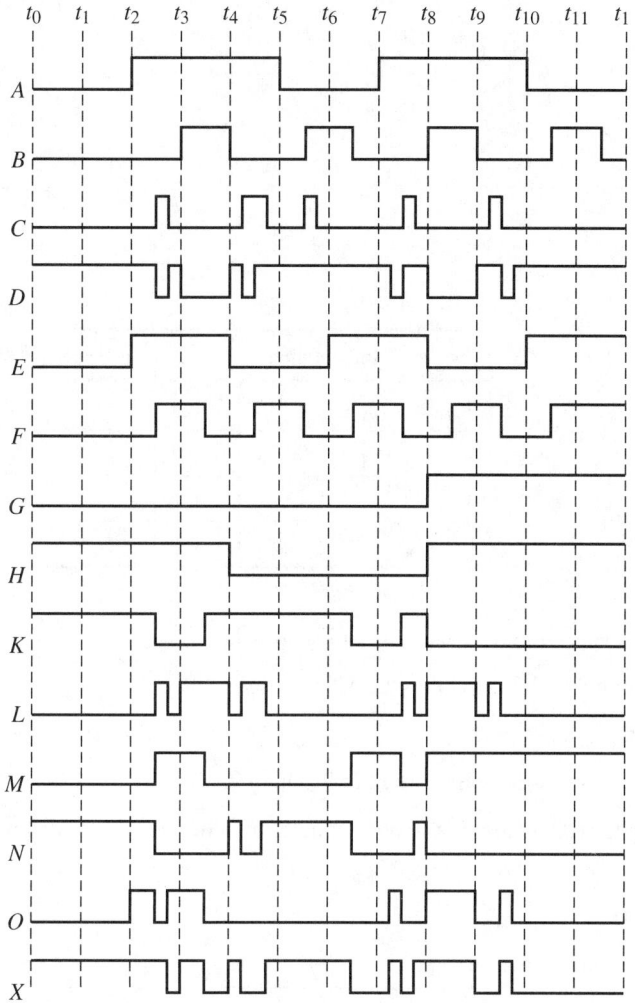

FIGURE 10.57 Normal timing diagrams.

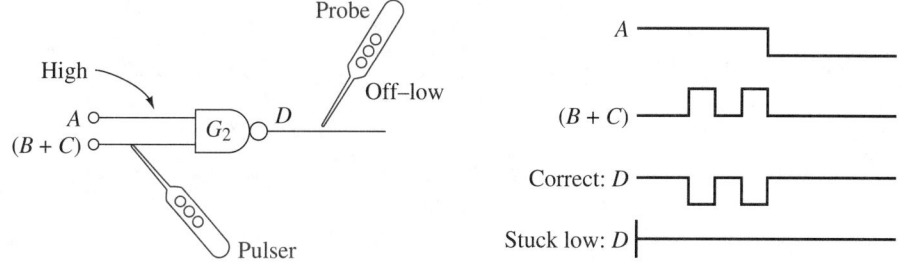

FIGURE 10.58 Troubleshooting Example 3.

4. Convert the following decimal numbers to binary:

    a. 5_{10}
    b. 14_{10}
    c. 248_{10}
    d. 0.625_{10}

5. Construct a truth table for the circuit in Figure 10.59.

6. You are analyzing the circuit in Figure 10.60. Determine the fault, if any (Table 10.28).

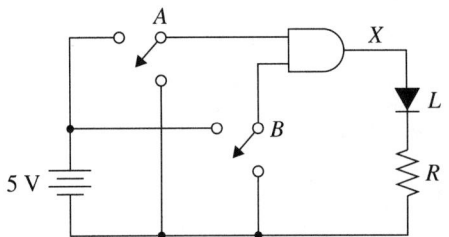

FIGURE 10.59 Problem 5.

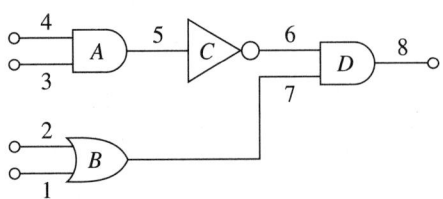

FIGURE 10.60 Problem 6.

TABLE 10.28

Pin Number								Fault
1	2	3	4	5	6	7	8	
0	1	0	1	0	1	0	0	
1	1	1	1	1	1	1	1	
0	1	1	1	0	1	1	1	

7. Write the Boolean expression for the circuit of Figure 10.61.

8. How would you define the IC chip in Figure 10.62?

9. What is the fault in Figure 10.63?

10. Write the Boolean expression for the circuit in Problem 9.

11. A logic probe (Figure 10.64) shows X to be an indeterminate level. What are the possible causes?

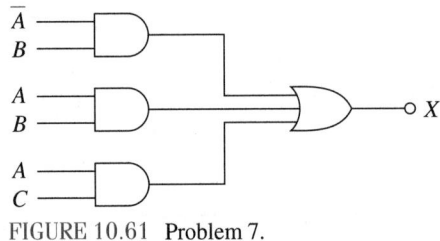

FIGURE 10.61 Problem 7.

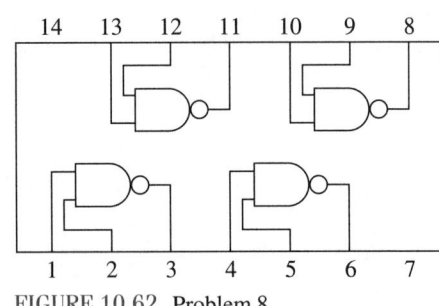

FIGURE 10.62 Problem 8.

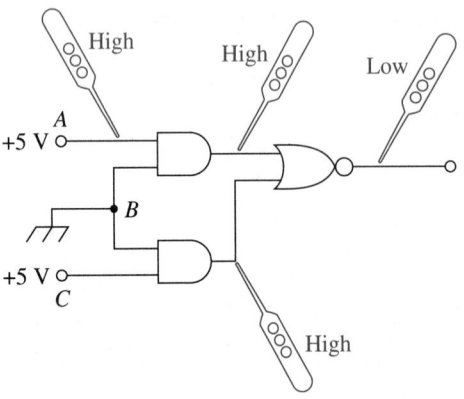

FIGURE 10.63 Problem 9.

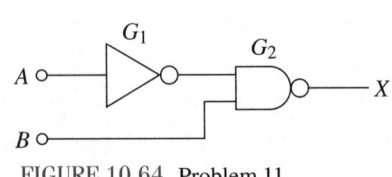

FIGURE 10.64 Problem 11.

10.26   TROUBLESHOOTING EXAMPLE 4

See Figure 10.65. For the given fault, complete Table 10.29.

10.27   TROUBLESHOOTING EXAMPLE 5

When the output of an *RS* flip-flop is stuck at 1, interesting patterns exist (Figure 10.66).
Since the 1 is fed back to NAND gate *B*, the *Q* output will always follow the *R* settings.

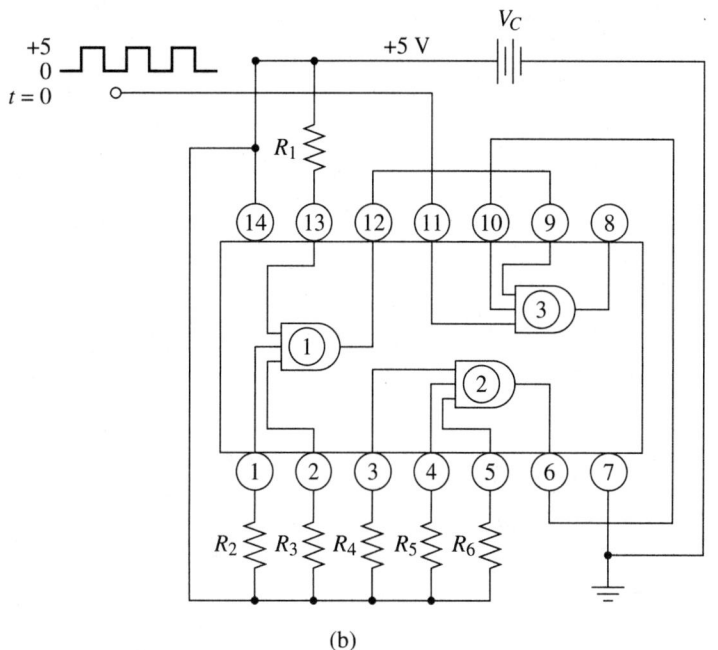

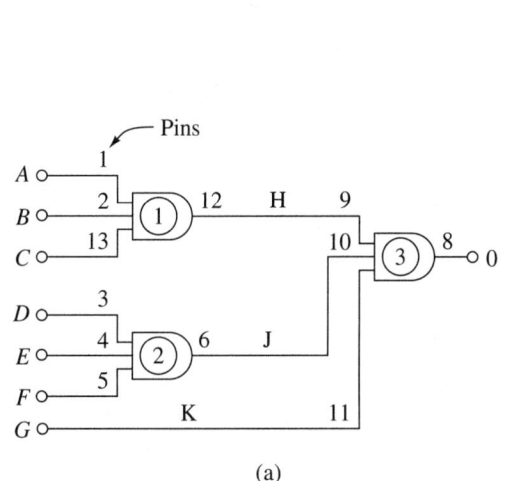

(a)   (b)

FIGURE 10.65 Troubleshooting Example 4.

TABLE 10.29

For the given fault, complete the table:	
R₁ Open	Waveform on Pin 8
Pin 11 stuck high.	
Pin 12 shorted to V_C.	
Pin 10 shorted to ground.	
Pin 11 shorted to Pin 12.	
Pin 11 shorted to Pin 12.	

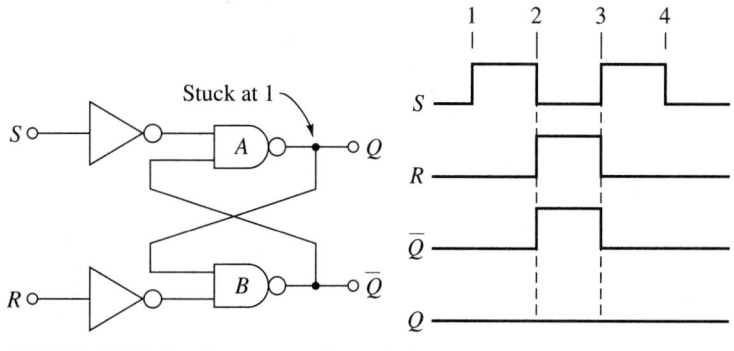

FIGURE 10.66 Troubleshooting Example 5.

Appendix A

COMPONENT IDENTIFICATION AND SYMBOLOGY

A.1 RESISTORS

Important Parameters

1. Resistance
2. Tolerance
3. Power rating
4. Type

Resistance Values

The standard preferred values are given in Table A.1.

Color Codes

The usual method of coding is by color bands on the body of the resistor, as shown in Figure A.1.

Some older resistors use a colored body, colored tip, and color spot. The code is read body-tip-spot.

Some resistors are not color-coded but use letters and numbers. The numbers indicate the numerical value and the letters are the multipliers:

$$R = \times 1$$
$$K = \times 100$$
$$M = \times 1,000,000$$

The tolerances are also designated by letters:

$$M = 20\%$$
$$K = 10\%$$
$$J = 5\%$$
$$H = 2.5\%$$
$$G = 2\%$$
$$F = 1\%$$

For example: The code 8K2K means 8.2 kΩ with 10% tolerance.

TABLE A.1 Resistor Preferred Values Available in Ohms, Kilohms, or Megohms

10^0	10^1	10^2
1.0	10	100
1.1	11	110
1.2	12	120
1.3	13	130
1.5	15	150
1.6	16	160
1.8	18	180
2.0	20	200
2.2	22	220
2.4	24	240
2.7	27	270
3.0	30	300
3.3	33	330
3.6	36	360
3.9	39	390
4.3	43	430
4.7	47	470
5.1	51	510
5.6	56	560
6.2	62	620
6.8	68	680
7.5	75	750
8.2	82	820
9.1	91	910

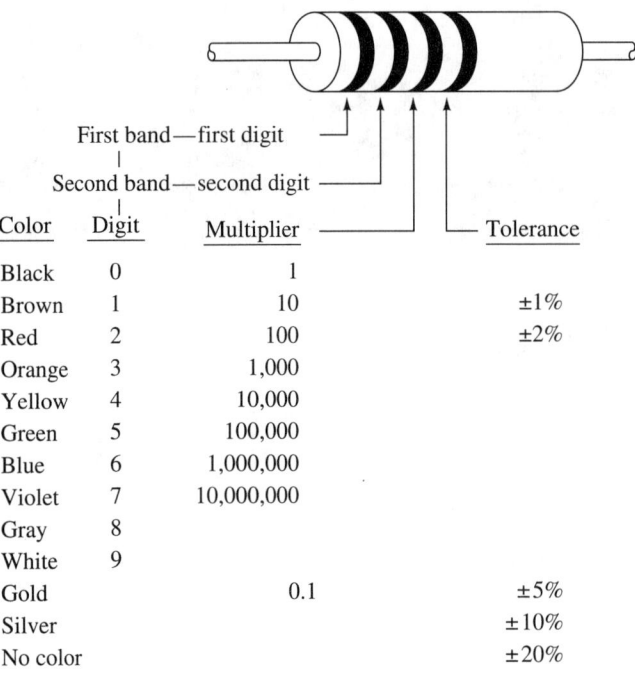

Color	Digit	Multiplier	Tolerance
Black	0	1	
Brown	1	10	±1%
Red	2	100	±2%
Orange	3	1,000	
Yellow	4	10,000	
Green	5	100,000	
Blue	6	1,000,000	
Violet	7	10,000,000	
Gray	8		
White	9		
Gold		0.1	±5%
Silver			±10%
No color			±20%

First band—first digit
Second band—second digit

FIGURE A.1 Color coding of resistors.

Types

There are 5 main types:

1. *Molded carbon.* These devices are constructed of a rod of carbon-binder material protected with a lacquer, paper, or ceramic coating.

2. *High-stability carbon.* These devices are made of a carbon film. They come in micro, miniature and subminiature size. They offer low noise, low cost, and good stability.
- *Micro:* Available in $\frac{1}{20}$-, $\frac{1}{10}$-, and $\frac{1}{8}$-W, 150-V ratings with $\pm 5\%$ tolerance.
- *Miniature:* Available in $\frac{1}{3}$-W, 250-V ratings with $\pm 5\%$ tolerance.
- *Standard:* $\frac{1}{2}$-W, 500-V rating with $\pm 5\%$ tolerance.
- *Power:* 1-W, 750-V rating with $\pm 5\%$ tolerance.
- *High power:* 3-W rating and wire-wound.

3. *Metal film.* The metal-film resistor is made by depositing a film of nickel-chromium on a ceramic body. The film is then cut to achieve the proper resistance value. The body is coated with a lacquer for protection. The metal film is more stable than carbon and costs more. Ratings are $\frac{1}{10}$ W and up. The most common is the thick-film resistor with $\frac{1}{2}$ W and $\pm 10\%$ tolerance.

4. *Metal-oxide.* This device is made of tin oxide film on a glass rod and is covered with a heat-resistant coating. Typical ratings are $\frac{1}{2}$ W with $\pm 2\%$ tolerance.

5. *Wire-wound.* These devices are for very low resistance values or very high currents. Ratings are 1 W and up.

Temperature Effects

As a component carries current, it heats up internally. This heat can cause a change in the component's resistance. In fact, most materials exhibit a positive temperature coefficient (ppm/°C) and experience an increase in resistance with a rise in temperature. Carbon, however, has a negative temperature coefficient, and experiences a decrease in resistance as the temperature rises. Carbon-film resistors have a negative temperature coefficient. Metal-film and metal-oxide resistors have a small positive temperature coefficient, which is usually less than 100 ppm/°C.

Effects of Aging

Carbon-composition resistors can change their values by as much as 20% with continuous use.

Effects of High Frequency

When used in high-frequency circuits, the common resistor can become an *RLC* network. The resistor leads can represent inductance and the spacing between the leads and the circuit board can look like capacitance. The effect is much worse for carbon-composition and wire-wound resistors.

Potentiometers

- *Carbon track.* These devices are suitable for general-purpose and low-power levels. Typical values are 100 Ω, 220 Ω, 470 Ω, 1 kΩ, 2.2 kΩ, 4.7 kΩ, 10 kΩ, 22 kΩ, 47 kΩ, 100 kΩ, 220 kΩ, 470 kΩ, 1 MΩ, 2.2 MΩ, and 4.7 MΩ.
- *Wire-wound.* These devices are typically 1 W to 3 W with tolerances of ± 10% or ± 20% and values as follows: 10 Ω, 22 Ω, 47 Ω, 100 Ω, 220 Ω, 470 Ω, 1 kΩ, 2.2 kΩ, 4.7 kΩ, 10 kΩ, 22 kΩ, and 47 kΩ.
- *Preset.* These devices are designed to be adjusted and then left alone. Typical values are 100 Ω to 1 MΩ.

Characteristics

The change in resistance can be linear or logarithmic or have an audio taper.

A.2 CAPACITORS

Categories

The two broad categories of capacitors are polarized and nonpolarized.

Types

Ceramic. Ceramic capacitors are inexpensive. They usually have a high working voltage rating, have a high value of leakage resistance, and are available in ranges of 1 pF to 1 μF. They are usually disk or tubular.

Disk. Disk capacitors may have a low (12-V) rating but large capacitance values, such as 0.01 μF to 0.1 μF. The general-purpose ones will have 50-V ratings and values of 0.01 μF to 0.1 μF. You can also find them with 500-V ratings with values of 10 pF to 10,000 pF. The temperature coefficients are typically ±15%/°C, and the insulation resistance is normally greater than 1010 MΩ.

Tubular. Tubular capacitors are available in ranges similar to the disk capacitors and have voltage ratings up to 350 V.

Ceramic Plate. These capacitors are used as couplers and decouplers in noncritical circuits. Their tolerance is ± 10%, the temperature coefficients are 150 to 750 ppm/°C and they normally have insulation resistance greater than 1000 MΩ.

Silvered Mica. These capacitors are more expensive because of their excellent high-frequency response, tighter tolerances, and very high working voltages. They are superior for critical RF applications. The values are 2.2 pF to 100,000 pF. Temperature coefficients are 35 to 75 ppm/°C, and their insulation resistance is greater than 50,000 MΩ.

Polystyrene. This class of capacitors normally exhibits low losses at high frequencies and good stability and reliability. Values range from 10 pF to 100,000 pF, with tolerance of ± 1%. These devices are recommended for tuned circuits and filter networks. Their temperature coefficient is ±160 ppm/°C. Insulation resistances are greater than 1011 MΩ.

Polycarbonate. These devices are low-loss capacitors used primarily on printed circuit boards with values up to 1 μF. The temperature coefficient is − 65 ppm/°C and insulation resistance is greater than 1010 MΩ.

Polyester film. These capacitors are used for printed circuit boards and are color-coded. Their values are 0.01 μF to 2.2 μF. The temperature coefficient is +350 ppm/°C. Insulation resistance is greater than 1010 MΩ.

Mylar Film. These capacitors are general-purpose devices with values from 0.001 μF to 0.22 μF. The working voltage is up to 100 V, and tolerances of ±10% are typical.

PETP Film. These capacitors are primarily for suppression of interference and are connected directly across the power-line mains. Values are from 0.01 μF to 0.47 μF. Working voltages are 250 V ac to 275 V ac.

Polyester Paper. These capacitors are also used directly across the power-line mains and have values from 0.001 μF to 1 μF.

Electrolytic. The value of electrolytic capacitors is usually not critical. It can range from 0.1 μF to 10,000 μF, with working voltages from 10 V dc to 500 V dc. The tolerances are usually quite wide. Due to their low dielectric resistance (500 kΩ to 10 MΩ), they have high values of leakage currents. The shelf life is short. They also have inherently high values of inductance.

Tantalum. These capacitors are more expensive due to their longer shelf life and higher dielectric resistance values. Typical values are 0.1 μF to 100 μF, and typical voltage ratings are 10 V dc to 35 V dc.

Variable. Variable capacitors are used for tuning circuits. Air dielectric is common, and values range from 10 pF to 1000 pF.

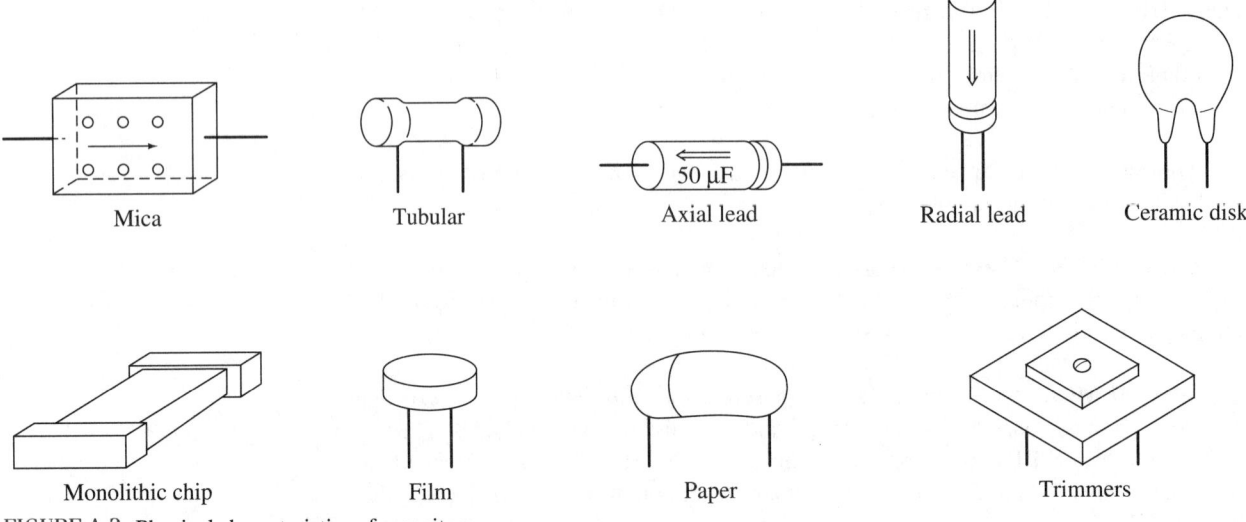

FIGURE A.2 Physical characteristics of capacitors.

Physical Characteristics

See Figure A.2.

Color Codes and Labeling

See Figure A.3 and Table A.2.

Important Parameters

Tolerance. The variation of capacitance expressed as a percentage of its specified value at 25°C.

Temperature Coefficient. Change in capacitance per degree change in temperature.

Working Voltage. Maximum voltage that can be used across the capacitor in continuous operation.

Breakdown Voltage. The voltage level that will cause damage to the dielectric.

DC Leakage. The small direct current in the capacitor when a small direct voltage is applied.

Insulation Resistance. The dielectric resistance.

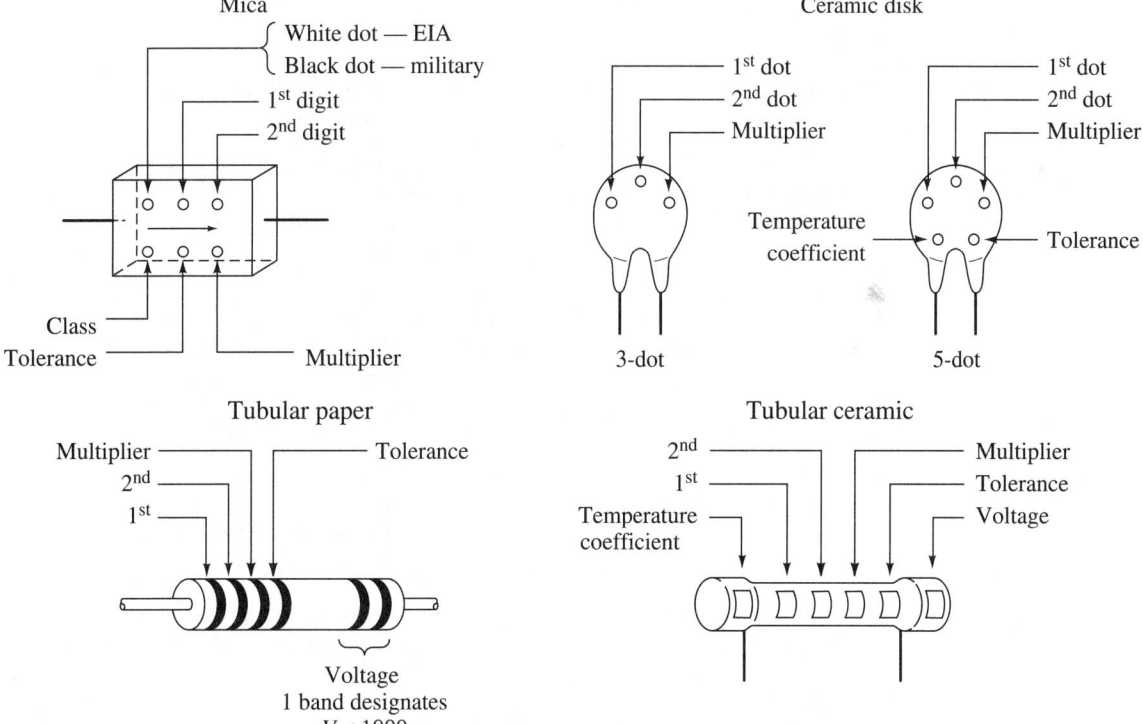

Common labeling: The capacitor body may or may not be marked to indicate pF or μF.
Microfarad may be marked as μF, mFd, MFD or MF.
Picofarad may be marked as pF, PF, pfD, PFD, μμF or MMF.
Whole numbers (5, 100, 500, etc.) designate picofarads.
Decimal numbers (0.002, 0.01, 0.047) designate microfarads.

FIGURE A.3 Common capacitor codes.

TABLE A.2 Color Coding of Resistors With Voltage Ratings

	Significant	Multiplier	Tolerance	Voltage Rating
Black	0	1	± 20%	—
Brown	1	10	± 1%	100
Red	2	100	± 2%	200
Orange	3	1,000	± 3%	300
Yellow	4	10,000	—	400
Green	5	100,000	± 5 EIA	500
Blue	6	1,000,000	± 6%	600
Violet	7	10,000,000		700
Gray	8		± 30%	800
White	9		± 10%	900
Gold	—	0.1	± 5 MIL	1000
Silver	—	0.01	± 10%	2000
No band	—		± 20%	500

TABLE A.3 Common Diode Parameters

Parameter	Symbol	Definition
Junction Diodes		
DC forward voltage	V_F	Anode-cathode voltage across a forward-biased diode.
DC forward current	I_F	Current in a forward-biased diode.
DC reverse current	I_R	Leakage current in a reverse-biased diode.
DC reverse voltage	V_R	Voltage across a reverse-biased diode.
Reverse breakdown voltage	V_{BR}, PRV, PIV	Maximum reverse voltage before breakdown.
Power dissipation	P_D	Maximum power dissipated by the junction.
Reverse recovery time	t_{RR}	Time for reverse current to reach a given level after being put in reverse-bias mode.
Forward recovery time	t_{FR}	Time to reach a level of forward current after being switched to forward-biased mode.
Zener Diodes		
Zener voltage	V_Z	Voltage at which zener regulates.
Knee current	I_{ZK}	Minimum current to give zener action.
Maximum zener current	I_{ZM}	Maximum permissible current.
Zener impedance	Z_Z	Designates a zener voltage change.

A.3 DIODES

Common Parameters

The parameters are listed in Table A.3.

Coding

The diode uses color bands for coding. See Figure A.4 for examples.

Symbols

Some common diode symbols are given in Figure A.5.

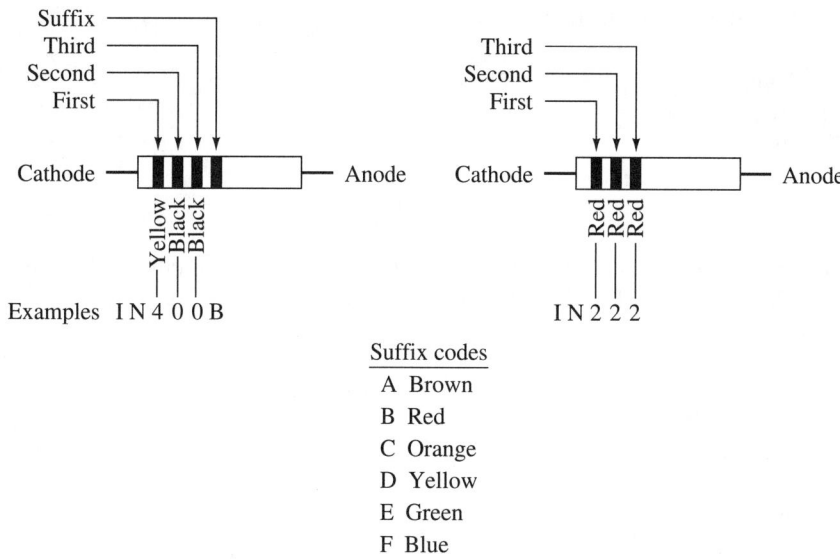

Suffix codes
A Brown
B Red
C Orange
D Yellow
E Green
F Blue

FIGURE A.4 Diode coding.

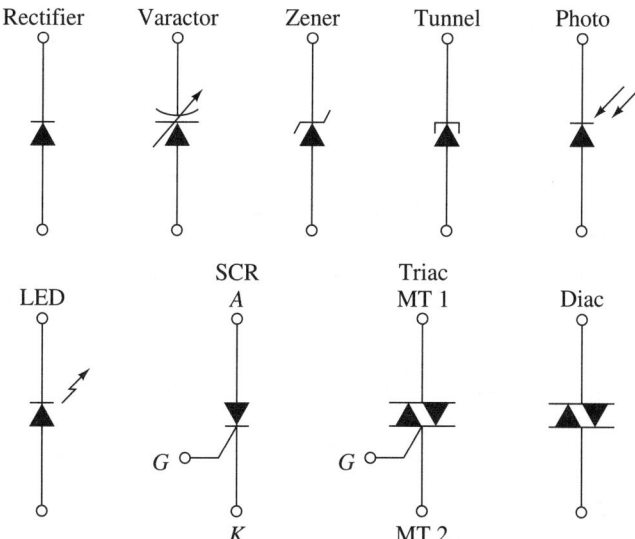

FIGURE A.5 Common diode symbols.

A.4 TRANSISTORS

Transistors are identified by code letters and numbers assigned by the manufacturer. If you cannot find the exact replacement,

1. Look for a transistor that is the specified equivalent. This transistor will be one with characteristics that are very close to those of the original.

2. If the circuit is not critical, you can use a replacement that is of the same functional group.

    a. Audio frequency—small signal
    b. Audio frequency—low level, low noise
    c. Small-signal amplifier
    d. RF amplifier
    e. RF oscillator
    f. Medium- to low-power switching
    g. High frequency, medium power
    h. General-purpose switching
    i. Complementary pairs
    j. Medium power
    k. High power

Coding

Pro-electron. The code is of the form XX where the first letter is either A (for germanium) or B (for silicon). The second letter is as follows:

A: Detector diode
B: Varactor
C: Audio-frequency transistor
D: Audio-frequency power transistor
F: Radio-frequency transistor
L: Radio-frequency power transistor
S: Switching transistor
V: Power-switching transistor
Y: Rectifier
Z: Zener diode

EIA. Type numbers are registered with the Electronics Industry Association (EIA). The letter *N* indicates a semiconductor device whose number of junctions is given by a prefix number. For example:

1N: Diodes
2N: Transistors
3N: FETs

In the component designator XNZZZZ, ZZZZ is used to indicate specific types.

Japanese Industrial Standard. A designator such as XSAZZZ is used, where

X: Number of junctions
S: Semiconductor

A gives polarity and applications such as

A: *pnp* high frequency
B: *pnp* low frequency
C: *npn* high frequency
D: *npn* low frequency
E: *p*-gate thyristor

G: *n*-gate thyristor
H: *n*-base UJT
J: *p*-channel FET
K: *n*-channel FET
M: triac

Thus, *2SA* means *pnp* and *2SC* means *npn*. For example, a 748 low-frequency transistor would be represented by 2SB748.

JEDEC (Joint Electronic Device Engineering Council). The standard packaging is given in Figure A.6.

TO stands for transistor outline.

TO92   Small plastic transistors
        Small signal
        200 – 800 mW
        I_C of 1 – 50 mA

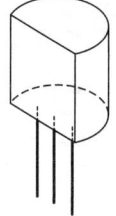

Bottom view

TO-5   Medium power
        Typical: 5 W
        I_C = 2 A

Can

Bottom view

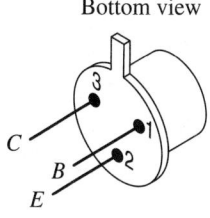

TO220   Power transistors
         Typical 10 W
         I_C = 4 A

Mounting hole

Heat sink mounted on chassis

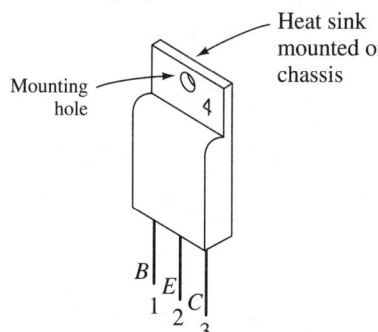

TO3   High power
       Typical 25 – 100 W
       I_C = 5 A

Bottom view

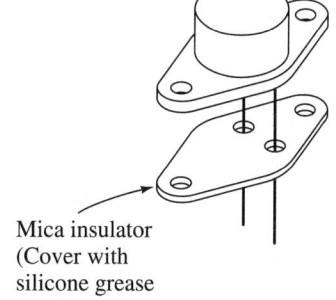

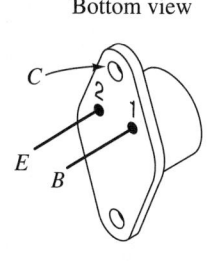

Mica insulator
(Cover with silicone grease for thermal transfer.)

FIGURE A.6 JEDEC (standard packaging).

Symbols. See Figure A.7.

Logic Symbols

The traditional digital logic symbols have been well-known, standard symbols for many years. However, as logic devices became more and more complex, these standard symbols were inadequate. To meet the needs, a new set of symbols was introduced in 1984 under IEEE/ANSI Standard 91-1984. Eventually, the traditional symbols will be completely replaced by the newer ones. See Figure A.8 for some examples.

Primary differences are as follows:

1. Rectangular symbols are used for all devices.
2. A small right triangle (▷) replaces the small bubble of traditional symbols to indicate
   • Inversion of the logic level
   • Active low (presence of triangle)
   • Active high (absence of triangle)
3. A numeral 1 inside the rectangle denotes one input.
4. An ampersand (&) denotes the AND condition.
5. Greater than or equal (≥) denotes the OR function.

Complete logic package can be represented by the new symbology. The 7475 QUAD latch in Figure A.8 is an example. Each logic gate is a separate block. The operation is indicated only in the top block but applies to all blocks.

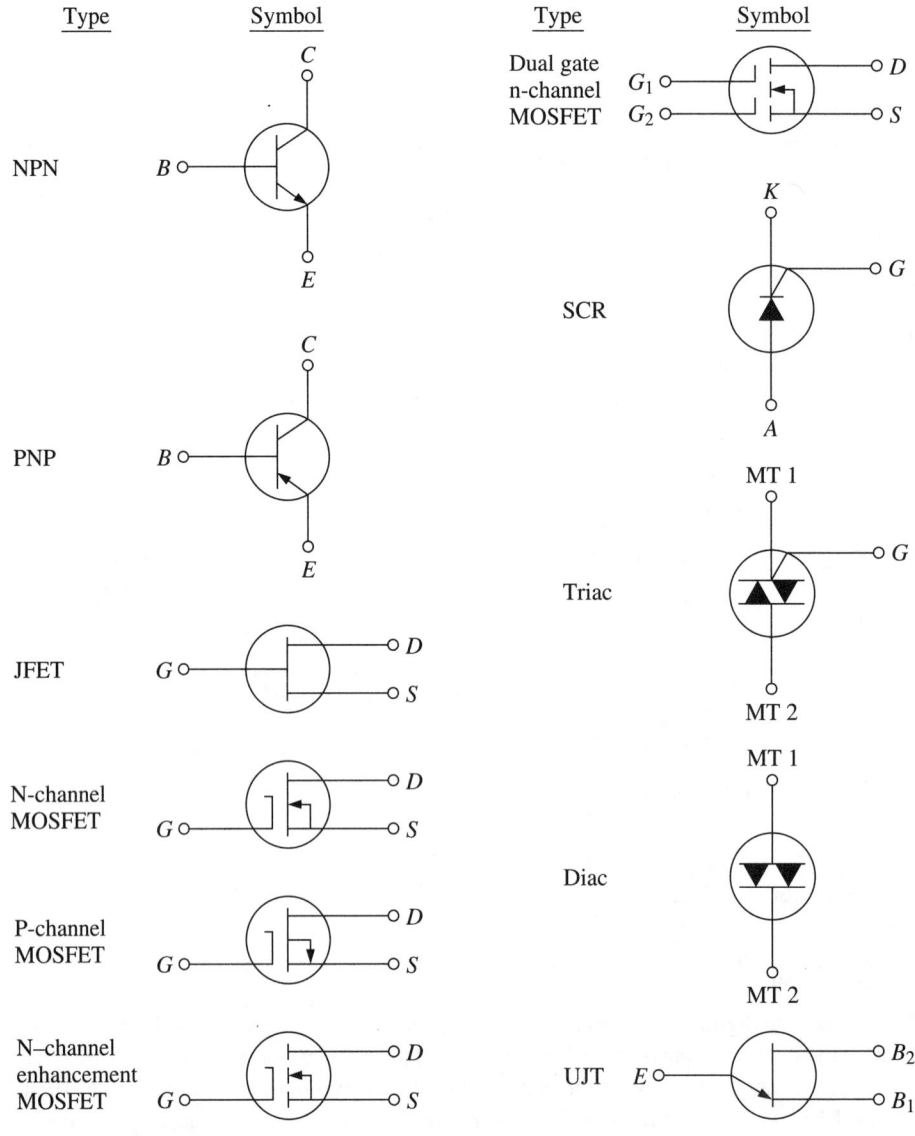

FIGURE A.7 Transistor symbols/Thyristor symbols.

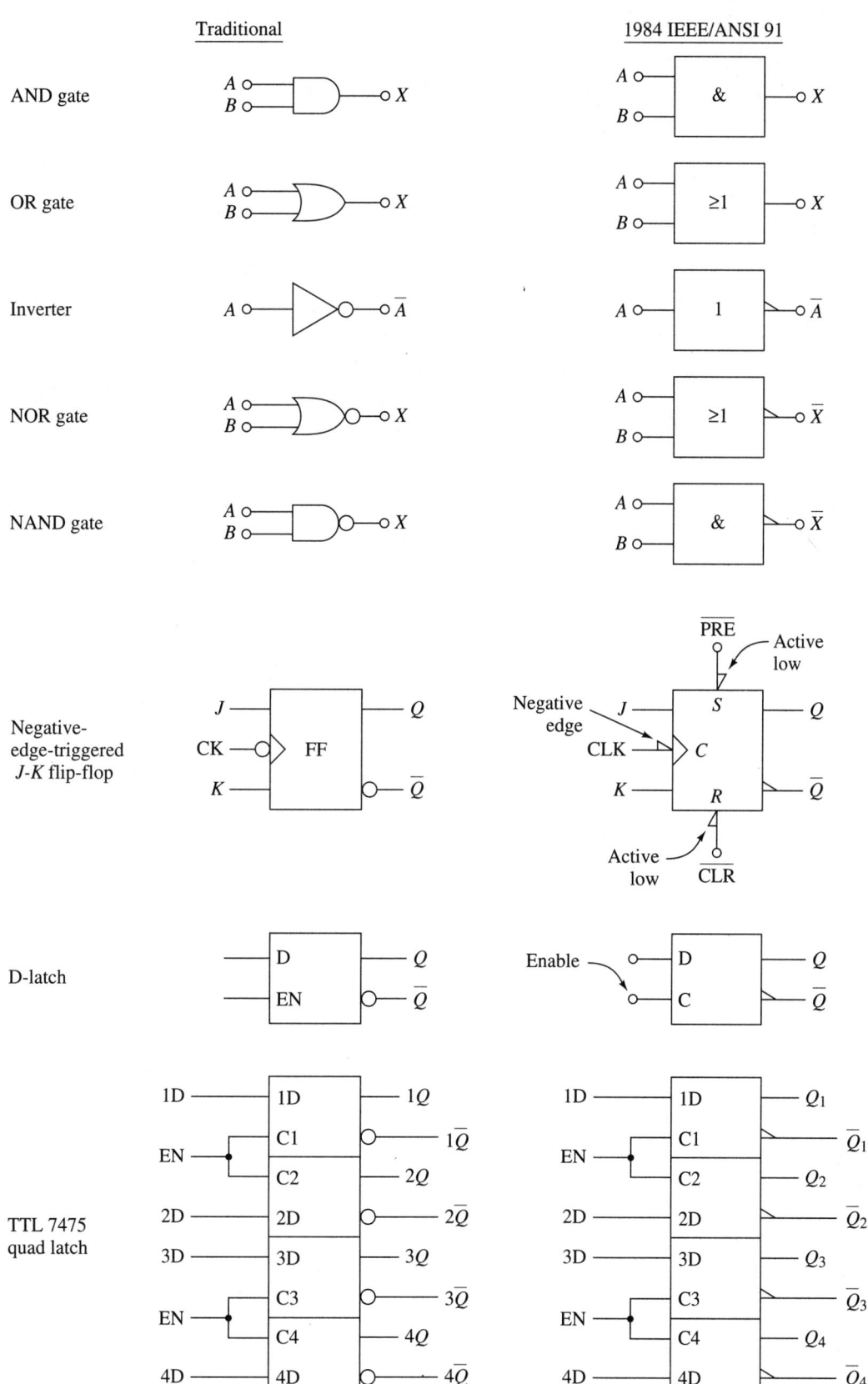

FIGURE A.8 Digital logic symbols.

Appendix B

INTEGRATED CIRCUITS PACKAGING

The packaging style is determined by

1. The number of components on the chip
2. The number of external connections
3. The type of environment expected
4. The method of mounting

The most common types are the DIP (dual in-line package), the ceramic flat pack, and the surface-mount package. These are shown in Figure B.1.

DIP
(Dual-in-line package)

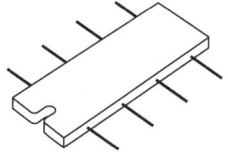

Ceramic flat pack
• Hermetically sealed
• Immune to humidity

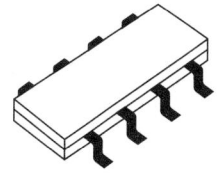

Surface-mount
• Pins soldered directly
  to PCB
• Smaller than DIPs
• Can be handled by automatic
  insertion machines

Linear integrated circuits

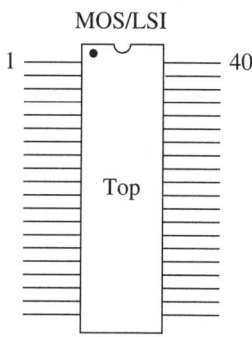

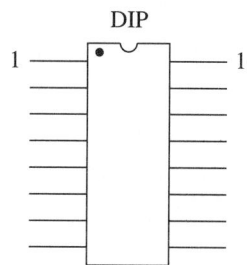

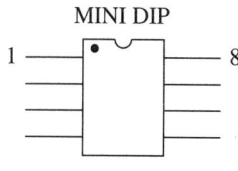

FIGURE B.1   IC packages.

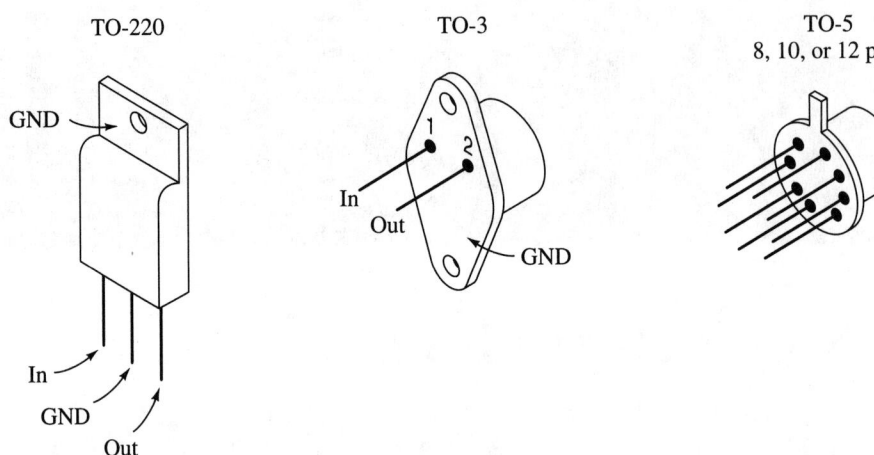

FIGURE B.1 (continued) IC packages.

Appendix C

IC TTL DEVICE FAMILIES AND CHARACTERISTICS

TTL: TRANSISTOR-TRANSISTOR LOGIC

TTL uses bipolar transistors with high conductivity, which means low heat loss. These devices are used extensively in digital logic circuits. They offer the following:

- Are inexpensive
- Use little power
- Fast
- Totem-pole outputs (see Figure C.1)

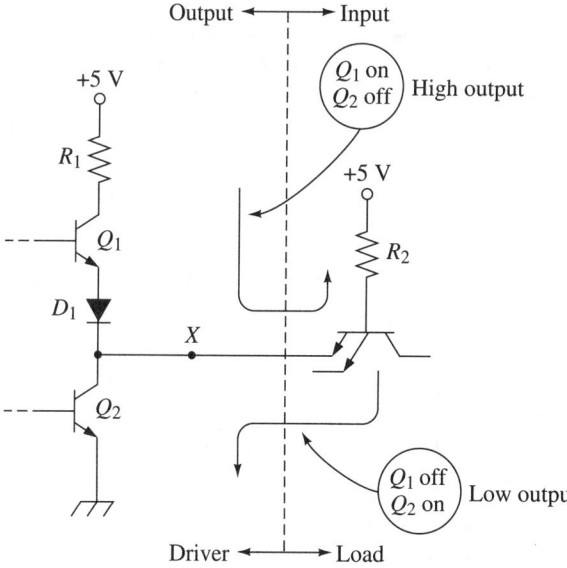

Low-output state (current sink):
   Q_2 is on.
   Q_1 is off and prevents current flow through R_1. This decreases total power dissipation.
   Pt. X is shorted ground.

High-output state (current source):
   Q_1 acts like an emitter follower with very low output impedance.

During the transition between on and off switching between Q_1 and Q_2, both transistors are conducting and larger currents limited by R_1 flow.

Q_1	Q_2
Current sourcing transistor or Pull-up transistor	Current sinking transistor or Pull-down transistor

FIGURE C.1 Totem-pole output circuit.

- Readily available
- Easy to troubleshoot
- Easy to interface with
- Multiple-emitter inputs (see Figure C.2)

Subfamilies

The various families are summarized in Table C.1.

TTL Series Characteristics

Series characteristics are summarized in Table C.2.

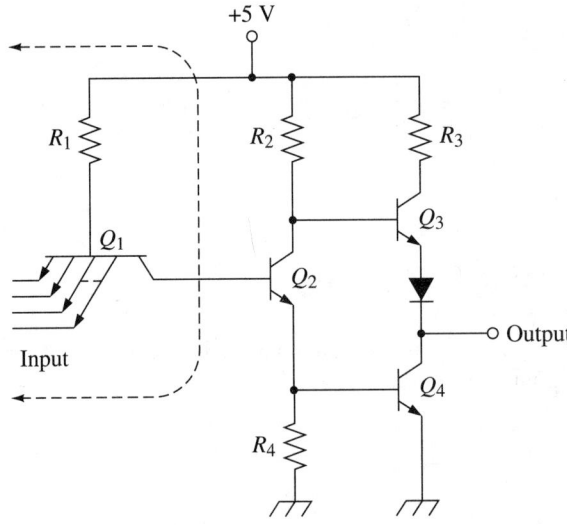

FIGURE C.2 Multiple-emitter input.

TABLE C.1 TTL Families

Standard	Developed in 1965. Rarely used in new designs. 7400 series (5400 series for military). Example: 7404 (5404) hex inverter
High speed	74HXX series. Double the switching speed and power consumption of standard. Example: 74H04 (54H04)
Low power	One-tenth power consumption of standard, but slower. Example: 74L04 (54L04)
Schottky	First version to include the Schottky clamp. (Schottky diode shunting the collector-base junction of the saturating transistor.) Three times the speed of the standard. Example: 74S04 (54S04)
Low-power Schottky	Same speed as standard. However, consumes only 20% power level. Example: 74LS04
Advanced Schottky	Twice the speed of standard and one-half the power consumption. Example: 74AS04
Advanced low-power Schottky	Example: 74ALS04
Fast	Trademark of Fairchild and Motorola. Example: 74F04

TABLE C.2 TTL Characteristics

	Standard	Low Power	Schottky	Low-Power Schottky
Supply voltage	5 V	5 V	5 V	5 V
Maximum supply voltage	5.5 V	5.5 V	5.5 V	5.5 V
Output voltage	Logic 0 = 0.1 typical Logic 1 = 3.4 typical	0.1 typical 3.4 typical	0.1 typical 3.4 typical	0.1 typical 3.4 typical
Power dissipation	10 mW	1 mW	19 mW	2 mW
Propagation delays	10 ns	33 ns	3 ns	9 ns
Fan-out	10	20	10	20
Clock frequency	35 MHz	3 MHz	125 MHz	45 MHz

Appendix D

COMPONENT-SELECTION PROCESS

The goal of troubleshooting is to locate the faulty circuit component, replace it, and return the circuit to service. All too often, troubleshooting and locating the bad component is the easiest part. The harder part is identifying the circuit component and locating a replacement. The information given to you in Appendixes A, B, and C will help you to identify the part. Now, we will consider the next step. Once you have identified the component, you must determine the operational characteristics and maximum ratings. Some of these are listed here.

Resistors
- Resistance value
- Wattage rating
- Tolerance requirements
- Type required

Capacitors
- Working voltage
- Capacitance value
- Polarized or nonpolarized

Inductors
- Inductance value
- Type of core material

Diodes
- Small signal
- Rectifier
- Germanium
- Silicon
- PIV rating

Transistors
- *npn* or *pnp*
- Audio or RF
- Switching
- Breakdown voltages
- Maximum collector current
- FET or CMOS

IC Circuits

- Device
- Family
- Type
- Linear
- Logic

Now that you know the component's function, part number, and maximum ratings, you are ready to look for a suitable part as a replacement. You will require access to a number of different catalog sources. Perhaps you will be fortunate enough to work for an employer who has provided some of the following:

- IC Master
- D.A.T.A. Books
- Manufacturers' data books
- SAM'S Cross Reference Guides
- Manufacturers' catalogs
- Distributors' catalogs (such as NTE)
- Computerized parts' locator guides

If you know the IC device number, start with the IC Master. If you know the particular category, you can look into the D.A.T.A. Books for help. If you know the manufacturer, check its data books.

Let's follow a step-by-step example.

Example 1

The failed component in a power supply was identified as a high-power transistor (used as a pass transistor). The part number is D44H11. The D.A.T.A. Book transistor replacement guide lists eight replacement part numbers along with the manufacturers. One of these is an RCA 2N6292. By using this number to check other cross-reference guides, we can find a Radio Shack 276 2020, which is immediately available.

Be sure to verify the replacement. Your application may be somewhat different than the typical, and the suggested replacement may not be adequate. Locate the technical data on the replacement number and compare with your own requirements. You will need to verify both the physical dimensions and the electrical parameters.

No available cross-reference part number is available. If your original part number contains a suffix code, try locating one without the suffix. If you are successful, however, you must still compare its characteristics with those needed. If you are unable to locate a replacement component, do the following:

1. Check the listed parameters of components that are comparable.
2. Select one that is closest to your needs.

INDEX